Reactive Oxygen Species and Antioxidant Systems in Plants: Role and Regulation under Abiotic Stress

M. Iqbal R. Khan • Nafees A. Khan
Editors

Reactive Oxygen Species and Antioxidant Systems in Plants: Role and Regulation under Abiotic Stress

Springer

Editors
M. Iqbal R. Khan
Plant Physiology Section,
Department of Botany
Aligarh Muslim University
Aligarh, India

Crop and Environmental
Sciences Division
International Rice Research Institute
Los Banos, Philippines

Nafees A. Khan
Department of Botany
Aligarh Muslim University
Aligarh, India

ISBN 978-981-10-5253-8 ISBN 978-981-10-5254-5 (eBook)
DOI 10.1007/978-981-10-5254-5

Library of Congress Control Number: 2017948012

Printed on acid-free paper

This Springer imprint is published by Springer Nature
The registered company is Springer Nature Singapore Pte Ltd.
The registered company address is: 152 Beach Road, #21-01/04 Gateway East, Singapore 189721, Singapore

Preface

The increase in abiotic stress factors has become a major threat to the sustainability of crop production. This situation has led to thinking of ways which can help to come up with potential measures. It is, therefore, necessary to understand the influence of abiotic stress factors on crop performance and the mechanisms by which these factors impact plants. It has now become evident that abiotic stress impacts negatively on plant growth and development at every stage of the plant's life. Plants adapt to the changing environment with the adjustment at physiological, biochemical and molecular levels. The primary possible mechanism involved in the negative effects of abiotic stress factors is the excess production of reactive oxygen species (ROS). They alter physiological and molecular mechanisms leading to poor performance of plants. Plants, however, are able to cope with these adverse effects by inducing antioxidant systems as the priority. Nevertheless, the dual role of ROS has now also been ascertained which provides an evidence for the regulation of plant metabolism positively on a concentration-dependent manner. Under conditions of high ROS production, the antioxidant system plays a major role in diminishing the ROS effects. Thus, ROS production and antioxidant system are interwoven with abiotic stress conditions. The antioxidants have the capacity to hold the stability in metabolism in order to avoid disruption due to environmental disturbances.

The present edited book is an attempt to update the state of art of the knowledge on metabolism of ROS and antioxidants and their relationship in plant adaptation to abiotic stresses involving physiological, biochemical and molecular processes. The chapters are also focused on the current climate issues and how ROS metabolism can interact with the antioxidant system to accelerate detoxification mechanism. It will enhance the mechanistic understanding on ROS and antioxidant systems and will pave the path for agricultural scientists in developing tolerant crops to achieve sustainability under the changing environmental conditions. Additionally, the present book could provide an excellent material for undergraduate, postgraduate and doctoral students to understand fundamentals of ROS metabolism and antioxidant system under both optimal and stressful environments.

Los Banos, Philippines
Aligarh, India

M. Iqbal R. Khan
Nafees A. Khan

Contents

Contributors

Asad Ahmad Department of Bioengineering, Integral University, Lucknow, India

Iffat Zareen Ahmad Department of Bioengineering, Integral University, Lucknow, India

Jubayer Al Mahmud Laboratory of Plant Stress Responses, Department of Applied Biological Science, Faculty of Agriculture, Kagawa University, Kita-gun, Kagawa, Japan

Department of Agroforestry and Environmental Science, Faculty of Agriculture, Sher-e-Bangla Agricultural University, Dhaka, Bangladesh

Taufika Islam Anee Department of Agronomy, Faculty of Agriculture, Sher-e-Bangla Agricultural University, Dhaka, Bangladesh

Laboratory of Plant Stress Responses, Department of Applied Biological Science, Faculty of Agriculture, Kagawa University, Kita-gun, Kagawa, Japan

Shagufta Anjum Centre for Biodiversity Studies, School of Biosciences and Biotechnology, Baba Ghulam Shah Badshah University, Rajouri, India

Abid Ali Ansari Department of Biology, Faculty of Science, University of Tabuk, Tabuk, Kingdom of Saudi Arabia

Saroj Arora Department of Botanical and Environment Sciences, Guru Nanak Dev University, Amritsar, Punjab, India

Mohd Asgher Centre for Biodiversity Studies, School of Biosciences and Biotechnology, Baba Ghulam Shah Badshah University, Rajouri, India

Shagun Bali Department of Botanical and Environment Sciences, Guru Nanak Dev University, Amritsar, Punjab, India

Renu Bhardwaj Department of Botanical and Environment Sciences, Guru Nanak Dev University, Amritsar, Punjab, India

Pooja Bisht Department of Botany, Sri Venkateswara College, University of Delhi, Delhi, India

Thammineni Chakradhar Plant Molecular Biology Group, International Centre for Genetic Engineering and Biotechnology (ICGEB), New Delhi, India

Giacomo Cocetta Department Agricultural and Environmental Science – Production, Landscape, Agroenergy, Università degli Studi di Milano, Milano, Italy

Mudasir Irfan Dar Environmental Botany Section, Department of Botany, Aligarh Muslim University, Aligarh, Uttar Pradesh, India

Sibel Dervis Department of Plant Protection, Faculty of Agriculture, Mustafa Kemal University, Hatay, Antakya, Turkey

Murat Dikilitas Department of Plant Protection, Faculty of Agriculture, Harran University, Sanliurfa, Turkey

Kummari Divya Cell, Molecular Biology & Genetic Engineering Group, International Crops Research Institute for the Semi-Arid Tropics (ICRISAT), Hyderabad, Telangana, India

Antonio Ferrante Department Agricultural and Environmental Science – Production, Landscape, Agroenergy, Università degli Studi di Milano, Milano, Italy

Alessandra Francini Institute of Life Sciences, Scuola Superiore Sant'Anna, Pisa, Italy

Masayuki Fujita Laboratory of Plant Stress Responses, Department of Applied Biological Science, Faculty of Agriculture, Kagawa University, Kita-gun, Kagawa, Japan

Vandana Gautam Department of Botanical and Environment Sciences, Guru Nanak Dev University, Amritsar, Punjab, India

Iain D. Green Department of Life and Environmental Science, The Faculty of Science and Technology, Bournemouth University, Dorset, UK

Neha Handa Department of Botanical and Environment Sciences, Guru Nanak Dev University, Amritsar, Punjab, India

Mirza Hasanuzzaman Department of Agronomy, Faculty of Agriculture, Sher-e-Bangla Agricultural University, Dhaka, Bangladesh

Molecular Biotechnology Group, Center of Molecular Biosciences (COMB), Tropical Biosphere Research Center, University of the Ryukyus, Nishihara, Okinawa, Japan

Masashi Inafuku Molecular Biotechnology Group, Center of Molecular Biosciences (COMB), Tropical Biosphere Research Center, University of the Ryukyus, Okinawa, Japan

Deepti Josula Life Sciences, Sri Venkateswara College, University of Delhi, Delhi, India

Rashmi Kalia BBK DAV College for Women, Amritsar, Punjab, India

Yuriy Victorovich Karpets Dokuchaev Kharkiv National Agrarian University, Kharkiv, Ukraine

Jatinder Katnoria Department of Botanical and Environment Sciences, Guru Nanak Dev University, Amritsar, Punjab, India

Parminder Kaur Department of Botanical and Environment Sciences, Guru Nanak Dev University, Amritsar, Punjab, India

Ravdeep Kaur Department of Botanical and Environment Sciences, Guru Nanak Dev University, Amritsar, Punjab, India

Fareed Ahmad Khan Environmental Botany Section, Department of Botany, Aligarh Muslim University, Aligarh, Uttar Pradesh, India

M. Iqbal R. Khan Plant Physiology Section, Department of Botany, Aligarh Muslim University, Aligarh, India

Crop and Environmental Sciences Division, International Rice Research Institute, Los Banos, Philippines

Nafees A. Khan Plant Physiology Section, Department of Botany, Aligarh Muslim University, Aligarh, India

Kanika Khanna Department of Botanical and Environment Sciences, Guru Nanak Dev University, Amritsar, Punjab, India

Tushar Khare Department of Biotechnology, Modern College (Savitribai Phule Pune University), Pune, India

Sukhmeen Kaur Kohli Department of Botanical and Environment Sciences, Guru Nanak Dev University, Amritsar, Punjab, India

Yuriy Evgenovich Kolupaev Dokuchaev Kharkiv National Agrarian University, Kharkiv, Ukraine

Vinay Kumar Department of Biotechnology, Modern College (Savitribai Phule Pune University), Pune, India

Abdul Mabood Department of Bioengineering, Integral University, Lucknow, India

Srikrishna Mahanty Plant Molecular Biology Group, International Centre for Genetic Engineering and Biotechnology (ICGEB), New Delhi, India

Asim Masood Plant Physiology Section, Department of Botany, Aligarh Muslim University, Aligarh, India

Neeti Mehla Department of Botany, Sri Venkateswara College, University of Delhi, Delhi, India

Vishnu Varthini Nachimuthu Centre of Excellence in Molecular Breeding, Tamil Nadu Agricultural University, Coimbatore, Tamil Nadu, India

Kamrun Nahar Laboratory of Plant Stress Responses, Department of Applied Biological Science, Faculty of Agriculture, Kagawa University, Kita-gun, Kagawa, Japan

Department of Agricultural Botany, Faculty of Agriculture, Sher-e-Bangla Agricultural University, Dhaka, Bangladesh

Avinash Nagpal Department of Botanical and Environment Sciences, Guru Nanak Dev University, Amritsar, Punjab, India

Mohd Irfan Naikoo Environmental Botany Section, Department of Botany, Aligarh Muslim University, Aligarh, Uttar Pradesh, India

Fauzia Naushin Department of Botany, Women's College, Aligarh Muslim University, Aligarh, Uttar Pradesh, India

Puja Ohri Department of Zoology, Guru Nanak Dev University, Amritsar, Punjab, India

Hirosuke Oku Molecular Biotechnology Group, Center of Molecular Biosciences (COMB), Tropical Biosphere Research Center, University of the Ryukyus, Nishihara, Okinawa, Japan

Balaji Aravindhan Pandian Centre of Excellence in Molecular Breeding, Tamil Nadu Agricultural University, Coimbatore, Tamil Nadu, India

Tasir S. Per Plant Physiology Section, Department of Botany, Aligarh Muslim University, Aligarh, India

Malireddy K. Reddy Plant Molecular Biology Group, International Centre for Genetic Engineering and Biotechnology (ICGEB), New Delhi, India

Palakolanu Sudhakar Reddy Plant Molecular Biology Group, International Centre for Genetic Engineering and Biotechnology (ICGEB), New Delhi India

Cell, Molecular Biology & Genetic Engineering Group, International Crops Research Institute for the Semi-Arid Tropics (ICRISAT), Hyderabad, Telangana, India

Ramesha A. Reddy Plant Molecular Biology Group, International Centre for Genetic Engineering and Biotechnology (ICGEB), New Delhi, India

Farha Rehman Department of Botany, Faculty of Science, Mohammad Ali Jauhar University, Rampur, Uttar Pradesh, India

S. Robin Centre of Excellence in Molecular Breeding, Tamil Nadu Agricultural University, Coimbatore, Tamil Nadu, India

Poonam Saini Department of Botanical and Environment Sciences, Guru Nanak Dev University, Amritsar, Punjab, India

Shelja Sareen BBK DAV College for Women, Amritsar, Punjab, India

Anket Sharma Department of Botanical and Environment Sciences, Guru Nanak Dev University, Amritsar, Punjab, India

Mansi Sharma Department of Biotechnology, Modern College (Savitribai Phule Pune University), Pune, India

Vinita Sindhi Biological Sciences, Sri Venkateswara College, University of Delhi, Delhi, India

Ravinder Singh Department of Botanical and Environment Sciences, Guru Nanak Dev University, Amritsar, Punjab, India

Heena Tabassum Department of Bioengineering, Integral University, Lucknow, India

Ashwani Kumar Thukral Department of Botanical and Environment Sciences, Guru Nanak Dev University, Amritsar, Punjab, India

Alice Trivellini Institute of Life Sciences, Scuola Superiore Sant'Anna, Pisa, Italy

Susheel Verma Centre for Biodiversity Studies, School of Biosciences and Biotechnology, Baba Ghulam Shah Badshah University, Rajouri, India

Vinod Verma Department of Botanical and Environment Sciences, Guru Nanak Dev University, Amritsar, Punjab, India

Shabir H. Wani Division of Genetics and Plant Breeding, Faculty of Agriculture WADURA, Sher-e-Kashmir University of Agricultural Sciences and Technology, Jammu, India

Department of Plant, Soil and Microbial Sciences, Michigan State University, East Lansing, MI, USA

Nurcan Yucel Department of Plant Protection, Faculty of Agriculture, Harran University, Sanliurfa, Turkey

An Introduction to Antioxidants and Their Roles in Plant Stress Tolerance

1

Neeti Mehla, Vinita Sindhi, Deepti Josula, Pooja Bisht, and Shabir H. Wani

Abstract

Various abiotic stresses lead to the formation of reactive oxygen species (ROS) in plants which are highly reactive and toxic in plant cell. The ROS comprises both free radical (superoxide radicals, O_2^-; hydroxyl radical, OH^-; perhydroxyl radical, HO_2^-; and alkoxy radicals, RO^-) and non-radical (molecular) forms such as hydrogen peroxide (H_2O_2) and singlet oxygen (O_2). Chloroplasts and mitochondria are the major sites for the generation of O_2^-. Plant's abiotic stress tolerance requires a number of physiological and biochemical mechanisms which includes enzymatic (superoxide dismutase, SOD; catalase, CAT; ascorbate peroxidase, APX; glutathione reductase, GR; monodehydroascorbate reductase, MDHAR; dehydroascorbate reductase, DHAR; glutathione peroxidase, GPX; guaiacol peroxidase, GPOX; and glutathione-S-transferase, GST) and non-enzymatic (ascorbic acid, ASH; glutathione, GSH; phenolic compounds; alkaloids; flavonoids; carotenoids; non-protein amino acids; and α-tocopherols). Increased environmental stresses imbalance the production of reactive oxygen species and thereby quench the activity of antioxidants and thus

N. Mehla • P. Bisht
Department of Botany, Sri Venkateswara College, University of Delhi, Delhi 110021, India

V. Sindhi
Biological Sciences, Sri Venkateswara College, University of Delhi, Delhi 110021, India

D. Josula
Life Sciences, Sri Venkateswara College, University of Delhi, Delhi 110021, India

S.H. Wani (✉)
Division of Genetics and Plant Breeding, Faculty of Agriculture WADURA, Sher-e-Kashmir University of Agricultural Sciences and Technology, Srinagar, Jammu and Kashmir 191121, India

Department of Plant, Soil and Microbial Sciences, Michigan State University, East Lansing, MI 48824-1325, USA
e-mail: shabirhussainwani@gmail.com; wanishab@msu.edu

M.I.R. Khan, N.A. Khan (eds.), *Reactive Oxygen Species and Antioxidant Systems in Plants: Role and Regulation under Abiotic Stress*,
DOI 10.1007/978-981-10-5254-5_1

resulting in oxidative damage. ROS can cause damage to cell structures, nucleic acids, lipids and proteins. Certain ROS like OH- ions are said to react with all components of DNA and damage the purines and pyrimidines. The increased production of antioxidants thus helps the plant to withstand the environmental stress. This chapter focuses on the description of antioxidant defence system under abiotic stress in plants and its involvement in the removal of reactive oxygen species. Hence various types of antioxidants, their types and role will be discussed in detail.

Keywords
Abiotic stress • Antioxidant defence system • Reactive oxygen species (ROS) • Oxidative stress

Abbreviations

$_1O^2$	Singlet oxygen
ABA	Abscisic acid
APX	Ascorbate peroxidase
AsA	Ascorbate
AsA-GSH	Ascorbate-glutathione
ASH	Ascorbic acid
ATP	Adenosine triphosphate
Ca2+	Calcium ions
Car	Carotene
Cars	Carotenoids
CAT	Catalase
Cd	Cadmium
Cu/ZnSOD	Copper/zinc superoxide dismutase
Cys	Cysteine
DHAR	Dehydroascorbate reductase
DNA	Deoxyribonucleic acid
FeSOD	Iron superoxide dismutase
GB	Glycine betaine
GDH	Glutamate dehydrogenase
GPOX	Guaiacol peroxidase
GPX	Glutathione peroxidase
GR	Glutathione reductase
GSH	Glutathione
GSSG	Oxidized glutathione
GST	Glutathione-S-transferase

H_2O_2	Hydrogen peroxide
HO_2^-	Perhydroxyl radical
LPO	Lipid peroxidation
LDL	Low density lipoprotein
MDA	Malondialdehyde
MDHAR	Monodehydroascorbate reductase
Mg^{2+}	Magnesium ion
mM	Millimolar
MnSOD	Manganese superoxide dismutase
NaCl	Sodium chloride
NADPH	Nicotinamide adenine dinucleotide phosphate
NO	Nitric oxide
NO	Reactive nitric oxide
O_2^-	Superoxide radical
OH	Hydroxyl radicals
POX	Peroxidases
Pro	Proline
PSI	Photosystem I
PSII	Photosystem II
ROS	Reactive oxygen species
SODs	Superoxide dismutases

1.1 Introduction

Plants produce excessive reactive oxygen radicals in response to stress caused as a result of environmental changes (Dabrowska et al. 2007; Khan and Khan 2014; Khan et al. 2014, 2015, 2016a, b). Reactive oxygen species (ROS) accumulate as a result of various abiotic stress factors such as salinity, UV radiations, heavy metals, extreme temperature changes, drought, air pollution, herbicides, nutrient deficiency, etc. (Wang et al. 2014; Feigl et al. 2015; Silveira et al. 2015; Thao et al. 2015; Farnese et al. 2016).

The ROS are free radical and non-radical molecules (Sharma et al. 2012) and are the key components of the signalling pathways' network, which act as primary regulators of cellular responses and cell physiology of plant in response to environmental factors (Das and Roychoudhury 2014). However, sudden rise in intracellular levels of ROS is caused due to the imbalance between production and scavenging of ROS under stress conditions (Mittler et al. 2004, Miller et al. 2010; Srivastava and Dubey 2011). In plant tissues, a variety of reactions which consume 1–2% oxygen can lead to excess ROS production which results in cell structure damage (Bhattacharjee 2005). The ROS are by-products of various metabolic activities which take place in the mitochondria, chloroplast and peroxisomes of the plant cell (Navrot et al. 2007; Luis 2015). Plants utilize ROS, which are both free radical like perhydroxyl radical (HO_2^-), superoxide radical (O_2^-), alkoxy radical (RO^-)

and hydroxyl radical (OH^-) and non-radical molecular forms like singlet oxygen (O_2) and hydrogen peroxide (H_2O_2), in various metabolic processes like lignin formation in cell wall (Inzé and Van Montagu 1995; Denness et al. 2011), the abscission of flower and leaf, cell senescence, ripening of fruit and flowering (Mehlhorn et al. 1996; Bhattacharjee 2005). Increased levels of ROS however result in loss of crop productivity (Mittler 2002; Apel and Hirt 2004; Mahajan and Tuteja 2005; Zlatev et al. 2006; Khan et al. 2007; Tuteja 2010; Gill et al. 2011). It is interesting to note that increased levels of ROS are highly reactive and affecting various cellular, biochemical and physiological functions such as disruption of the cell membrane through lipid peroxidation; damage to DNA, pigments and enzymes; carbohydrate deoxidation; and protein denaturation (Noctor and Foyer 1998; Bose et al. 2013). In addition, ROS have the ability to initiate new gene expression and at times can even damage a cell (oxidize proteins, damage nucleic acids or cause lipid peroxidation (LPO) (Foyer and Noctor 2005).

The role of ROS as a protective signalling factor or a damaging factor is dependent on the equilibrium between ROS production and scavenging (Gratão et al. 2005). Plants have their own defence mechanism (antioxidant system) that includes both enzymatic and non-enzymatic systems to regulate the oxidative stress. The enzymatic system includes CAT, peroxidase (POX), SOD, glutathione reductase (GR), ascorbate peroxidase (APX), polyphenol oxidase (PPO), etc., and non-enzymatic system includes carotenes, ascorbic acid (vitamin C), α-tocopherols, etc. When plants become senesced, the antioxidant system does not work well, with some antioxidants increasing and others decreasing as a result of which the ROS may accumulate in a large amount. This is often indicated by enhanced lipid peroxidation and decreased levels of antioxidant enzymes leading to programmed cell death (PCD), particularly under severe abiotic stress conditions (Dhindsa and Matowe 1981; Hodges and Forney 2000; Gill et al. 2011; Duarte et al. 2013). There are some natural antioxidants like vitamins, phenols, flavonoids, carotenoids, glutathione and endogenous metabolites which also help in scavenging ROS (Larson 1988) (Fig. 1.1). The antioxidant system scavenges toxic radicals by functioning as singlet and triplet oxygen quenchers, synergists, enzyme inhibitors and peroxide decomposers (Manach et al. 1998). In this chapter we will try to highlight the antioxidant defence system during abiotic stress in plants and its association with the removal of reactive oxygen species.

1.1.1 Antioxidant Defence Mechanism

The atoms or molecules having unpaired electrons are known as free radicals and are considered as a major source of ageing. These free radicals are stabilized by special category of substances known as antioxidants which prevent and protect the cell from the damage caused by free radicals. Few examples of such substances include beta-carotene; lycopene; vitamins C, E and A; catalase; etc. which help in sustaining the rate of oxidation reactions in a cell (Sies 1997). The process of transfer of electrons from one substance to another with low reduction potential is

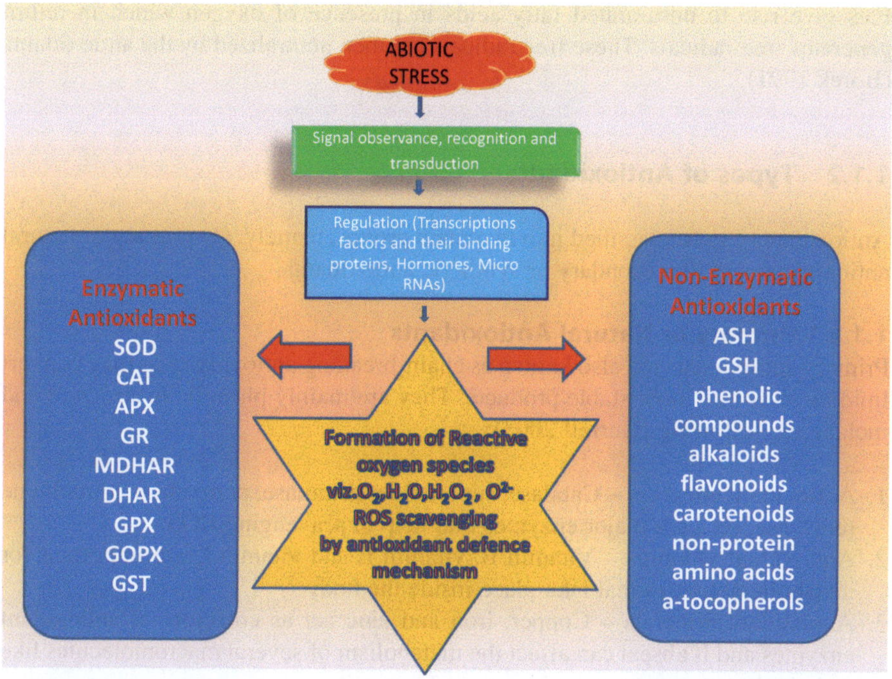

Fig. 1.1 Abiotic stress responses in plants

termed as oxidation, and these oxidation reactions are responsible for the generation of free radicals, which can cause injury to the cells by an order of chain reactions. Antioxidants play a very crucial role in mitigating or preventing the process of oxidation of other molecules (Ahamadi et al. 2015).

In order to prevent damage by oxidation, plants and animals have a complex systems of antioxidants like catalase, superoxide dismutase, peroxidises, glutathione, vitamin C; etc. which act as reducing agents that terminate oxidative reactions by removing free radicals. The production of these antioxidant molecules may, at first, be inhibited to allow the occurrence of the oxidative burst (Luis 2015), but once the signal is initiated, these mechanisms are activated and perform function in a united way (Viehweger 2014; Farnese et al. 2016).

There are two mechanisms by which the antioxidants act on ROS:

1. Chain-breaking mechanism – Electrons are donated to the free radicals by primary antioxidants (e.g. lipid radical)
2. Removal of ROS – Quenching chain initiator catalyst removes ROS and RNS (reactive nitrogen species)

Fats give rise to unsaturated fatty acids in presence of oxygen which in return generates free radicals. These free radicals are then neutralized by the antioxidants (Borek 1991).

1.1.2 Types of Antioxidants

Antioxidants can be classified into two main groups, namely, (1) primary or natural antioxidants and (2) secondary or synthetic antioxidants.

1.1.2.1 Primary or Natural Antioxidants

Primary antioxidants are also known as chain-breaking antioxidants as they convert lipid radicals into more stable products. They are mainly phenolic in structure and include the following (Hurrell 2003):

1. Antioxidant enzymes – Catalase, superoxide dismutase, peroxidase, glutathione reductase, etc., are major enzymes that help in scavenging ROS.
2. Antioxidant vitamins – Vitamin B, vitamin E and vitamin C are important for metabolic activities that take place inside the body.
3. Antioxidant minerals – Copper, iron and zinc act as cofactors of antioxidant enzymes and if absent can affect the metabolism of several macromolecules like carbohydrates.
4. Phytochemicals – These are phenolic compounds and include flavonoids.

1.1.2.2 Secondary or Synthetic Antioxidants

Phenolic compounds that capture free radicals and stop the chain reactions are called secondary antioxidants. These include (Hurrell 2003):

1. Nordihydroguaiaretic acid (NDGA)
2. Butylated hydroxyl anisole (BHA)
3. Tertiary butyl hydroquinone (TBHQ)
4. Butylated hydroxytoluene (BHT)
5. Propyl gallate (PG) and metal-chelating agent (EDTA)

1.2 Natural Antioxidants

1.2.1 Antioxidant Enzymes

Plants have a unique and incredible defence mechanism for detoxification of the free radicals. Basically these mechanisms include catalysis of various antioxidant enzymes like superoxide dismutase, glutathione peroxidase and catalase followed by a metastasis of superoxide to hydrogen peroxide and further conversion into water and oxygen. The ratio of GSSG: GSH regulates the catalysis reactions of glutathione peroxidase where GSH is required as a substrate. The detoxification of

Table 1.1 Antioxidant defence mechanism (antioxidant enzymes) in various plant species in response to abiotic stress

Antioxidant enzymes	Type of stress	Plant species	References
Superoxide dismutase	Salt	*Mulberry*	Harinasut et al. (2003)
	Salt	*C. arietinum*	Kukreja et al. (2005) and Eyidogan and Öz (2007)
	Salt and drought	*Glycyrrhiza uralensis*	Pan et al. (2006)
	Metal (Cd)	*Brassica juncea*	Mobin and Khan (2007)
		Vigna mungo	Singh et al. (2008)
	Aluminium	*H. vulgare*	Šimonovičová et al. (2004)
	Drought	*O. sativa*	Sharma and Dubey (2005)
		P. vulgaris	Zlatev et al. (2006)
		Alternanthera philoxeroides	Wang et al. (2008b)
	Water	*Lycopersicum esculentum*	Sánchez-Rodríguez et al. (2016)
	Salt	Maize	Hamada AbdElgawad et al. (2016)
Catalase	Salt+copper	*A. dolium*	Srivastava et al. (2005)
	Salt+drought	*Glycyrrhiza uralensis*	Pan et al. (2006)
	Ultraviolet-B rays	*C. auriculata*	Agarwal (2007)
	Drought	Wheat	Simova-Stoilova et al. (2010)
	Salt	Maize	Hamada AbdElgawad et al. (2016)
	Water	*Lycopersicum esculentum*	Sánchez-Rodríguez et al. (2016)
Peroxidase	Metal (Fe)	Bean	Pekker et al. (2002)
	Metal (Cd)	*A. thaliana*	Cho and Seo (2005)
	Water	*P. vulgaris*	Zlatev et al. (2006)
		P. asperata	Yang et al. (2008)
	Heat +drought	*Arabidopsis thaliana*	Koussevitzky et al. (2008)
	Salt+osmotic stress	*O. sativa*	Cunha et al. (2016)
	Salt	*Eurya emarginata*	Zheng et al. (2016)
Glutathione reductase	Light	*P. asperata*	Yang et al. (2008)
	Cold	*W. somnifera*	(Mir et al. 2015)
	Water	*Arachis hypogaea*	Chakraborty et al. (2015)
	Salt	Chickpea	Shankar et al. (2016)
	Water	*O. sativa*	Kundur et al. (2016)
	Drought	Maize	Ahmad et al. (2016)

ROS largely depends on the concentration of these reactants and their ratio. The concentration of redox-active metals such as iron should be monitored by binding to storage and transport proteins (e.g. ferritin, transferrin, lactoferrin) so as to control the formation of reactive oxygen species (ROS).

Different antioxidant enzymes play different roles (Table 1.1). Some enzymes like superoxide dismutase, catalase and peroxidase eliminate ROS, while others like glutathione peroxidase eliminate internal lipid peroxidation products. Glutathione S-transferase plays an important role in eliminating toxic secondary oxidation radicals (Timofeyev and Steinberg 2006).

1.2.1.1 Superoxide Dismutase

Superoxide dismutase (SOD, EC 1.15.1.1) catalyses the destruction of the O^{2-} free radical. They act as the first line of defence against harm caused by ROS (Das and Roychoudhury 2014):

$$2O^{2-} + 2H^+ \rightarrow O_2 + H_2O_2$$

It protects oxygen-metabolizing cells against harmful effects of superoxide free radicals. Three types of superoxide dismutase are characterized by different metal contents (Worthington-biochem):

1. A blue-green Cu(II)-Zn(II) enzyme comes from human and bovine erythrocytes.
2. A wine-red Mn(III) protein is found in *E. coli* and in chicken and rat (Peeters-Joris et al. 1975) liver mitochondria (Tyler 1975).
3. A yellow Fe (III) enzyme from *E. coli* (Villafranca et al. 1974).

In higher plants, SOD isozymes are localized in various cell organelles (Mn-SOD in mitochondria and peroxisomes, Fe-SOD in chloroplasts and peroxisomes and CuZn-SOD in chloroplasts, cytosol, apoplast and peroxisomes) and protect them from being oxidized by reactive oxygen species (ROS) (Corpas et al. 2001, 2006; Alscher et al. 2002). A substantial increase in SOD activity in response to salt stress has been observed in various plants, viz. *C. arietinum* (Kukreja et al. 2005), mulberry (Harinasut et al. 2003) and *Lycopersicon esculentum* (Gapińska et al. 2007). Eyidogan and Oz found three SOD activity bands (MnSOD, FeSOD and Cu/ZnSOD) in *C. arietinum* with significantly high activities of Cu/ZnSOD and MnSOD isozymes under salt stress (Eyidogan and Öz 2007). Pan et al. studied drought and salt stress effects on *Glycyrrhiza uralensis* and found a significant increase in SOD activity and presence of an additional MnSOD at the time of salt stress (Pan et al. 2006). On treating *C. arietinum* (Hasan et al. 2008), *Hordeum vulgare* (Guo et al. 2004), *O. sativa* (Hsu and Kao 2004), *Triticum aestivum* (Khan et al. 2007), *Brassica juncea* (Mobin and Khan 2007), *Vigna mungo* (Singh et al. 2008) and *A. thaliana* (Skórzyńska-Polit et al. 2003) with cadmium, SOD activity was found to be high. Increase in SOD activity following drought stress was noted in three cultivars of *O. sativa* (Sharma and Dubey 2005), *P. vulgaris* (Zlatev et al. 2006) and *Alternanthera philoxeroides* (Gao et al. 2008).

The effects of water stress on the activities of total leaf SOD and chloroplast SOD in *Trifolium repens* were studied, and it was found that with the increase in water stress, the activity of SOD also increased (Wang et al. 2008a). An increase in SOD activity in *H. vulgare* root tips was seen under Al stress at 72 h (Šimonovičová et al. 2004). Yang et al. (2008) demonstrated the effect on *Picea asperata* when a combination of drought and low light were used.

1.2.1.2 Catalase

Catalase (H_2O_2:H_2O_2 oxidoreductase, EC 1.11.1.6; CAT) is a tetrameric haem-containing enzyme which acts on H_2O_2 and degrades it at a rapid rate. It is found in all aerobic organisms and is unique as it does not require a reducing equivalent (Das and Roychoudhury 2014). It has been reported that one molecule of CAT can convert 26 million molecules of H_2O_2 to H_2O and O_2 per minute and depending on the concentration of H_2O_2, it exerts a dual function (Deisseroth and Dounce 1970). At low concentrations ($<10^{-6}$ M) of H_2O_2, it acts "peroxidatically", where a variety of hydrogen donors (e.g. ethanol, ascorbic acid) can be oxidized in the following manner:

$$RH_2 + H_2O_2 \rightarrow R + 2H_2O$$

When the substrate concentrations are high, catalase decomposes toxic H_2O_2 at an extremely rapid rate using the "catalytic" reaction in which H_2O_2 acts as both acceptor and donor of hydrogen molecules (Tausz et al. 2004):

$$2H_2O_2 \rightarrow 2H_2O + O_2$$

A decrease in CAT activity was seen in *A. doliolum* under NaCl and Cu stress (Srivastava et al. 2005). In wheat plants, CAT activity was seen to be enhanced under drought stress, and its levels were higher especially in sensitive varieties (Simova-Stoilova et al. 2010). However, Sharma and Dubey reported a decrease in CAT activity under drought stress in rice seedlings (Sharma and Dubey 2005). Increased levels of CAT activity were seen in *P. asperata* under drought stress and high light conditions (Yang et al. 2008). The UV-B stress increased CAT activity in *C. auriculata* seedlings (Agarwal 2007). The combined effect of salt and drought stress decreased the CAT act in *Glycyrrhiza uralensis* seedlings (Pan et al. 2006).

1.2.1.3 Peroxidases

Guaiacol Peroxidase

Guaiacol peroxidise (GPOX) comprises of 40–50 kDa monomers and uses aromatic electron donors such as guaiacol and pyrogallol and decomposes indole-3-acetic acid (IAA) (Das and Roychoudhury 2014). It consumes H_2O_2 and is important for the biosynthesis of lignin. The activity of GPOX is different in different plant species and with varying stress conditions. It increases in Cd-exposed plants of *T. aestivum* (Milone et al. 2003), *A. thaliana* (Cho and Seo 2005) and *C. demersum*

(Aravind and Prasad 2003). On subjecting spruce needles to Cd stress, it was observed that GPOX activity initially increased and later decreased as the stress was further increased (Radotić et al. 2000). The leaf and root tissues of *Vigna radiata* (Panda 2001) and *O. sativa* (Yamane et al. 2009) showed an increase in the levels of CAT under salinity stress.

Ascorbate Peroxidase (APX)

Ascorbate peroxidases (APXs) are haem-containing proteins with iron present at the catalytic site. The substrate used for reducing hydrogen peroxide is the ascorbate, and its concentration defines the sensitivity of the enzyme. Lower concentrations (lower than 20 μM) make the enzymes become less stable with decreased activity (Shigeoka et al. 2002). They are mainly found in higher plants, algae and some cyanobacteria (Miyake et al. 1991; Sano et al. 2001; Sharma and Dubey 2004; Shigeoka et al. 1980a, b; Takeda et al. 1992, 1998, 2000) but are also found in tissues rich in ascorbate (Boveris et al. 1980; Wada et al. 1998), in insects (Mathews et al. 1997), in *Trypanosoma cruzi* (Mathews et al. 1997) and in the choroid and iris epithelium of the bovine eye. Plants contain five isoforms of APX: mitochondria isoforms, cytosolic isoforms, chloroplastic isoforms and peroxisomal/glyoxysomal isoforms. They function as scavengers of H_2O_2 (Miyake and Asada 1996). When plants are exposed to various stress conditions, APX activity generally increases. Enhanced leaf APX activity under cadmium stress has been reported in *B. juncea* (Mobin and Khan 2007), *V. mungo* (Singh et al. 2008), *T. aestivum* (Khan et al. 2007) and *Ceratophyllum demersum* (Aravind and Prasad 2003). Rice seedlings showed increased APX activity with pretreatment of H_2O_2 under non-heat shock conditions (Hsu and Kao 2007). Salt-stressed *A. doliolum* showed high APX activity (Srivastava et al. 2005). Significant increase in APX activity was noted under water stress in three cultivars of *P. vulgaris* (Zlatev et al. 2006) and *P. asperata* (Yang et al. 2008). Sharma and Dubey reported higher chloroplast APX activity in plants under mild drought stress than control-grown plants, but the activity later decreased with the increasing drought stress (Sharma and Dubey 2005). Pekker et al. found an increase in cAPX levels in the leaves of de-rooted bean plants in response to Fe stress (Pekker et al. 2002). Cytosolic APX1 is important in protecting plants against heat and drought stress (Koussevitzky et al. 2008). Simonovicova et al. found an increase in APX activity in *H. vulgare* L. cv. Alfor root tips under aluminium stress at 72 h (Šimonovičová et al. 2004).

1.3 Halliwell-Asada Cycle

Halliwell-Asada cycle (ascorbate-glutathione cycle) is one of the most important antioxidant systems in plants. It is a set of reactions catalysed by glutathione reductase (GR), dehydroascorbate reductase (DHAR) and monodehydroascorbate reductase (MDHAR). ATP and NAPH are utilized to convert H_2O_2 with the help of glutathione and ascorbate (Foyer and Noctor 2005). The highly energetic reactions of photosynthesis that take place in chloroplasts are responsible for the organelle to act as a major source of superoxide and H_2O_2. The ascorbate-glutathione cycle plays an important role in chloroplast redox reactions because of non-residence of catalase enzyme.

1.3.1 Glutathione Reductase

Glutathione reductase (GR, EC 1.6.4.2) is a ubiquitous enzyme, which catalyses the reduction of oxidized glutathione (GSSG) to glutathione (GSH). Glutathione redox cycle is needed to maintain the required amount of reduced GSH in the cell and GSH functions as an antioxidant as it acts upon organic peroxides and free radicals in amino acid transport. Glutathione reductase is needed for glutathione redox cycle, and it also serves as a substrate for glutathione S-transferases and glutathione peroxidises. It is mainly found in chloroplasts, but small amount is also present in cytosol and mitochondria (Creissen et al. 1994; Edwards et al. 1990). GR and GSH determine the tolerance of a plant under various stresses (Rao and Reddy 2008). GR activity showed an increase in the presence of Cd in *C. annuum* (León et al. 2002), *A. thaliana* (Skórzyńska-Polit et al. 2003), *V. mungo* (Singh et al. 2008), *T. aestivum* (Khan et al. 2007) and *B. juncea* (Mobin and Khan 2007). Eyidogan and Oz observed that under salt stress, the GR activity increased in the leaf tissue of *C. arietinum* L. cv. Gokce (Eyidogan and Öz 2007). An increased GR activity was seen in *C. arietinum* roots following salt stress (Kukreja et al. 2005). Decrease in GR activity has been observed on exposing *A. doliolum* to Cu stress but showed to be escalated when subjected under salt stress (Srivastava et al. 2005). A significant increase in GR activity is reported when *O. sativa* seedlings were exposed to drought stress (Sharma and Dubey 2005).GR activity increases in *P. asperata* seedlings under high light intensity, but no change was observed in low light (Yang et al. 2008).

1.3.2 Monodehydroascorbate Reductase (MDHAR)

MDHAR is a flavin adenine dinucleotide (FAD) enzyme having chloroplastic and cytosolic isozymes. Monodehydroascorbate (MDHA) is an electron acceptor specific to MDHAR. It uses NADH rather than NADPH as the electron donor (Asada 1999). APX and MDHAR scavenge H_2O_2 in the mitochondria and peroxisomes. A decrease in MDHAR activity in Cd-exposed poplar hybrids (*Populus* and *Canescens*) and an increase in Cd-exposed *Pinus sylvestris* were observed by Schutzendubel et al. (2001). The enzymes responsible for regeneration of ASH were found by Sharma and Dubey (2005). DHAR, GR and MDHAR were higher in rice seedlings under drought stress. Increased MDHAR activity develops a tolerance in tomato plant against chill stress (Stevens et al. 2008). Overexpression of MDHAR in transgenic tobacco resulted with an increase in tolerance against salt and osmotic stresses (Eltayeb et al. 2007).

1.3.3 Dehydroascorbate Reductase (DHAR)

DHAR plays an important role in regulating and regenerating ASH. This role of DHAR has been studied in transgenic tobacco plants overexpressing *A. thaliana* cytosolic DHAR when exposed to Al stress. It was seen that plants overexpressing DHAR, by maintaining high ASH level, showed tolerance to aluminium stress (Yin et al. 2010). Higher levels of DHAR help in protecting tobacco plants against ozone toxicity (Chen and Gallie 2005) and drought and ozone stress tolerance (Eltayeb et al. 2006) and increased salt tolerance in *Arabidopsis* (Ushimaru et al. 2006).

1.3.4 Glutathione Peroxidase

GPXs (EC 1.11.1.9) are a family of selenium-containing heterogeneous isozymes that shield plant cells against oxidative damage by reducing H_2O_2 and organic and lipid hydroperoxides utilizing GSH (Noctor et al. 2002). Selenium acts as scavenger of peroxides from cytosol and cell membrane (Mahapatra et al. 2013). Millar et al. identified a family of seven related proteins named from GPX1 to GPX7 in various cellular organelles including mitochondria cytosol, endoplasmic reticulum and chloroplast *Arabidopsis thaliana* (Millar et al. 2003). Stress increases GPX activity in cultivars of *C. annuum* plants (León et al. 2002) but is reduced in roots and shows no significant change in the leaves of cadmium-exposed *P. sativum* plants (Dixit et al. 2001). Gapinska et al. (2007) reported that 150 mM NaCl stress resulted in higher activity of GPX in *L. esculentum* roots.

1.3.5 Glutathione S-Transferases (GST)

Glutathione S-transferases (GST, EC 2.5.1.18) or the plant glutathione transferases are large group of enzymes that catalyse tripeptide glutathione and electrophilic xenobiotic substrate amalgamation (GSH; g-gluecysegly). Plant GSTs help in herbicide detoxification, vacuolar sequestration of anthocyanin, hormone homeostasis, hydroxyperoxide detoxification, tyrosine metabolism and regulation of apoptosis and affect plant responses to biotic and abiotic stresses (Dixon et al. 2010). It has been observed that GSTs can eliminate cytotoxic or genotoxic compounds, which can damage the DNA, RNA and proteins (Noctor et al. 2002). They use GSH to decrease peroxides. GSTs represent one of the exuberant protein and in few instances exemplifying more than 1% of soluble protein in plant cells (Edwards et al. 2000). An increased GST activity was observed in roots of *Phragmites australis* plants (Moons 2003) and leaves and roots of cadmium-exposed *P. sativum* plants (Dixit et al. 2001). Gapinska et al. (2007) observed GST activity to be enhanced in *L. esculentum* roots when subjected under salinity stress. When two contrasting sorghum varieties, drought sensitive (SPV-839) and drought tolerant (M35-1), were exposed to 150 mM NaCl for 72 h, it was observed that the tolerant variety showed better response in terms of efficient H_2O_2 scavenging mechanisms with high GST and CAT activities (Jogeswar et al. 2006).

1.3.6 Low Molecular Weight Antioxidants

These antioxidants can be lipid soluble and membrane associated like α-tocopherol and β-carotene or water-soluble reductants like ascorbate and glutathione (Ahmad et al. 2008). Non-enzymatic antioxidants can also be divided into nutrient antioxidants and metabolic antioxidants. Metabolic antioxidants are produced by the body like bilirubin, metal-chelating proteins and uric acid. Nutrient antioxidants are provided through diet or supplements like vitamins E and C and trace metals (Madhavi et al. 1995). Radical-scavenging antioxidants (e.g. vitamin E) capture the radical, thus interrupting chain reactions; the vitamin E radical is relatively stable and can be converted back to its non-radical form. These ROS defence mechanisms are valuable for the whole body and subcellular distribution of the different components. Concentrations of the enzymes (SOD, catalase and GSH peroxidase), vitamin E and substrates (GSH) are higher in areas where ROS damage is more likely (ROS 1997).

1.3.6.1 Ascorbic Acid (ASH)

Ascorbic acid commonly known as vitamin C has been reported to be the most opulent, strong and water-soluble antioxidant which checks or neutralizes the damage done by the ROS in plants (Smirnoff 2000; Khan and Ashraf 2008). It is mostly found in all plant tissues but is abundant in meristematic regions, photosynthetic cells and some fruits. ASH is considered as an efficient scavenger of reactive oxygen species as it has the potential to donate electrons through enzymatic or non-enzymatic reactions.

It either regenerates a-tocopherol from tocopheroxyl radical or directly scavenges the OH^- and O_2 radicals. In chloroplast, the excess excitation energy is dissipated by ASH acting as a cofactor of violaxanthin de-epoxidase (Smirnoff 2000). In addition, it also plays a vital role in preserving the activities of enzymes that contain transition metal ions as a prosthetic group (Noctor and Foyer 1998). The ASH redox system comprises of L-ascorbic acid, MDHA and DHA. Both the oxidized forms of ASH are comparatively unstable in aqueous environments, while DHA can be chemically reduced by GSH to ASH (Foyer and Halliwell 1976).

1.3.6.2 Glutathione (GSH)

Glutathione acts as an intracellular defence in plants against oxidative damage. It is localized in reduced state in the cell compartments like mitochondria, peroxisomes, cytosol, vacuole, apoplasts and even endoplasmic reticulum (Mittler and Zilinskas 1992; Jiménez et al. 1998) where it is involved in several physiological processes, including signal transduction, regulation of sulphate transport, detoxification of xenobiotics, conjugation of metabolites (Xiang et al. 2001) and the expression of genes responsive during stress conditions (Mullineaux and Rausch 2005). Growth and development-related events in plants, including cell senescence and death, cell differentiation, enzymatic regulation and pathogen resistance, also require GSH (Rausch and Wachter 2005). In many cellular reactions, GSH acts as a substrate and produces GSSG, which are two glutathione molecules linked by a disulphide bond. The redox state of a cell can be maintained with the help of this equilibrium between GSH and GSSG (Zlatev et al. 2006). GSH counteracts the inhibitory effects of ROS-induced oxidative stress by retaining the reduced state of the cells (Meyer 2008). It acts as an efficient scavenger of O_2, H_2O_2 (Briviba et al. 1997; Noctor and Foyer 1998) and most treacherous ROS like $OH^{\cdot}$ (Larson 1988). Furthermore, GSH plays an important role in the anti-oxidative defence system by regenerating another adequate water-soluble antioxidant like ASH, through the ASH-GSH cycle (Foyer and Halliwell 1976). GSH also detoxifies many inhaled pollutants like ozone, NO_2, etc. (Mahapatra et al. 2013). Glutathione, a major non-protein thiol, has long been reported to be essential for recycling of antioxidants like vitamin C and vitamin E (Kayang 2007).

1.3.6.3 Proline (Pro)

Besides being an osmolyte, proline is a non-enzymatic antioxidant that is useful for microbes, animals and plants to reduce the adverse effects of ROS. Pro accumulates under salt, drought and metal stress as a result of increase in synthesis or decrease in degradation. Free Pro acts as a protein stabilizer, an osmo-protectant, a metal chelator, an OH^- and O_2 scavenger and an inhibitor of LPO (Ashraf and Foolad 2007; Trovato et al. 2008). Proline, mannitol, sorbitol and myo-inositol have been tested, and proline has been found to possess excellent OH^--scavenging ability (Smirnoff and Cumbes 1989). Hence in case of the reactive oxygen species produced as a result of metal, salt and dehydrated stress, proline acts as an effective quencher along with a redox-signalling molecule (Alia and Sardhi 1991).

1.3.6.4 α-Tocopherols (Vitamin E)

Tocopherols commonly known as vitamin E are known to be powerful scavengers of ROS and lipid radicals. In plants they are found in thylakoid membrane and are lipid soluble in nature. There are four different isomers of tocopherols which are present in the plant system (α, β, γ, δ). However, α-tocopherol shows the maximum antioxidant activity due to presence of three methyl groups in its molecular structure. They also stabilize polysaturated fatty acids within lipid bilayers and low density lipoprotein (LDL) by protecting them from lipoxygenase attack (Sathishkumar et al. 2010; Mahapatra et al. 2013).

1.3.6.5 Carotenoids (Car)

In every photosynthetic organisms, b-carotene, zeaxanthin, carotenoids and tocopherols play a vital photoprotective role, by dissipating excess excitation energy as heat or by scavenging ROS and suppressing LPO. B-carotene is an excellent singlet oxygen scavenger (Mahapatra et al. 2013). Carotenoids are pigments that are soluble in lipid and are generally found in plants and microorganisms. There are more than 600 carotenoids found in plants that have several metabolic functions like providing plant tolerance to oxidative stress. There are predominantly three major functions that are carried out by Car:

1. An accessory light-harvesting role – they absorb light (wavelength between 400 and 550 nm) and transfer to the chlorophyll (Siefermann-Harms 1987).
2. An antioxidant – they quench naturally formed triplet sensitizer (Chl3), $O_2^{\cdot}$ and other harmful free radicals during photosynthesis (Collins 2001).
3. Structural role – they provide stability to thylakoid membrane and light-harvesting complex proteins and important for the PSI assembly (Niyogi et al. 2001).

1.3.6.6 Flavonoids

Flavonoids are mostly found in the floral parts, leaves and pollens. They can also be seen accumulated in aerial parts of the plant, in leaf surface and in the plant vacuole as glycosides. Flavonoid concentration in a plant cell is predominantly maintained over 1 Mm (Vierstra et al. 1982). Flavonoids can be categorized into different types depending on their structural forms like flavonols, flavones, isoflavones and anthocyanin. Flavonoids have many functions like acting as signal molecules in plant-microbe interactions; protection against UV light; flowers, fruits and seed pigmentation; role in plant fertility; germination of pollen; and defence against pathogenic microorganisms, insects and animals (phytopathogens) (Olsen et al. 2010). They are considered to be the most bioactive plant secondary metabolites and can perform better than certain well-known antioxidants, like ASH and α-tocopherol (Hernandez et al. 2009). They prevent cell damage in plants under environmental stress by locating and neutralizing radicals (Løvdal et al. 2010). Flavonoids inhibit lipid peroxidation and lipoxygenases and are functionally dependent on the number and alignment of their hydroxyl groups attached to ring structures (Mahapatra et al. 2013). The antioxidant behaviour of flavonoids gets influenced by the reduction potentials and the accessibility of the radicals. Plants that produce flavonoids and other phenolic compounds develop a tolerance to high UV irradiation as these compounds can absorb UV light (Cle et al. 2008). Stress conditions may induce various flavonoid biosynthetic genes. The biotic and abiotic stresses such as wounding, drought, metal toxicity and nutrient deprivation can increase the levels of flavonoids in the plants as a part of their defence strategy (Winkel-Shirley 2002). Phenolic compounds also act as antioxidants because of their ability to act as reducing agents, hydrogen donors and singlet oxygen quenchers. They are classified into two groups – polyphenols and simple phenols (Prakash et al. 2001). They are generally known to be present in both edible and non-edible plants. Phenolic compounds are also used as metal chelators (Oboh and Rocha 2007).

1.3.6.7 Secondary Antioxidants

Secondary antioxidants are preventives that lower the rate of chain initiation through various mechanisms. They slow down the rate of auto-oxidation of lipids by various processes like decomposing hydro peroxides to non-radical products, scavenging oxygen and binding metal ions (Gordon 1990). Primary antioxidants are generated by secondary antioxidants by donating electron or hydrogen to primary antioxidant radicals. Chelating agents remove pro-oxidant metals and prevent metal-catalysed oxidation by stabilizing the oxide form of metals which have reduced redox potential. Metal chelating is an example of secondary antioxidant mechanism (Madhavi et al. 1995).

1.4 Conclusion

Oxidative stress is caused due to overproduction of ROS generated by different types of abiotic stresses in plants. These ROS or free radicals, dwelling in either extracellular or intracellular environment, impose their detrimental effects to the cells. The cells have antioxidant defence mechanisms which prevent oxidative damage to proteins, lipids and nucleic acids. The antioxidant defence mechanisms could be enzymatic such as superoxide dismutase, glutathione peroxidise, catalase and glutathione reductase or non-enzymatic including proline, a-tocopherols, glutathione, carotenoids and flavonoids. The ROS, though being key regulatory molecules, can harm cells when generated in abundance or when the antioxidant defence system is not working properly. The free radicals have the ability to interact with each other and with antioxidant systems. ROS, as previously explained in pathogenesis, is known to play a dual role in its mechanism of action. Plant cell and organelles have cellular antioxidant machinery for controlling the concentration of ROS in the cell. ROS-scavenging enzymes like catalase, peroxidases, superoxide dismutase, glutathione reductases, etc. are useful in helping plants develop tolerance against various abiotic stresses. Therefore, to tolerate extreme environmental conditions, plants with higher ability to scavenge and/or control the level of cellular ROS may be useful in future.

Acknowledgment Authors are grateful to the University of Delhi for providing financial support under the DU Innovation Project SVC 103/2012. We also acknowledge Dr. (Mrs.) P. Hemalatha Reddy, Principal, Sri Venkateswara College, for providing the institutional support.

Conflict of Interest Authors have no confliction.

References

Agarwal S (2007) Increased antioxidant activity in cassia seedlings under UV-B radiation. Biol Plant 51:157–160. doi:10.1007/s10535-007-0030-z

Ahamadi SAK, Ebadi A, Jahanbakhsh S et al (2015) Changes in enzymatic and nonenzymatic antioxidant defense mechanisms of canola seedlings at different drought stress and nitrogen levels. Turkish J Agric For 39:601–612

Ahmad P, John R, Sarwat M, Umar S (2008) Responses of proline, lipid peroxidation and antioxidative enzymes in two varieties of *Pisum sativum* L. under salt stress. Int J Plant Prod 2:353–365

Ahmad N, Malagoli M, Wirtz M, Hell R (2016) Drought stress in maize causes differential acclimation responses of glutathione and sulfur metabolism in leaves and roots. BMC Plant Biol 16:247

Alia P, Sardhi P (1991) Proline accumulation under heavy metal stress. J Plant Physiol 138:554–558

Alscher RG, Erturk N, Heath LS (2002) Role of superoxide dismutases (SODs) in controlling oxidative stress in plants. J Exp Bot 53:1331–1341

Apel K, Hirt H (2004) Reactive oxygen species: metabolism, oxidative stress, and signal transduction. Annu Rev Plant Biol 55:373–399. doi:10.1146/annurev.arplant.55.031903.141701

Aravind P, Prasad MNV (2003) Zinc alleviates cadmium-induced oxidative stress in *Ceratophyllum demersum* L.: a free floating freshwater macrophyte. Plant Physiol Biochem 41:391–397. doi:10.1016/S0981-9428(03)00035-4

Asada K (1999) The water-water cycle in chloroplasts: scavenging of active oxygens and dissipation of excess photons. Annu Rev Plant Physiol Plant Mol Biol 50:601–639. doi:10.1146/annurev.arplant.50.1.601

Ashraf M, Foolad M (2007) Roles of glycinebetaine and proline in improving plant abiotic stress resistance. Environ Exp Bot 59:206–211

Bhattacharjee S (2005) Reactive oxygen species and oxidative burst: roles in stress, senescence and signal transduction in plants. Curr Sci 89:1113–1121. doi:10.1080/15216540252774694

Borek C (1991) Antioxidants and cancer, science and medicine: the baby-boomer's guide. New Canaan Connect keats Publ 4:51–61

Bose J, Rodrigo-Moreno A, Shabala S (2013) ROS homeostasis in halophytes in the context of salinity stress tolerance. J Exp Bot ert 430

Boveris A, Sies H, Martino E et al (1980) Deficient metabolic utilization of hydrogen peroxide in Trypanosoma cruzi. Biochem J 188:643–648

Briviba K, Klotz LO, Sies H (1997) Toxic and signaling effects of photochemically or chemically generated singlet oxygen in biological systems. Biol Chem 378:1259–1265

Chakraborty K, Singh AL, Kalariya KA et al (2015) Physiological responses of peanut (Arachis hypogaea L.) cultivars to water deficit stress: status of oxidative stress and antioxidant enzyme activities. Acta Bot Croat 74:123–142

Chen Z, Gallie DR (2005) Increasing tolerance to ozone by elevating foliar ascorbic acid confers greater protection against ozone than increasing avoidance. Plant Physiol 138:1673–1689. doi:10.1104/pp.105.062000

Cho U, Seo N (2005) Oxidative stress in Arabidopsis thaliana exposed to cadmium is due to hydrogen peroxide accumulation. Plant Sci 168:113–120

Cle C, Hill LM, Niggeweg R et al (2008) Modulation of chlorogenic acid biosynthesis in *Solanum lycopersicum*; consequences for phenolic accumulation and UV-tolerance. Phytochemistry 69:2149–2156. doi:10.1016/j.phytochem.2008.04.024

Collins A (2001) Carotenoids and genomic stability. Mutat Res 475:21–28

Corpas F, Barroso J, del Rio L (2001) Peroxisomes as a source of reactive oxygen species and nitric oxide signal molecules in plant cells. Trends Plant Sci 6:145–150

Corpas FJ, Fernandez-Ocana A, Carreras A et al (2006) The expression of different superoxide dismutase forms is cell-type dependent in olive (*Olea europaea* L.) leaves. Plant Cell Physiol 47:984–994. doi:10.1093/pcp/pcj071

Creissen GP, Broadbent P, Kular B et al (1994) Manipulation of glutathione reductase in transgenic plants: implications for plant responses to environmental stress. Proc R Soc Edinburgh 102B:167–175

Cunha JR, Neto MCL, Carvalho FEL et al (2016) Salinity and osmotic stress trigger different antioxidant responses related to cytosolic ascorbate peroxidase knockdown in rice roots. Environ Exp Bot 131:58–67

Dabrowska G, Kata A, Goc A et al (2007) Characteristics of the plant ascorbate. Acta Biol Cracov Ser Bot 49:7–17

Das K, Roychoudhury A (2014) Reactive oxygen species (ROS) and response of antioxidants as ROS-scavengers during environmental stress in plants. Front Environ Sci 2:53

Deisseroth A, Dounce A (1970) Catalase: physical and chemical properties, mechanism of catalysis, and physiological role. Physiol Rev 50:319–375

Denness L, McKenna JF, Segonzac C et al (2011) Cell wall damage-induced lignin biosynthesis is regulated by a reactive oxygen species-and jasmonic acid-dependent process in Arabidopsis. Plant Physiol 156:1364–1374

Dhindsa RS, Matowe W (1981) Drought tolerance in two mosses: correlated with enzymatic defence against lipid peroxidation. J Exp Bot 32:79–91

Dixit V, Pandey V, Shyam R (2001) Differential antioxidative responses to cadmium in roots and leaves of pea (*Pisum sativum* L. cv. Azad). J Exp Bot 52:1101–1109

Dixon DP, Skipsey M, Edwards R (2010) Roles for glutathione transferases in plant secondary metabolism. Phytochemistry 71:338–350. doi:10.1016/j.phytochem.2009.12.012

Duarte B, Santos D, Marques JC, Cacador I (2013) Ecophysiological adaptations of two halophytes to salt stress: photosynthesis, PS II photochemistry and anti-oxidant feedback – implications for resilience in climate change. Plant Physiol Biochem PPB 67:178–188. doi:10.1016/j.plaphy.2013.03.004

Edwards EA, Rawsthorne S, Mullineaux PM (1990) Subcellular distribution of multiple forms of glutathione reductase in leaves of pea (*Pisum sativum* L.) Planta 180:278–284. doi:10.1007/BF00194008

Edwards R, Dixon DP, Walbot V (2000) Plant glutathione S-transferases: enzymes with multiple functions in sickness and in health. Trends Plant Sci 5:193–198. doi:10.1016/S1360-1385(00)01601-0

Eltayeb AE, Kawano N, Badawi GH et al (2006) Enhanced tolerance to ozone and drought stresses in transgenic tobacco overexpressing dehydroascorbate reductase in cytosol. Physiol Plant 127:57–65. doi:10.1111/j.1399-3054.2006.00624.x

Eltayeb AE, Kawano N, Badawi GH et al (2007) Overexpression of monodehydroascorbate reductase in transgenic tobacco confers enhanced tolerance to ozone, salt and polyethylene glycol stresses. Planta 225:1255–1264. doi:10.1007/s00425-006-0417-7

Eyidogan F, Öz MT (2007) Effect of salinity on antioxidant responses of chickpea seedlings. Acta Physiol Plant 29:485–493. doi:10.1007/s11738-007-0059-9

Farnese FS, Menezes-Silva PE, Gusman GS, Oliveira JA (2016) When bad guys become good ones: the key role of reactive oxygen species and nitric oxide in the plant responses to abiotic stress. Front Plant Sci 7:471. doi:10.3389/fpls.2016.00471

Feigl G, Lehotai N, Molnár Á et al (2015) Zinc induces distinct changes in the metabolism of reactive oxygen and nitrogen species (ROS and RNS) in the roots of two Brassica species with different sensitivity to zinc stress. Ann Bot 116:613–625. doi:10.1093/aob/mcu246

Foyer CH, Halliwell B (1976) The presence of glutathione and glutathione reductase in chloroplasts: a proposed role in ascorbic acid metabolism. Planta 133:21–25. doi:10.1007/BF00386001

Foyer CH, Noctor G (2005) Redox homeostasis and antioxidant signaling: a metabolic interface between stress perception and physiological responses. Plant Cell 17:1866–1875. doi:10.1105/tpc.105.033589

Gao J, Xiao Q, Ding L, et al (2008) Differential responses of lipid peroxidation and antioxidants in Alternanthera philoxeroides and Oryza sativa subjected to drought stress. 56:89. Plant Growth Regul. doi:10.1007/s10725-008-9291-6J

Gapińska M, Skłodowska M, Gabara B (2007) Effect of short- and long-term salinity on the activities of antioxidative enzymes and lipid peroxidation in tomato roots. Acta Physiol Plant 30:11. doi:10.1007/s11738-007-0072-z

Gill SS, Khan NA, Anjum NA, Tuteja N (2011) Amelioration of cadmium stress in crop plants by nutrient management: morphological. Physiol Biochem Asp Plant Stress:1–23

Gordon MH (1990) The mechanism of antioxidant action in vitro. In: Food antioxidants. Springer, pp 1–18

Gratão PL, Polle A, Lea PJ, Azevedo RA (2005) Making the life of heavy metal-stressed plants a little easier. Funct Plant Biol 32:481–494

Guo T, Zhang G, Zhou M et al (2004) Effects of aluminum and cadmium toxicity on growth and antioxidant enzyme activities of two barley genotypes with different Al resistance. Plant Soil 258:241–248

Hamada AbdElgawad GZ, Hegab MM, Pandey R, et al (2016) High salinity induces different oxidative stress and antioxidant responses in maize seedlings organs

Harinasut P, Poonsopa D, Roengmongkol K et al (2003) Salinity effects on antioxidant enzymes in mulberry cultivar. ScienceAsia 29:109–113

Hasan SA, Hayat S, Ali B, Ahmad A (2008) 28-Homobrassinolide protects chickpea (*Cicer arietinum*) from cadmium toxicity by stimulating antioxidants. Environ Pollut 151:60–66

Hernandez I, Alegre L, Van Breusegem F, Munne-Bosch S (2009) How relevant are flavonoids as antioxidants in plants? Trends Plant Sci 14:125–132. doi:10.1016/j.tplants.2008.12.003

Hodges DM, Forney CF (2000) The effects of ethylene, depressed oxygen and elevated carbon dioxide on antioxidant profiles of senescing spinach leaves. J Exp Bot 51:645–655

Hsu YT, Kao CH (2004) Cadmium toxicity is reduced by nitric oxide in rice leaves. Plant Growth Regul 42:227–238

Hsu YT, Kao CH (2007) Heat shock-mediated H2O2 accumulation and protection against Cd toxicity in rice seedlings. Plant Soil 300:137–147

Hurrell RF (2003) Influence of vegetable protein sources on trace element and mineral bioavailability. J Nutr 133:2973S–2977S

Inzé D, Van Montagu M (1995) Oxidative stress in plants. Curr Opin Biotechnol 6:153–158

Jiménez A, Hernández JA, Pastori G et al (1998) Role of the ascorbate-glutathione cycle of mitochondria and peroxisomes in the senescence of pea leaves. Plant Physiol 118:1327–1335

Jogeswar G, Pallela R, Jakka NM et al (2006) Antioxidative response in different sorghum species under short-term salinity stress. Acta Physiol Plant 28:465–475

Kayang H (2007) Tribal knowledge on wild edible plants of Meghalaya, Northeast India. Indian J Tradit Knowl 6:177–181

Khan A, Ashraf M (2008) Exogenously applied ascorbic acid alleviates salt-induced oxidative stress in wheat. Environ Exp Bot 63:224–231

Khan MIR, Khan NA (2014) Ethylene reverses photosynthetic inhibition by nickel and zinc in mustard through changes in PS II activity, photosynthetic-nitrogen use efficiency and antioxidant metabolism. Protoplasma 251:1007–1019

Khan NA, Singh S, Nazar R (2007) Activities of antioxidative enzymes, sulphur assimilation, photosynthetic activity and growth of wheat (*Triticum aestivum*) cultivars differing in yield potential under cadmium stress. J Agron Crop Sci 193:435–444

Khan MIR, Asgher M, Khan NA (2014) Alleviation of salt-induced photosynthesis and growth inhibition by salicylic acid involves glycinebetaine and ethylene in mungbean (*Vigna radiata* L.) Plant Physiol Biochem 80:67–74

Khan MIR, Nazir F, Asgher M, Per TS, Khan NA (2015) Selenium and sulfur influence ethylene formation and alleviate cadmium-induced oxidative stress by improving proline and glutathione production in wheat. J Plant Physiol 178:9–18

Khan MIR, Iqbal N, Masood A, Mobin M, Anjum NA, Khan NA (2016a) Modulation and significance of nitrogen and sulfur metabolism in cadmium challenged plants. Plant Growth Regul 78:1–11

Khan MIR, Khan NA, Masood A, Per TS, Asgher M (2016b) Hydrogen peroxide alleviates nickel-inhibited photosynthetic responses through increase in use-efficiency of nitrogen and sulfur, and glutathione production in mustard. Front Plant Sci 7:44

Koussevitzky S, Suzuki N, Huntington S et al (2008) Ascorbate peroxidase 1 plays a key role in the response of Arabidopsis thaliana to stress combination. J Biol Chem 283:34197–34203. doi:10.1074/jbc.M806337200

Kukreja S, Nandwal AS, Kumar N et al (2005) Plant water status, H2O2 scavenging enzymes, ethylene evolution and membrane integrity of *Cicer arietinum* roots as affected by salinity. Biol Plant 49:305–308

Kundur R, Reddy PT, Rao MD (2016) Effect of PEG mediated water stress on solute accumulation, relative water content, biomass and antioxidant enzymes in rice. Indian J Agric Res 50:398–405

Larson RA (1988) The antioxidants of higher plants. Phytochemistry 27:969–978

León AM, Palma JM, Corpas FJ et al (2002) Antioxidative enzymes in cultivars of pepper plants with different sensitivity to cadmium. Plant Physiol Biochem 40:813–820

Løvdal T, Olsen KM, Slimestad R et al (2010) Synergetic effects of nitrogen depletion, temperature, and light on the content of phenolic compounds and gene expression in leaves of tomato. Phytochemistry 71:605–613

Luis A (2015) ROS and RNS in plant physiology: an overview. J Exp Bot 66:2827–2837

Madhavi DL, Singhal RS, Kulkarni PR (1995) Technological aspects of food antioxidants. Food Sci Technol York-Marcel Dekker:159–266

Mahajan S, Tuteja N (2005) Cold, salinity and drought stresses: an overview. Arch Biochem Biophys 444:139–158

Mahapatra S, Mohanta YK, Panda S (2013) Methods to study antioxidant properties with special reference to medicinal plants. Int J Pharm North Orissa Univ Baripada-757003 Odisha, India 3:91–97

Manach C, Morand C, Crespy V et al (1998) Quercetin is recovered in human plasma as conjugated derivatives which retain antioxidant properties. FEBS Lett 426:331–336

Mathews MC, Summers CB, Felton GW (1997) Ascorbate peroxidase: a novel antioxidant enzyme in insects. Arch Insect Biochem Physiol 34:57–68

Mehlhorn H, Lelandais M, Korth HG, Foyer CH (1996) Ascorbate is the natural substrate for plant peroxidases. FEBS Lett 378:203–206

Meyer AJ (2008) The integration of glutathione homeostasis and redox signaling. J Plant Physiol 165:1390–1403

Millar AH, Mittova V, Kiddle G et al (2003) Control of ascorbate synthesis by respiration and its implications for stress responses. Plant Physiol 133:443–447

Miller G, Suzuki N, Ciftci-Yilmaz S, Ron M (2010) Reactive oxygen species homeostasis and signalling during drought and salinity stresses. Plant Cell Environ 33:453–467

Milone MT, Sgherri C, Clijsters H, Navari-Izzo F (2003) Antioxidative responses of wheat treated with realistic concentration of cadmium. Environ Exp Bot 50:265–276

Mir BA, Mir SA, Khazir J et al (2015) Cold stress affects antioxidative response and accumulation of medicinally important withanolides in *Withania somnifera* (L.) Dunal. Ind Crop Prod 74:1008–1016

Mittler R (2002) Oxidative stress, antioxidants and stress tolerance. Trends Plant Sci 7:405–410

Mittler R, Zilinskas BA (1992) Molecular cloning and characterization of a gene encoding pea cytosolic ascorbate peroxidase. J Biol Chem 267:21802–21807

Mittler R, Vanderauwera S, Gollery M, Van Breusegem F (2004) Reactive oxygen gene network of plants. Trends Plant Sci 9:490–498

Miyake C, Asada K (1996) Inactivation mechanism of ascorbate peroxidase at low concentrations of ascorbate; hydrogen peroxide decomposes compound I of ascorbate peroxidase. Plant Cell Physiol 37:423–430

Miyake C, Michihata F, Asada K (1991) Scavenging of hydrogen peroxide in prokaryotic and eukaryotic algae: acquisition of ascorbate peroxidase during the evolution of cyanobacteria. Plant Cell Physiol 32:33–43

Mobin M, Khan NA (2007) Photosynthetic activity, pigment composition and antioxidative response of two mustard (Brassica juncea) cultivars differing in photosynthetic capacity subjected to cadmium stress. J Plant Physiol 164:601–610

Moons A (2003) Osgstu3 and osgtu4, encoding tau class glutathione S-transferases, are heavy metal- and hypoxic stress-induced and differentially salt stress-responsive in rice roots. FEBS Lett 553:427–432. doi:10.1016/S0014-5793(03)01077-9

Mullineaux PM, Rausch T (2005) Glutathione, photosynthesis and the redox regulation of stress-responsive gene expression. Photosynth Res 86:459–474

Navrot N, Rouhier N, Gelhaye E, Jacquot J (2007) Reactive oxygen species generation and antioxidant systems in plant mitochondria. Physiol Plant 129:185–195

Niyogi KK, Shih C, Chow WS et al (2001) Photoprotection in a zeaxanthin-and lutein-deficient double mutant of Arabidopsis. Photosynth Res 67:139–145

Noctor G, Foyer CH (1998) A re-evaluation of the ATP: NADPH budget during C3 photosynthesis: a contribution from nitrate assimilation and its associated respiratory activity? J Exp Bot 49:1895–1908

Noctor G, Gomez L, Vanacker H, Foyer CH (2002) Interactions between biosynthesis, compartmentation and transport in the control of glutathione homeostasis and signalling. J Exp Bot 53:1283–1304

Oboh G, Rocha JBT (2007) Polyphenols in red pepper [*Capsicum annuum var. aviculare* (Tepin)] and their protective effect on some pro-oxidants induced lipid peroxidation in brain and liver. Eur Food Res Technol 225:239–247

Olsen KM, Hehn A, Jugdé H et al (2010) Identification and characterisation of CYP75A31, a new flavonoid 3′5 ′-hydroxylase, isolated from *Solanum lycopersicum*. BMC Plant Biol 10:21. doi:10.1186/1471-2229-10-21

Pan Y, Wu LJ, Yu ZL (2006) Effect of salt and drought stress on antioxidant enzymes activities and SOD isoenzymes of liquorice (Glycyrrhiza uralensis Fisch). Plant Growth Regul 49:157–165

Panda SK (2001) Oxidative response of green gram seeds under salinity stress. Indian J Plant Physiol 6:438–440

Peeters-Joris C, Vandevoorde A-M, Baudhuin P (1975) Subcellular localization of superoxide dismutase in rat liver. Biochem J 150:31–39

Pekker I, Tel-Or E, Mittler R (2002) Reactive oxygen intermediates and glutathione regulate the expression of cytosolic ascorbate peroxidase during iron-mediated oxidative stress in bean. Plant Mol Biol 49:429–438

Prakash A, Rigelhof F, MIller E (2001) Antioxidant activity. Medallion Lab Anal Prog 19:1–4

Radotić K, Dučić T, Mutavdžić D (2000) Changes in peroxidase activity and isoenzymes in spruce needles after exposure to different concentrations of cadmium. Environ Exp Bot 44:105–113

Rao ASVC, Reddy AR (2008) Glutathione reductase: a putative redox regulatory system in plant cells. In: Khan NA, Singh S, Umar S (eds) Sulfur assimilation and abiotic stress in plants. Springer, Berlin, pp 111–147

Rausch T, Wachter A (2005) Sulfur metabolism: a versatile platform for launching defence operations. Trends Plant Sci 10:503–509

ROS (1997) Reactive oxygen species (ROS): R&D systems. https://www.rndsystems.com/resources/articles/reactive-oxygen-species-ros

Sánchez-Rodríguez E, Romero L, Ruiz JM (2016) Accumulation of free polyamines enhances the antioxidant response in fruits of grafted tomato plants under water stress. J Plant Physiol 190:72–78. http://dx.doi.org/10.1016/j.jplph.2015.10.010

Sano S, Ueda M, Kitajima S et al (2001) Characterization of ascorbate peroxidases from unicellular red alga Galdieria partita. Plant Cell Physiol 42:433–440

Sathishkumar R, Lakshmi PTV, Annamalai A (2010) Comparative analyses of non enzymatic and enzymatic antioxidants of Enicostemma littorale blume. Int J Pharm Bio Sci 1:1–16

Schützendübel A, Schwanz P, Teichmann T et al (2001) Cadmium-induced changes in antioxidative systems, hydrogen peroxide content, and differentiation in Scots pine roots. Plant Physiol 127:887–898

Shankar V, Kumar D, Agrawal V (2016) Assessment of antioxidant enzyme activity and mineral nutrients in response to NaCl stress and its amelioration through glutathione in chickpea. Appl Biochem Biotechnol 178:267–284

Sharma P, Dubey RS (2004) Ascorbate peroxidase from rice seedlings: properties of enzyme isoforms, effects of stresses and protective roles of osmolytes. Plant Sci 167:541–550

Sharma P, Dubey RS (2005) Modulation of nitrate reductase activity in rice seedlings under aluminium toxicity and water stress: role of osmolytes as enzyme protectant. J Plant Physiol 162:854–864

Sharma P, Jha AB, Dubey RS, Pessarakli M (2012) Reactive oxygen species, oxidative damage, and antioxidative defense mechanism in plants under stressful conditions. J Bot:1–26. doi:10.1155/2012/217037

Shigeoka S, Nakano Y, Kitaoka S (1980a) Metabolism of hydrogen peroxide in *Euglena gracilis* Z by L-ascorbic acid peroxidase. Biochem J 186:377–380

Shigeoka S, Nakano Y, Kitaoka S (1980b) Purification and some properties of L-ascorbic acid-specific peroxidase in *Euglena gracilis* Z. Arch Biochem Biophys 201:121–127

Shigeoka S, Ishikawa T, Tamoi M et al (2002) Regulation and function of ascorbate peroxidase isoenzymes. J Exp Bot 53:1305–1319

Siefermann-Harms D (1987) The light-harvesting and protective functions of carotenoids in photosynthetic membranes. Physiol Plant 69:561–568. doi:10.1111/j.1399-3054.1987.tb09240.x

Sies H (1997) Oxidative stress: oxidants and antioxidants. Exp Physiol 82:291–295

Silveira NM, de Oliveira JA, Ribeiro C et al (2015) Nitric oxide attenuates oxidative stress induced by arsenic in lettuce (*Lactuca sativa*) leaves. Water Air Soil Pollut 226:1–9

Šimonovičová M, Tamás L, Huttová J, Mistrík I (2004) Effect of aluminium on oxidative stress related enzymes activities in barley roots. Biol Plant 48:261–266

Simova-Stoilova L, Vaseva I, Grigorova B et al (2010) Proteolytic activity and cysteine protease expression in wheat leaves under severe soil drought and recovery. Plant Physiol Biochem 48:200–206

Singh S, Khan NA, Nazar R, Anjum NA (2008) Photosynthetic traits and activities of antioxidant enzymes in blackgram (Vigna mungo L. Hepper) under cadmium stress. Am J Plant Physiol 3:25–32

Skórzyńska-Polit E, Dra M, Krupa Z (2003) The activity of the antioxidative system in cadmium-treated Arabidopsis thaliana. Biol Plant 47:71–78

Smirnoff N (2000) Ascorbic acid: metabolism and functions of a multi-facetted molecule. Curr Opin Plant Biol 3:229–235

Smirnoff N, Cumbes QJ (1989) Hydroxyl radical scavenging activity of compatible solutes. Phytochemistry 28:1057–1060

Srivastava S, Dubey RS (2011) Manganese-excess induces oxidative stress, lowers the pool of antioxidants and elevates activities of key antioxidative enzymes in rice seedlings. Plant Growth Regul 64:1–16

Srivastava AK, Bhargava P, Rai LC (2005) Salinity and copper-induced oxidative damage and changes in the antioxidative defence systems of Anabaena doliolum. World J Microbiol Biotechnol 21:1291–1298

Stevens R, Page D, Gouble B et al (2008) Tomato fruit ascorbic acid content is linked with monodehydroascorbate reductase activity and tolerance to chilling stress. Plant Cell Environ 31:1086–1096

Takeda T, Shigeoka S, Hirayama O, Mitsunaga T (1992) The presence of enzymes related to glutathione metabolism and oxygen metabolism in *Chlamydomonas reinhardtii*. Biosci Biotechnol Biochem 56:1662–1663

Takeda T, Yoshimura K, Ishikawa T, Shigeoka S (1998) Purification and characterization of ascorbate peroxidase in *Chlorella vulgaris*. Biochimie 80:295–301

Takeda T, Yoshimura K, Yoshii M et al (2000) Molecular characterization and physiological role of ascorbate peroxidase from halotolerant Chlamydomonas sp. W80 strain. Arch Biochem Biophys 376:82–90

Tausz M, Šircelj H, Grill D (2004) The glutathione system as a stress marker in plant ecophysiology: is a stress-response concept valid? J Exp Bot 55:1955–1962

Thao NP, Khan MIR, Thu NBA et al (2015) Role of ethylene and its cross talk with other signaling molecules in plant responses to heavy metal stress. Plant Physiol 169:73–84

Timofeyev MA, Steinberg CEW (2006) Antioxidant response to natural organic matter (NOM) exposure in three Baikalean amphipod species from contrasting habitats. Comp Biochem Physiol Part B Biochem Mol Biol 145:197–203

Trovato M, Mattioli R, Costantino P (2008) Multiple roles of proline in plant stress tolerance and development. Rend Lincei 19:325–346

Tuteja N (2010) In: Hirt H (ed) Cold, salt and drought stress. Plant stress biology: from genomics towards system biology. Wiley-Blackwell, Weinheim, pp 137–159

Tyler DD (1975) Polarographic assay and intracellular distribution of superoxide dismutase in rat liver. Biochem J 147:493–504

Ushimaru T, Nakagawa T, Fujioka Y et al (2006) Transgenic Arabidopsis plants expressing the rice dehydroascorbate reductase gene are resistant to salt stress. J Plant Physiol 163:1179–1184

Viehweger K (2014) How plants cope with heavy metals. Bot Stud 55:1–12

Vierstra RD, John TR, Poff KL (1982) Kaempferol 3-o-galactoside, 7-o-rhamnoside is the major green fluorescing compound in the epidermis of *Vicia faba*. Plant Physiol 69:522–525

Villafranca JJ, Yost FJ, Fridovich I (1974) Magnetic resonance studies of manganese (III) and iron (III) superoxide dismutases temperature and frequency dependence of proton relaxation rates of water. J Biol Chem 249:3532–3536

Wada N, Kinoshita S, Matsuo M et al (1998) Purification and molecular properties of ascorbate peroxidase from bovine eye. Biochem Biophys Res Commun 242:256–261

Wang C-Q, Zhang Y-F, Zhang Y-B (2008a) Scavenger enzyme activities in subcellular fractions of white clover (*Trifolium repens* L.) under PEG-induced water stress. J Plant Growth Regul 27:387–393

Wang L, Zhou Q, Ding L, Sun Y (2008b) Effect of cadmium toxicity on nitrogen metabolism in leaves of *Solanum nigrum* L. as a newly found cadmium hyperaccumulator. J Hazard Mater 154:818–825

Wang L, Su H, Han L et al (2014) Differential expression profiles of poplar MAP kinase kinases in response to abiotic stresses and plant hormones, and overexpression of PtMKK4 improves the drought tolerance of poplar. Gene 545:141–148. doi:10.1016/j.gene.2014.04.058

Winkel-Shirley B (2002) Biosynthesis of flavonoids and effects of stress. Curr Opin Plant Biol 5:218–223

Xiang C, Werner BL, E'Lise MC, Oliver DJ (2001) The biological functions of glutathione revisited in Arabidopsis transgenic plants with altered glutathione levels. Plant Physiol 126:564–574

Yamane K, Mitsuya S, Kawasaki M et al (2009) Antioxidant capacity and damages caused by salinity stress in apical and basal regions of rice leaf. Plant Prod Sci 12:319–326

Yang Y, Han C, Liu Q et al (2008) Effect of drought and low light on growth and enzymatic antioxidant system of Picea asperata seedlings. Acta Physiol Plant 30:433–440

Yin L, Wang S, Eltayeb AE et al (2010) Overexpression of dehydroascorbate reductase, but not monodehydroascorbate reductase, confers tolerance to aluminum stress in transgenic tobacco. Planta 231:609–621

Zheng J-L, Zhao L-Y, Shen B et al (2016) Effects of salinity on activity and expression of enzymes involved in ionic, osmotic, and antioxidant responses in Eurya emarginata. Acta Physiol Plant 38:1–9

Zlatev ZS, Lidon FC, Ramalho JC, Yordanov IT (2006) Comparison of resistance to drought of three bean cultivars. Biol Plant 50:389–394

An Introduction to Reactive Oxygen Species Metabolism Under Changing Climate in Plants

2

Mudasir Irfan Dar, Mohd Irfan Naikoo, Fareed Ahmad Khan, Farha Rehman, Iain D. Green, Fauzia Naushin, and Abid Ali Ansari

Abstract

Plants are subjected to various kinds of abiotic stresses throughout their life cycles which include salinity, drought, temperature (heat and cold), heavy metal, nutrient deficiency and UV radiation. Such types of stresses in plants lead the overproduction of certain chemical entities known as reactive oxygen species (ROS) which are highly reactive and cause damage to proteins, lipids, carbohydrates and macromolecules. The ROS include hydrogen peroxide (H_2O_2), superoxide radical ($O_2{\bullet}^-$), hydroxyl radical ($OH^{\bullet}$), perhydroxyl radical ($HO_2^{\bullet}$), etc. But, nature has gifted the plants with very efficient enzymatic (superoxide dismutase, SOD; catalase, CAT; ascorbate peroxidase, APX; glutathione reductase, GR; monodehydroascorbate reductase, MDHAR; dehydroascorbate reductase, DHAR; glutathione peroxidase, GPX; guaiacol peroxidase, GOPX;

M.I. Dar (✉) • M.I. Naikoo • F.A. Khan
Environmental Botany Section, Department of Botany, Aligarh Muslim University, Aligarh, Uttar Pradesh 202002, India
e-mail: irfanmudasir@gmail.com

F. Rehman
Department of Botany, Faculty of Science, Mohammad Ali Jauhar University, Rampur, Uttar Pradesh 24490, India

I.D. Green
Department of Life and Environmental Science, The Faculty of Science and Technology, Bournemouth University, Talbot Campus, Poole, Dorset BH12 5BB, UK

F. Naushin
Department of Botany, Women's College, Aligarh Muslim University, Aligarh, Uttar Pradesh 202002, India

A.A. Ansari
Department of Biology, Faculty of Science, University of Tabuk, Tabuk, Kingdom of Saudi Arabia

M.I.R. Khan, N.A. Khan (eds.), *Reactive Oxygen Species and Antioxidant Systems in Plants: Role and Regulation under Abiotic Stress*,
DOI 10.1007/978-981-10-5254-5_2

and glutathione-S-transferase, GST) and non-enzymatic (ascorbic acid, AA; glutathione, GSH; tocopherols; flavonoids; proline; etc.) antioxidant defence systems which work in coordination to detoxify and protect plant cells from oxidative damage. Moreover, ROS play a role as secondary messenger which induces cell to undergo programmed cell death.

Keywords
Antioxidants • Biomolecules • Climate change • Plants • ROS

2.1 Introduction

Molecular oxygen was brought into our atmosphere about 2.7 billion years ago by the O_2 evolving photosynthetic organisms, and reactive oxygen species (ROS) have been the unavoidable byproducts of aerobic metabolism (Halliwell 2006). The ROS are highly reactive and are restrained to the different cellular compartments of plant, such as chloroplasts, mitochondria and peroxisomes (del Rio et al. 2006; Navrot et al. 2007). Apoplast has also been reported as a site for ROS production (Jubany-Mari et al. 2009; Roychoudhury and Basu 2012). Under normal environmental conditions, ROS molecules are incapable to cause any damage as they are constantly being scavenged by a range of antioxidative mechanisms (Foyer and Noctor 2005). But, the delicate equilibrium between the ROS production and their scavenging by antioxidants is disturbed by different biotic and abiotic stress factors such as salinity, water scarcity, UV radiation, heavy metals, temperature extremes, deficiency of nutrients, pesticides and infection by pathogens (Gill and Tuteja 2010). These disturbances lead an increased intracellular production of ROS which may cause substantial damage to the cellular components (Fig. 2.1). It has been reported that about 2% of O_2 consumption by plant cells leads to the ROS generation in these cells (Bhattachrjee 2005).

The ROS include both free radical ($O_2{\bullet}^-$, superoxide radical; $OH^{\bullet}$, hydroxyl radical; $HO_2^{\bullet}$, perhydroxyl radical; and $RO^{\bullet}$, alkoxy radicals) and nonradical (molecular) forms (H_2O_2, hydrogen peroxide, and $^1O_2$, singlet oxygen). These reactive intermediates are very toxic and cause extensive damage to biomolecules like proteins, DNA and lipids and eventually lead to cell death (Apel and Hirt 2004; Foyer and Noctor 2005). During stressful environments, the increased levels of ROS get scavenged by the plant's antioxidative machinery comprising of the enzymatic components that include superoxide dismutase (SOD), ascorbate peroxidase (APX), guaiacol peroxidase (GOPX), glutathione-S-transferase (GST) and catalase (CAT) and the non-enzymatic metabolites such as ascorbic acid (AA), reduced glutathione (GSH), α-tocopherol, carotenoids, phenolics, proline and flavonoids (Gill and Tuteja 2010; Miller et al. 2010; Gill et al. 2011). The presence of the antioxidative compounds in majority of the cellular compartments reveals the importance of ROS detoxification for continuous cellular existence (Gill et al. 2011; Khan and Khan 2014; Khan et al. 2014, 2015; 2016a, b). The present chapter

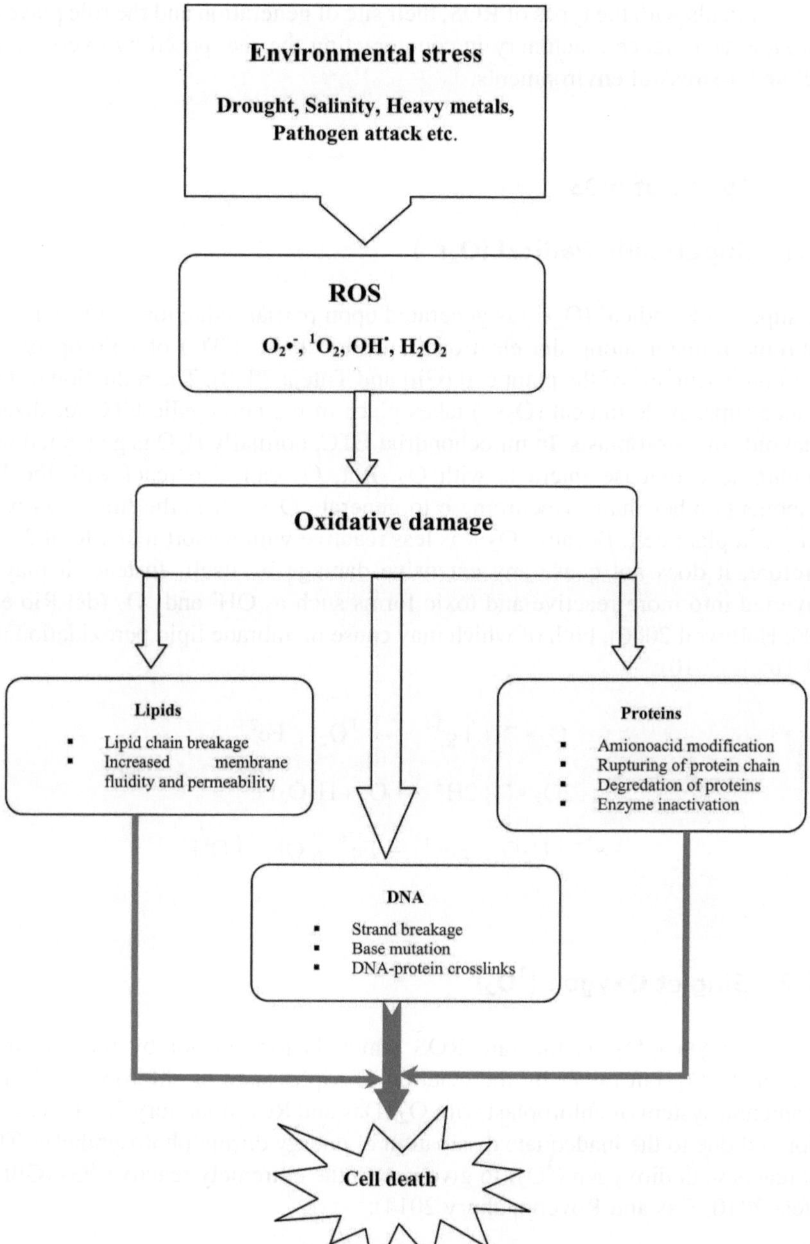

Fig. 2.1 Environmental stress (biotic and abiotic)-induced ROS generation and cell death

primarily deals with the types of ROS, their site of generation and the role played by antioxidative defence machinery in counteracting the risk posed by overproduced ROS under stressful environments.

2.2 Types of ROS

2.2.1 Superoxide Radical ($O_2{\bullet}^-$)

The superoxide radical ($O_2{\bullet}^-$) is generated upon partial reduction of O_2 during the electronic transfer along the electron transport chain (ETC) of chloroplasts and other compartments of the plant cell (Gill and Tuteja 2010). The reduction of O_2 to produce superoxide radical ($O_2{\bullet}^-$) takes place in the non-cyclic ETC localized in thylakoids of chloroplasts. In mitochondrial ETC, normally H_2O is generated when cytochrome c oxidase interacts with O_2. But, O_2 can also react with the ETC components other than cytochrome c to generate $O_2{\bullet}^-$. It is the first ROS which forms in a plant cell. Because $O_2{\bullet}^-$ is less reactive with a short half-life of 2–4 μs, therefore, it does not cause any extensive damage by itself. Instead, it may get converted into more reactive and toxic forms such as $OH^{\bullet}$ and $^1O_2$ (del Rio et al. 2006; Halliwell 2006), both of which may cause membrane lipid peroxidation (Gill and Tuteja 2010):

$$O_2{\bullet}^- + Fe^{3+} \rightarrow {}^1O_2 + Fe^{2+}$$

$$2O_2{\bullet}^- + 2H^+ \rightarrow O_2 + H_2O_2Fe^{3+}$$

$$Fe^{2+} + H_2O_2 + Fe^{3+} \rightarrow Fe^{3+} + OH^- + OH^{\bullet}$$

2.2.2 Singlet Oxygen ($^1O_2$)

Singlet oxygen ($^1O_2$) is the rare ROS which is formed not by the transfer of electrons to O_2, but rather by the reaction of triplet state of chlorophyll (^{3}Chl) in the antenna system of chloroplast with O_2 (Das and Roychoudhury 2014). The ^{3}Chl is formed due to the inadequate dissipation of energy during photosynthesis. This ^{3} Chl reacts with dioxygen ($^3O_2$) to give rise to the extremely reactive $^1O_2$ (Gill and Tuteja 2010; Das and Roychoudhury 2014):

$$Chl \rightarrow {}^3Chl$$

$$^3Chl + {}^3O_2 \rightarrow Chl + {}^1O_2$$

The formation of $^1O_2$ has a potent negative effect on both PSI and PSII and puts the whole photosynthetic machinery under threat (Das and Roychoudhury 2014).

Moreover, different abiotic stresses such as salinity, heavy metals, water scarcity and water logging result into closing of stomata which ultimately lead to lower amounts of CO_2 in the chloroplast, favouring the generation of $^1O_2$ (Gill and Tuteja 2010). Although the lifetime of $^1O_2$ in a cell is only about 3 μs (Hatz et al. 2007), some amounts of it can diffuse up to hundred nanometres (nm) and may damage a wide range of biomolecules such as proteins, pigments, nucleic acids and lipids. Singlet oxygen ($^1O_2$) is believed to be the major ROS involved in light-induced loss of PSII activity, which sets off cell death (Wagner et al. 2004; Krieger-Liszkay et al. 2008).

2.2.3 Hydrogen Peroxide (H_2O_2)

Hydrogen peroxide, a moderately reactive oxygen species, is generated on univalent reduction or protonation of $O_2{\bullet}^-$. It can also occur non-enzymatically by being dismutated to H_2O_2 under low pH conditions or mostly by a reaction catalysed by SOD (Das and Roychoudhury 2014):

$$2O_2{\bullet}^- + 2H^+ \rightarrow H_2O_2 + O_2$$

$$2O_2{\bullet}^- + 2H^+ \overset{SOD}{\rightarrow} H_2O_2 + O_2$$

The main sources of H_2O_2 generation in plant cells are the ETC in the chloroplast, mitochondria, endoplasmic reticulum (ER), cell membranes, β-oxidation of fatty acids and photorespiration. The reactions involving photo-oxidation by NADPH oxidase and xanthine oxidase (XOD) can also produce H_2O_2 (Sharma et al. 2012). The production of H_2O_2 in plant cells doesn't take place only under normal environmental conditions but also when plants are under stress conditions like drought, chilling, intense light, UV radiation, wounding and pathogen infection (Sharma et al. 2012; Das and Roychoudhury 2014). Under stressful conditions, stomata are closed, resulting in low availability of CO_2 for fixation. This will in turn favour the oxygenation of ribulose1,5-bisphosphate (RuBP) and thus the enhancement of photorespiration. It has been reported that above 70% of the H_2O_2 was produced as a result of drought stress (Noctor et al. 2002a, b).

The accumulation of H_2O_2 in plant cells has both good and bad effects. It is advantageous at low concentrations but can be detrimental at higher concentrations in the cell (Gill and Tuteja 2010). At lower amounts, it behaves as a regulatory signalling molecule for crucial physiological processes like senescence (Peng et al. 2005), photorespiration and photosynthesis (Noctor et al. 2002a, b), stomatal movement (Bright et al. 2006), cell cycle, growth and development (Tanou et al. 2009) and also involved in providing tolerance to various stresses (Gill and Tuteja 2010). At high concentrations, H_2O_2 can go over considerable distances and cross membranes through aquaporins due to its longer half-life of about 1 ms (Bienert et al. 2007). At these concentrations, H_2O_2 has the potential to oxidize both cysteine (-SH) and methionine (-SCH_3) residues and also inactivates the Cu/Zn SOD and

Fe-SOD enzymes of Calvin cycle by oxidizing their thiol groups (Halliwell 2006). It has been reported that enzymes like fructose-1,6-bisphosphatase, sedoheptulose-1,7-bisphosphatase and phosphoribulokinase were inactivated at concentrations of 10 μM H_2O_2. H_2O_2 at higher concentrations also causes programmed cell death (Dat et al. 2000). However, like $O_2{\bullet}^-$, H_2O_2 is moderately reactive, and its damage is only recognized when it is transformed into more reactive intermediates (Das and Roychoudhury 2014).

2.2.4 Hydroxyl Radical ($OH^{\bullet}$)

Hydroxyl radical ($OH^{\bullet}$) is the most reactive and the most deadly ROS amongst its family members (Das and Roychoudhury 2014). It is produced by Fenton reaction between H_2O_2 and $O_2{\bullet}^-$ catalysed by transition metals like Fe (Fe^{2+}, Fe^{3+}) at neutral pH:

$$H_2O_2 + O_2{\bullet}^- \rightarrow OH^- + O_2 + OH^{\bullet}$$

It has the potential to react and damage biomolecules like DNA, proteins and lipids (Das and Roychoudhury 2014). There is an absence of any enzymatic system for scavenging this highly toxic ROS; therefore, its increased accumulation leads to cell death (Pinto et al. 2003).

2.3 ROS Generation Sites in Plant Cells

ROS are generated in both normal and stressed cells at several subcellular locations like chloroplasts, mitochondria, ER, plasma membranes, peroxisomes, apoplast and cell walls (Sharma et al. 2012). The production of ROS occurs due to the unavoidable leakage of electrons onto O_2 from the electron transport pathways of chloroplasts, mitochondria and plasma membranes or as a byproduct of various other metabolic processes in different cellular compartments. Chloroplasts and peroxisomes are the major sources of ROS production in presence of light, whilst the mitochondrion is the chief ROS generator under dark conditions (Choudhury et al. 2013).

2.3.1 Chloroplast

The chloroplast comprises of a highly organized system of thylakoids which harbours the efficient light-capturing photosynthetic machinery (Pfannschmidt 2003). The PSI and PSII which form the core of the light-harvesting systems in the thylakoids are the major sources of ROS production (Das and Roychoudhury 2014). Oxygen evolved through photolysis of water in the chloroplasts accepts electrons which pass through the photosystems, resulting in the formation of $O_2^{\bullet -}$.

Thus, the ROS-generating centres like ^{3}Chl and ETC in PSI and PSII render chloroplasts a leading site of ROS production (Gill and Tuteja 2010). Abiotic stress factors such as high light intensity, drought, saline conditions and limiting conditions of CO_2 enhance the production of ROS in chloroplasts (Gill and Tuteja 2010). Generally, the electrons excited in photosystems are accepted by $NADP^+$, which gets reduced to NADPH. This energy in the form of NADPH is then used in Calvin cycle to reduce the final electron acceptor CO_2. But, stressful conditions in plants lead to poor NADP supply, which overload the ETC of photosystems and cause leakage of electron from ferredoxin to O_2, reducing it to $O_2^{\bullet -}$ via Mehler reaction (Elstner 1991):

$$2O_2 + 2Fd_{red} \rightarrow 2O_2^{\bullet -} + 2Fd_{ox}$$

Leakage of electrons also occurs through the two electron acceptors (Q_A and Q_B) of PSII which contribute to generation of $O^{\bullet -}{}_2$ (Cleland and Grace 1999). Once $O_2^{\bullet -}$ is formed, it may either get protonated on the internal 'lumen' membrane surface into more noxious ROS like $HO_2^{\bullet}$ or dismutated spontaneously into H_2O_2 on the external 'stromal' membrane surface by a membrane-bound Cu/Zn-SOD at the PSI (Miller et al. 2010). At Fe-S centres where Fe^{2+} is available, H_2O_2 may be transformed through the Fenton reaction into much more toxic $OH^{\bullet}$ (Sharma et al. 2012; Das and Roychoudhury 2014). Therefore, chloroplasts are major source of ROS production.

2.3.2 Mitochondria

Although partially, mitochondria are also the important sites of generation of harmful ROS, like H_2O_2 and $O_2^{\bullet -}$ (Navrot et al. 2007). The mitochondrial ETC harbours electrons with ample amounts of free energy to reduce O_2 to form ROS, thus becoming the main source of ROS generation (Rhoads et al. 2006). Several sites in the mitochondrial ETC like NADH dehydrogenase (Complex I) and ubiquinone-cytochrome region (Complex III) are involved in the generation of ROS (Moller et al. 2007; Noctor et al. 2007). The Complex I spontaneously reduces O_2 to $O_2^{\bullet -}$ in its flavoprotein region. The ROS production at Complex I further increases when the electrons start to flow in a reverse direction from Complex III to Complex I due to the shortage of NAD^+-linked substrates. This reversal flow of electrons is controlled by ATP hydrolysis (Turrens 2003).

Ubiquinone cytochrome region of Complex III of ETC in its reduced form gives away its electron to cytochrome c_1 leaving an unstable ubisemiquinone which promotes seepage of electrons to O_2 and converts it into $O_2^{\bullet -}$ (Murphy 2009). Various enzymes present in the mitochondrial matrix are also the sources of ROS production. Some enzymes such as aconitase directly generate ROS, whilst others like 1-galactono-γ-lactone dehydrogenase (GAL) indirectly produce ROS by passing electrons to the ETC (Rasmusson et al. 2008). Although $O_2^{\bullet -}$ is the chief ROS formed inside the mitochondria, it is converted to relatively stable H_2O_2 by the

Mn-SOD (a mitochondrial form of SOD) or the APX (Sharma et al. 2012). H_2O_2 can then be converted to extremely active hydroxyl radical ($OH^•$) via the Fenton reaction. These uncharged $OH^•$ have the potential to infiltrate membranes and move out of the mitochondrion (Rhoads et al. 2006). Peroxidation of mitochondrial membranous polyunsaturated fatty acids (PUFAs) is initiated by the abstraction of a hydrogen atom by ROS, especially by $OH^•$. This results in the formation of cytotoxic lipid aldehydes, alkenals and hydroxyl alkenals, notably 4-hydroxy-2-nonenal and malondialdehyde (Gill and Tuteja 2010). Lipid peroxidation products can cause cellular injury by interacting with proteins, other lipids and nucleic acids (Gill and Tuteja 2010). Generally, ROS are produced in mitochondrion under normal conditions, but their production increases to a great extent during abiotic stress conditions (Pastore et al. 2007). Under such harsh environments, the coupling of ETC and ATP synthesis gets affected, resulting in over reduction of electron carriers like ubiquinone (UQ) pool, thereby producing greater amounts of ROS (Rhoads et al. 2006; Blokhina and Fagerstedt 2010). To compensate the lower rate of ATP synthesis in chloroplast during the conditions of water scarcity, the mitochondrial ATP synthesis gets boosted up, hence further increasing the mitochondrial ROS generation (Atkin and Macherel 2009).

2.3.3 Endoplasmic Reticulum

The generation of ROS like $O_2^{•-}$ also occurs in ER mediated by NADPH-dependent electron transport involving Cyt P_{450} (Mittler 2002). After the interaction of organic substrate RH with the Cyt P_{450}, reduction occurs by a flavoprotein to form a free radical intermediate (Cyt P_{450} R^-). This intermediate spontaneously reacts with triplet oxygen ($^3O_2$) to form an oxygenated complex (Cyt P_{450}-ROO^-). The complex may get reduced by Cyt-b or occasionally get decomposed to Cyt P_{450}-Rh generating $O_2^{•-}$ (Sharma et al. 2012; Das and Roychoudhury 2014).

2.3.4 Peroxisomes

Peroxisomes are small, usually spherical, subcellular membrane-bound organelles (Gill and Tuteja 2010) which act as the major sites of intracellular ROS production due to their essential oxidative metabolism (Palma et al. 2009). The main metabolic processes involved in the generation of H_2O_2 in peroxisomes are the photorespiratory glycolate oxidase reaction, the fatty acid β-oxidation, the enzymatic reaction of flavin oxidases and the disproportionation of $O_2^{•-}$ radicals (Baker and Graham 2002). During drought conditions, when the water availability is low and stomata remain closed, the ratio of CO_2 to O_2 is reduced to a great extent, resulting in increased photorespiration, ultimately leading to glycolate formation. This glycolate gets oxidized by the enzyme glycolate oxidase in peroxisomes to release H_2O_2, rendering it the leading producer of H_2O_2 (Noctor et al. 2002a, b). The generation of $O_2^{•-}$ in peroxisomes takes place at two different locations (del Rio et al. 2002; Sharma et al. 2012). One is in the peroxisomal matrix, where $O_2^{\bullet-}$ is produced during the oxidation of xanthine or hypoxanthine to uric acid by XOD enzyme (Corpas et al. 2001). Second site of $O_2^{\bullet-}$ generation lies in the peroxisome membranes dependent on NAD(P)H with a small ETC composed of a flavoprotein NADH and cytochrome b (Gill and Tuteja 2010). Additionally, three integral membrane proteins of molecular masses 18, 29 and 132 kDa are involved in $O_2^{\bullet-}$ production. The NADH works as electron donor of the 18 and 32 kDa integral proteins, whereas the 29 kDa membrane protein uses the NADPH as the electron donor to reduce cytochrome c (Das and Roychoudhury 2014).

Although increased production of H_2O_2 and $O_2^{\bullet-}$ in the peroxisomes may cause oxidative damage and possibly cell death (del Rio et al. 2002), their role as signal molecules, mediating pathogen-induced programmed cell death in plants, can't be ignored when present in small amounts (Grant and Loake 2000; McDowell and Dangl 2000).

2.3.5 Apoplast

ROS such as H_2O_2 production in plants also occurs extracellularly in apoplasts (Das and Roychoudhury 2014). During unfavourable environmental conditions, stress signals combined with the abscisic acid (ABA) render the apoplast an important location for H_2O_2 production (Hu et al. 2006). Moreover, there are also some ROS-producing enzymes, consisting of pH-dependent peroxidases (POXs), cell wall-linked oxidases, germin-like oxalate oxidases and polyamine oxidases, which produce H_2O_2 (Wojtaszek 1997; Apel and Hirt 2004). Amine oxidases in the apoplast may contribute to defence responses in plants against biotic stress, mainly through the enhanced production of H_2O_2 either spontaneously or by dismutation of superoxide into H_2O_2 (Cona et al. 2006). On the contrary to superoxide, H_2O_2 diffuse into cells and trigger myriad of the plant defences, including programmed cell death (Dangl and Jones 2001).

Induction of programmed cell death prevents the spread of disease from the infection site. Moreover, apoplastic H_2O_2 generated during the 'oxidative burst' elicited by some invading microorganisms is supposed to provide an oxidizing environment, unfavourable for the multiplication of the microbial cells (Grant and Loake 2000). The introduction of oxalate oxidase activity through genetic engineering in plants which normally lack H_2O_2-generating oxalate oxidase activity has been reported to provide resistance to oxalate-secreting pathogens (Donaldson et al. 2001). This confirms the important physiological role of apoplastic H_2O_2 in plants against pathogen infections.

2.3.6 Plasma Membranes

Plasma membrane plays a key role in interacting with the changing environmental conditions. Electron-transporting oxidoreductases found at plasma membranes lead to ROS production over there (Sharma et al. 2012). The NADPH-dependent oxidases localized in the plasma membrane catalyse transfer of electrons from cytoplasmic NADPH to O_2 to form $O_2^{\bullet -}$ which either directly dismutates to H_2O_2 or is catalysed by SOD activity (Sharma et al. 2012; Das and Roychoudhury 2014). The role of NADPH oxidase in plants against pathogenic infection and abiotic stresses is well documented (Kwak et al. 2003).

2.3.7 Cell Wall

Cell wall is also considered as an active site for ROS generation (Sharma et al. 2012). The generation of ROS such as H_2O_2 by cell wall-associated peroxidases has been revealed during hypersensitive response generated in cotton infected by bacterium *Xanthomonas campestris* pv. *malvacearum* (Martinez et al. 1998) and in *Arabidopsis* during potassium deficiency (Kim et al. 2010). During stress conditions, the cell wall-localized lipoxygenase (LOX) causes hydroperoxidation of PUFA rendering it an active source of ROS (Das and Roychoudhury 2014). The cell wall-localized diamine oxidases also produce activated O_2 in the cell wall using diamine or polyamines to reduce a quinone that autooxidizes to form peroxides (Elstner 1991). During pathogen attack, lignin precursors experience an extensive cross-linking through H_2O_2-mediated pathways to make the cell wall stronger with lignin (Higuchi 2006).

2.4 Oxidative Damage to Biomolecules

ROS are known to cause damages to biomolecules such as lipids, proteins and DNA (Fig. 2.1).

2.4.1 Lipids

Lipids are the main components of plasma membrane enveloping the cell (Das and Roychoudhury 2014). However, the oxidative damage to lipids also known as lipid peroxidation is known to occur in every organism and is often considered as an indicator to determine the extent of lipid damage under stressful conditions (Gill and Tuteja 2010; Das and Roychoudhury 2014). It is now well documented that products formed during lipid peroxidation are actually polyunsaturated precursors such as ketones, MDA, etc. (Garg and Manchanda 2009). Once lipid peroxidation occurs in cellular or organelle membranes, it triggers a chain reaction, thus further aggravating the oxidative stress by generating lipid radicals which damage biomolecules like proteins and DNA (Montillet et al. 2005). The two core targets of the ROS attack in membrane lipids are the double bond between two carbon atoms and the ester linkage between glycerol and fatty acids (Sharma et al. 2012). PUFAs like linoleic and linolenic acid, which are essential components of plasma membrane, are particularly more susceptible to damage by ROS like $^1O_2$ and $OH^•$, giving rise to complex mixtures of lipid hydroperoxides (Das and Roychoudhury 2014).

The resulting lipid hydroperoxides are converted into various reactive species like lipid alkoxy radicals, aldehydes like MDA, alkanes, lipid epoxides and alcohols (Davies 2001; Fam and Morrow 2003). The breakdown products of aldehydes potentially form conjugates with DNA and proteins. A single episode of initiation thus produces several peroxide molecules by a chain reaction (Gill and Tuteja 2010). This increased PUFA peroxidation diminishes membrane fluidity, increases its leakiness to materials that otherwise can cross it only through specific channels and results in the secondary damage to proteins of membrane (Moller et al. 2007). Several studies have reported an increase in lipid peroxidation due to the production of ROS in plants exposed to various abiotic stresses (Vanaja et al. 2000; Hernandez et al. 2000; Dar et al. 2015).

2.4.2 Proteins

Various types of direct or indirect alterations may take place in a protein of plant cell during stressful conditions (Sharma and Dubey 2005; Maheshwari and Dubey 2009; Tanou et al. 2009). The activity of the protein is altered due to modifications such as nitrosylation, carbonylation, disulphide bond formation and glutathionylation. Carbonylation of protein is often used as an indicator for determining the extent of protein oxidation (Moller et al. 2007). Protein carbonylation may be due to direct oxidation of amino acid side chains (e.g. proline and arginine to γ-glutamyl semialdehyde, lysine to amino adipic semialdehyde and threonine to aminoketobutyrate) (Shringarpure and Davies 2002). The protein carbonylation has been reportedly higher in the mitochondria than other subcellular organelles like chloroplasts and peroxisomes (Bartoli et al. 2004), suggesting that mitochondria are more vulnerable to damage by ROS attack.

2.4.3 DNA

The nuclear DNA of plants is well protected by histones and other linked proteins, but mitochondrial and chloroplastic DNA are susceptible to ROS attack due to lack of protective histones as well as their propinquity with ROS generation system (Das and Roychoudhury 2014). DNA is the genetic material, and any injury to the DNA may lead the alterations in the proteins encoded by it, which may ultimately result in malfunction or complete failure of the encoded proteins (Sharma et al. 2012). ROS may damage the DNA by oxidizing its deoxyribose sugar residue, altering the nucleotide base sequence, abstraction of a nucleotide, breaking and cross-linking the DNA and reacting with double bonds of the purine and pyrimidine bases (Halliwell 2006). The ROS abstraction of the C-4 H-atom of the deoxyribose sugar due to ROS attack results in the formation of deoxyribose radical which reacts further to produce single-strand breaks in the DNA (Evans et al. 2004). The base oxidation generates various damaged products including 8-hydroxyquinine, hydroxyl methyl urea, dehydro-2′-deoxyguanosine, thymine glycol and thymine and adenine ring opened (Sharma et al. 2012; Das and Roychoudhury 2014). It is reported that $OH^•$ is the most reactive intermediate and causes damage to all components of the DNA molecule. The $OH^•$ can potentially damage both the purine and pyrimidine bases and also the deoxyribose backbone (Halliwell and Gutteridge 1999), and it also causes DNA-protein cross-links when it reacts with either DNA or its linked proteins. These cross-links may prove lethal for the plant cell, if they don't get repaired in time before the initiation of crucial cellular processes like replication or transcription (Das and Roychoudhury 2014).

2.5 Antioxidative Defence System in Plants

Under normal conditions, there is complete equilibrium between the generation and scavenging of reactive oxygen metabolites. This equilibrium between production and quenching of ROS gets disturbed by various unfavourable environmental factors such as temperature extremes, drought, heavy metals, air pollutants and high doses of pesticides, resulting in the increased intracellular ROS production levels (Noctor et al. 2002a, b; Sharma et al. 2010). To evade the oxidative damage caused by these toxic ROS, the level of endogenous antioxidant defence system is raised in higher plants (Sharma et al. 2010). A great deal of research has established that the induction of the cellular antioxidant machinery is important for protection against various stresses (Tuteja 2007). The ROS defence mechanism in plants consists of the antioxidative defence system which helps to alleviate the oxidative stress-induced damages. This antioxidant machinery consists of non-enzymatic and enzymatic components to scavenge ROS and is found in different subcellular compartments such as chloroplasts, mitochondria and peroxisomes (Sharma et al. 2012). Enzymatic antioxidants include the enzymes such as superoxide dismutase (SOD), catalase (CAT), ascorbate peroxidase (APX), monodehydroascorbate reductase (MDHAR), dehydroascorbate reductase (DHAR), glutathione reductase (GR) and non-enzymatic antioxidants which are glutathione (GSH), ascorbic acid (AA), carotenoids, proline, tocopherols and flavonoids (Mittler et al. 2004; Singh et al. 2008; Dar et al. 2015).

2.5.1 Enzymatic Antioxidants

2.5.1.1 Superoxide Dismutase (SOD)

Superoxide dismutase (E.C.1.15.1.1) belongs to the metalloenzyme family and occurs universally in all aerobic organisms. Under stress conditions, SOD plays a vital role, forming the primary line of defence against ROS-induced damages (Liang et al. 2003). The SOD catalyses the elimination of $O_2^{•-}$ by dismutating it into O_2 and H_2O_2, therefore reducing the threat of $OH^•$ formation by the metal-catalysed Haber-Weiss reaction. This reaction is 10,000 times faster than the spontaneous dismutation (Gill and Tuteja 2010; Das and Roychoudhury 2014). SOD is found in most of the subcellular compartments that generate activated oxygen and are vulnerable to ROS-mediated oxidative stress (Sharma et al. 2012). On the basis of presence of metal cofactors, SODs are grouped into three isozymes, i.e. Mn-SOD (localized in mitochondria and peroxisomes), Fe-SOD (often non-detectable in plants, but when present are found in chloroplasts) and Cu/Zn-SOD (localized in cytosol, peroxisomes and chloroplasts) (Racchi et al. 2001; Alscher et al. 2002). The genes which code all the three isoforms of SOD are present in the nucleus of the plant cell, and all these isoforms after forming in the nucleus are sent to their relevant subcellular compartments by an amino terminal targeting sequence (Bowler et al. 1992).

The upregulation of SODs caused due to various types of stresses plays a crucial role in alleviating oxidative stress for the survival of the plant (Boguszewska et al. 2010). SOD processing is substrate inducible and ROS act as its substrate. An increase in the SOD activity means increased ROS generation (Abedi and Pakniyat 2010). SOD activity has been reported to increase in plants exposed to various environmental stresses such as salt stress (Harinasut et al. 2003; Kukreja et al. 2005; Gapinska et al. 2008), pesticide stress (Dar et al. 2015), drought (Sharma and Dubey 2005) and metal toxicity (Mishra et al. 2011). Increased activity of SOD is often attributed to enhanced tolerance of the plant against various environmental stresses (Gupta et al. 1993). Transgenic plants overexpressing SODs have been shown to be very tolerant against various abiotic stresses (Wang et al. 2004; Yang et al. 2009).

2.5.1.2 Catalase (CAT)

Catalase (E.C.1.11.1.6) was the first antioxidative enzyme to be discovered and characterized (Sharma et al. 2012). It is a heme-containing enzyme that catalyses the conversion of H_2O_2 into H_2O and O_2 (Das and Roychoudhury 2014). It has high affinity for H_2O_2 but lesser specificity for organic peroxides (R-O-O-R). It has a very high turnover rate and is exceptional amongst antioxidant enzymes as it can function without a cellular reducing equivalent (Sharma et al. 2012). Peroxisomes are the major sites of H_2O_2 generation due to β-oxidation of fatty acids, photorespiratory oxidation and purine catabolism (Mittler 2002). However, there are reports that CAT is also present in other subcellular organelles like cytosol, chloroplast and the mitochondria, but the CAT activity in these compartments has not been reported till date (Mhamdi et al. 2010). Angiosperms possess various isoforms of CAT enzyme (Polidoros and Scandalias 1999; Azpilicueta et al. 2007). For example, in *Zea mays* three genes are found to code the three isoforms of this enzyme (Scandalias 1990). *CAT1* found in peroxisomes and cytosol is expressed in pollens and seeds; *CAT2* is also localized in cytosol and peroxisomes and mostly expressed in photosynthetic tissues such as leaves, but also in tissues like roots and seeds; and *CAT3* which is localized in the mitochondria is highly expressed in seeds and seedlings (Scandalias 1990; Sharma et al. 2012). Under the stressful environmental conditions, cells need lot of energy, and this also generates H_2O_2 by increased catabolism rate. CAT eliminates the H_2O_2 in an energy-efficient manner:

$$H_2O_2 \overset{CAT}{\rightarrow} H_2O + (1/2)O_2$$

2.5.1.3 Ascorbate Peroxidase (APX)

Ascorbate peroxidase (E.C.1.1.11.1) is an important component of the ascorbate-glutathione (ASC-GSH) cycle and eliminates H_2O_2 produced in cytosol and chloroplast (Gill and Tuteja 2010. The APX dismutates H_2O_2 into H_2O and DHA by using AA as a reducing agent (Das and Roychodhury 2014):

$$H_2O_2 + AA \overset{APX}{\rightarrow} 2H_2O + DHA$$

On the basis of amino acid sequences, the enzyme APX is divided into five distinctive isoforms which are located in different subcellular organelles in higher plants. These locations are cytosol, mitochondria, peroxisomes and chloroplasts, and the APX enzymes found in these locations are, respectively, known as cytosolic, mitochondrial, peroxisomal and chloroplastidal APX (Noctor and Foyer 1998; Sharma and Dubey 2004). APX is more commonly distributed and has a superior specificity for H_2O_2 compared to CAT; therefore, it is more efficient in eliminating H_2O_2 during the stressful conditions (Sharma et al. 2012). Increased activity of APX in response to various abiotic stresses such as drought, salinity, chilling, metal toxicity, higher rates of pesticides and UV irradiation has been reported by many workers (Boo and Jung 1999; Sharma and Dubey 2005, 2007; Maheshwari and Dubey 2009; Hefny and Abdel-Kader 2009; Dar et al. 2015).

2.5.1.4 Monodehydroascorbate Reductase (MDHAR)

Monodehydroascorbate reductase (E.C.1.6.5.4) is a flavin adenine dinucleotide (FAD) enzyme which catalyses the rejuvenation of AA from the short-lived MDHA radical by using NADPH as a reducing agent/electron donor, eventually replenishing AA pool in the cell (Sharma et al. 2012). Because it regenerates AA, it is also found in the peroxisomes and mitochondria, where APX scavenges H_2O_2 and oxidizes AA in the process (Mittler 2002). The isozymes of MDHAR are also found in chloroplasts, mitochondria, peroxisomes, cytosol and glyoxysomes (Hossain et al. 1984; Dalton et al. 1993; Jimenez et al. 1998; Sharma et al. 2012; Das and Roychoudhury 2014):

$$MDHA + NADPH \rightarrow AA + NADP^+$$

2.5.1.5 Dehydroascorbate Reductase (DHAR)

Dehydroascorbate reductase (E.C.1.8.5.1) catalyses the reduction of dehydroascorbate (DHA) to AA using reduced glutathione (GSH) as an electron donor/reducing agent (Eltayeb et al. 2007). Thus, like MDHAR, DHAR is another agent that regenerates the AA pool in the cell (Das and Roychoudhury 2014). The role of DHAR is very crucial in maintaining the cellular AA pool in both symplast and apoplast, hence sustaining the redox state of the plant cell (Chen and Gallie

2006). DHAR is found profusely in seeds and roots and also in green and etiolated shoots (Sharma et al. 2012):

$$\text{DHA} + 2\text{GSH} \rightarrow \text{AA} + \text{GSSG}$$

2.5.1.6 Glutathione Reductase (GR)

Glutathione reductase (E.C.1.6.4.2) is a flavoprotein oxidoreductase which uses NADPH as an electron donor to reduce GSSG to GSH. Reduced glutathione (GSH) is then used to regenerate AA from MDHA and DHA, and it gets transformed to its oxidized form (GSSG). Generally GR is present in the chloroplasts in leaf tissues. However, small amounts of it are also found in cytosol, mitochondria, glyoxysomes and peroxisomes (Jimenez et al. 1998). Like APX, GR is one of the main components in the ascorbate-glutathione cycle, by which the efficient recycling of glutathione is ensured by GR and plays a key role in the protection of chloroplasts during the oxidative stress by maintaining a high reduced/oxidized glutathione (GSH/GSSG) ratio in a cell (Noctor and Foyer 1998):

$$\text{GSSG} + \text{NADPH} \rightarrow 2\text{GSH} + \text{NADP}^+$$

2.5.1.7 Guaiacol Peroxidase (GPX)

Guaiacol peroxidase (E.C.1.11.1.7) is a heme-containing enzyme that removes surplus amounts of H_2O_2 during normal as well as stress conditions (Das and Roychoudhury 2014). It helps in the lignin synthesis and also protects plants against biotic stress by degrading indoleacetic acid (IAA) and consuming H_2O_2 in the process. GPX uses aromatic compounds like guaiacol and pyrogallol (Asada 1999) as reducing agents. Because GPX plays active roles in the cytosol and vacuoles and also in the cell walls, it is considered as an essential enzyme in the elimination of H_2O_2 (Das and Roychoudhury 2014).

$$H_2O_2 + \text{GSH} \rightarrow H_2O + \text{GSSG}$$

2.5.2 Non-enzymatic Antioxidants

The non-enzymatic antioxidant machinery includes ascorbic acid, reduced glutathione, tocopherol, carotenoids, flavonoids and proline. These non-enzymatic antioxidants in addition to the protection of cellular components from damage play a very important role in plant growth and development by interfering with the essential processes like mitosis, cell elongation, senescence and cell death (de Pinto and DeGara 2004).

2.5.2.1 Ascorbic Acid (AA)

Ascorbic acid (AA) is the ample, powerful, low molecular weight and water-soluble antioxidant with a key defence role against oxidative stress caused due to enhanced ROS level, potentially due to its ability to donate electrons in a number of enzymatic and non-enzymatic reactions (Smirnoff 2005; Gill and Tuteja 2010). Ascorbic acid is produced in plants in mitochondria via Smirnoff-Wheeler pathway catalysed by L-galactono-γ-lactone dehydrogenase. Some ascorbates are also generated from D-galacturonic acid (Das and Roychoudhury 2014). It has a wide distribution in the plant tissues and is usually higher in meristematic and photosynthetic cells and play important roles in various physiological processes like plant growth, differentiation and metabolism. Majority of the AA is concentrated in cytosol with a significant amount of it being found in apoplast, thus making it the first line of defence against the dangerous ROS (Barnes et al. 2002). AA is oxidized in two sequential steps, first producing monodehydroascorbate (MDHA), which, if not reduced immediately to ascorbate, disproportionates to AA and DHA. Ascorbic acid protects the membranes from oxidative damage by reacting with H_2O_2, $OH^{\bullet}$ and $O_2^{\bullet -}$ and regenerating α-tocopherol from tocopheroxyl radical (Shao et al. 2005) and preserves the activities of metal-binding enzymes that contain prosthetic transition metal ions (Zaefyzadeh et al. 2009). Ascorbic acid also prevents photo-oxidation of PSII activity and its downregulation, associated with zeaxanthin formation.

2.5.2.2 Reduced Glutathione (GSH)

Glutathione is essentially a low molecular weight thiol tripeptide (γ-glutamyl-cysteinyl-glycine) with wide distribution in all cellular compartments like cytosol, chloroplasts, endoplasmic reticulum, mitochondria, vacuoles, peroxisomes and apoplast which is the prime site for intracellular defence against ROS-induced oxidative damage. Two ATP-dependent steps determine the synthesis of glutathione; firstly, γ-glutamylcysteine is synthesised from Cys and Glu in the presence of glutamate-cysteine ligase (GCL), and then glutathione synthetase (GS) aids Gly to γ-glutamylcysteine to yield GSH (Foyer and Noctor 2005). GSH plays an integral part in extensive biological processes like cell division, cell differentiation, cell death and senescence, pathogen resistance, regulation of enzymatic activity, regulation of sulphate transport, detoxification of xenobiotics, conjugation of metabolites, synthesis of proteins and nucleotides, synthesis of phytochelatins and ultimately expression of stress-responsive genes (Mullineaux and Rausch 2005; Rausch and Wachter 2005). The central cysteine residue with strong nucleophilic character makes GSH a potent reducing power. GSH scavenges H_2O_2, $^1O_2$, $OH^{\bullet}$ and $O_2^{\bullet -}$ and protects the different biomolecules by forming adducts directly with reactive electrophiles (glutathiolation) or by reducing them in the presence of ROS or organic free radicals, yielding GSSG as a byproduct. The generated GSSG is converted back to GSH by de novo synthesis or enzymatically by glutathione reductase (GR), thus maintaining the cellular GSH pool (Asada 1994).

2.5.2.3 Tocopherols

Tocopherols (α, β, γ and δ) are a group of lipophilic antioxidants with α-tocopherol having the highest antioxidant capability of all the four isomers. The α-tocopherol efficiently scavenges ROS and lipid radicals thereby making it an indispensable protector and essential component of biomembranes (Hollander-Czytko et al. 2005; Kiffin et al. 2006). The tocopherols are found only in green parts of plants as these are synthesized only by photosynthetic organisms. The α-tocopherol is synthesized from γ-tocopherol by γ-tocopherol-methyltransferase (γ-TMT encoded by VTE4). Tocopherols protect PSII, both structurally and functionally via quenching its excess energy by reacting with O_2, and thus protect the lipids and other membrane constituents of the chloroplasts. Tocopherols inhibit the chain propagation step of the lipid peroxidation cycle and serve as a potent free radical trap. α-Tocopherol reduces the lipid radicals ($RO^{\bullet}$, $ROO^{\bullet}$ and RO^{*}) present at the membrane-water interface and gets converted into $TOH^{\bullet}$ which interacts with GSH and AA and is recycled to its reduced form (Igamberdiev et al. 2004).

2.5.2.4 Carotenoids

Carotenoids, the lipophilic antioxidant pigments found in plants and microorganisms, possess the ability to detoxify ROS. Carotenoids carry out three-fold functions in plants:

(a) Absorb light in the 450–570 nm and transfer the energy to the chlorophyll molecule (Sieferman-Harms 1987)
(b) As an antioxidant, protect the photosynthetic machinery by reacting with lipid peroxidation products to terminate chain reactions, scavenge $^1O_2$, generate heat as a byproduct, preventing the formation of $^1O_2$ by reacting with ^{3}Chl and excited chlorophyll (Chl*), and dissipate the excess excitation energy in the xanthophyll cycle
(c) Help in structural stability of PSI assembly, light-harvesting complex and thylakoid membranes (Niyogi et al. 2001)

2.5.2.5 Proline

Proline, essentially an osmolyte, is considered as a powerful antioxidant and potent inhibitor of programmed cell death. It is a potential non-enzymatic antioxidant which has found its application to counteract the adverse effects of different ROS. In plants, proline is synthesized either via the glutamate or the ornithine pathway (Sekhar et al. 2007; Dar et al. 2016). The ornithine pathway occurs via deamination of ornithine which is then transaminated to P5C by ornithine-delta-aminotransferase (OAT) (Verbruggen and Hermans 2008). Proline acts as an osmoprotectant, a protein stabilizer, a metal chelator (Dar et al. 2016), a lipid peroxidation inhibitor and an efficient scavenger of $OH^{\bullet}$ and $^1O_2$. During stressful conditions, proline accumulation speeds up in plants either due to enhanced synthesis or reduced degradation (Verbruggen and Hermans 2008; Dar et al. 2016).

2.5.2.6 Flavonoids

Flavonoids are prevalent in the plant kingdom and are predominantly found in leaves, floral parts and pollens. Flavonoids are classified as flavonols, flavones, isoflavones and anthocyanins on the basis of their structure. Besides providing flower, fruit and seed pigmentation, UV light protection, defence against phytopathogens and role in fertility and seed germination, flavonoids act as the important secondary ROS scavengers, thus preventing the damage to the photosynthetic apparatus, by locating and neutralizing radicals before they damage the cell, and thus are important for plants under adverse environmental conditions (Fini et al. 2011). They scavenge $^1O_2$ and reduce the damages to the outer envelope of the chloroplastic membranes (Agati et al. 2012).

2.6 Detoxification by Antioxidative Machinery Under Stressful Environments

Under normal environmental conditions, the generation of ROS in plants is low. However, under stress conditions, ROS production considerably increases in plants, disbalancing the normal production of $O_2{\bullet}^-$, $OH^{\bullet}$ and H_2O_2 in the intracellular environment (Sharma et al. 2010). Under drought stress, SOD activity increased in different cultivars of *Phaseolus vulgaris* (Zlatev et al. 2006) and *Oryza sativa* (Sharma and Dubey 2005) and in the leaves of *Trifolium repens* (Chang-Quan and Rui-Chang 2008). The CAT activity was also enhanced in wheat under drought stress (Simova-Stoilova et al. 2010). During the drought stress conditions, the enhancement in ROS generation may occur in many ways. Under drought stress, there is restricted fixation of CO_2 due to closing of stomata, leading to reduced $NADP^+$ regeneration via C_3 cycle. Due to shortage of electron acceptors, overloading of the electron transfer occurs in photosynthetic electron transport chain, which results in higher seepage of electrons to O_2 by the Mehler reaction (Sharma et al. 2012). Biehler and Fock (1996) reported about 50% more leakage of electrons in photosynthetic electron transport chain to the Mehler reaction in wheat plants under drought stress as compared to control plants. Drought stress also causes enhanced activity of photorespiration, particularly, when RuBP oxygenation is higher due to less availability of CO_2 (Noctor et al. 2002a, b). Noctor et al. (2002a, b) reported that more than 70% of total H_2O_2 generation via photorespiration occurs under drought stress environments. The drought-tolerant genotype C306 showed elevated APX and CAT activity, with lesser H_2O_2 and MDA content as compared to the drought-susceptible genotype, HD2329 (Sairam et al.1998). The overexpression of APX in the chloroplasts of *Nicotiana tabacum* reduced the toxicity of H_2O_2, resulting in the generation of drought tolerance in a plant (Badawi et al. 2004). The greater vulnerability of the sensitive varieties IR-29 and Pusa Basmati to water deficiency was due to significant decline in GSH/GSSG ratio, as compared to the tolerant variety Pokkali (Basu et al. 2010b). Exogenous application of AA to wheat cultivars increased the level of photosynthetic pigments, net photosynthesis and growth, compared to the untreated plants grown under drought conditions (Malik and Ashraf 2012).

The SOD activity also enhanced under salt stress conditions in chickpea (Kukreja et al. 2005) and tomato (Gapinska et al. 2008). Transgenic *Arabidopsis* overexpressing Mn-SOD was witnessed to have increased tolerance against salt stress (Wang et al. 2004). High CAT activity was noticed in *Cicer arietinum* under salt stress (Kukreja et al. 2005). Under saline conditions, the activities of APX and GR were reportedly higher in salt-tolerant cultivars of potato, whereas significantly reduced in salt-sensitive varieties (Aghaei et al. 2009). Overexpression of MDHAR in *Nicotiana tabacum* and DHAR in *Arabidopsis* improved their tolerance against salt stress (Ushimaru et al. 2006; Eltayeb et al. 2007). Both AA and GSH contents were found to be higher in salt-tolerant cultivar Pokkali as compared to sensitive cultivar Pusa Basmati (Vaidyanathan et al. 2003). Exogenous application of AA to the salt stressed tomato plants accelerated their recovery process and certified their long-lasting survival (Shalata and Neumann 2001). Ascorbic acid also helped to mitigate oxidative stress in wheat, by improving its photosynthetic capability (Athar et al. 2008). GSH was also reported to relieve the oxidative damage in rice chloroplasts caused due to saline conditions (Wang et al. 2014). It was later found that heat stress brought about the accumulation of GSH levels and enhanced the enzymatic activity of enzymes implicated in GSH synthesis and the GSH/GSSG ratio (Hasanuzzaman et al. 2013).

Salt stress causes closure of stomata, which limits CO_2 fixation in the leaves, and finally results in the overreduction of photosynthetic electron transport system, leading to increased ROS production (Sharma et al. 2012). Low CO_2/O_2 ratio in chloroplasts also promotes photorespiration resulting into enhanced generation of H_2O_2 (Hernandez et al. 2000).

It is also reported that heavy metals like cadmium (Cd) stress enhanced the CAT activity in *Phaseolus aureus*, *Pisum sativum* and *Lemna minor* (Sreedevi and Krishnan 2012). There have been also reports that UV-B radiation and lead (Pb) treatments enhanced APX activity in *Arabidopsis thaliana* and *Eichhornia crassipes* (Rao et al. 1996; Malar et al. 2014). The plant species *Vaccinium myrtillus* is well known as a colonist of heavy metal-contaminated soil, and contents of GSH, GPX, nonprotein thiols and proline in this plant were found to be enhanced when grown in heavy metal-contaminated sites (Kandziora-Ciupa et al. 2013). The activity of peroxidase and CAT in *Vigna radiata* elevated with the increase in $CdCl_2$ concentration (Roychoudhury and Ghosh 2013).

It is well evidenced that α-tocopherol plays a key role in maintaining the cellular Na^+/K^+ homeostasis and hormonal balance in plants (Ellouzi et al. 2013). Transgenic tobacco overexpressing *Arabidopsis VTE1* (encoding tocopherol biosynthesis enzyme) showed diminished lipid peroxidation, electrolyte leakage and H_2O_2 content but had higher chlorophyll contents as compared to its wild species (Liu et al. 2008). *Arabidopsis vte1* and *vte4* mutants lacking α-tocopherol were found to have reduced growth and vigour under saline environment (Ellouzi et al. 2013). High exposure of plants to UV-B radiation resulted decline in their α-tocopherol contents, probably due to reactions with lipid radicals (Jain et al. 2003). Carotenoid, an essential component of the defence mechanism of plants, protects chlorophyll and photosynthetic membrane from photo-oxidative damage (Mishra et al. 2008). The enhanced levels of carotenoids at low doses of dimethoate (Mishra et al. 2008) *and* chlorpyrifos (Parween et al. 2011) exposure were observed. Moreover, the carotenoid content increased in drought-resistant plants when they were grown under drought-stressed conditions, thus protecting photosynthetic machinery from photo-oxidative damages (Munne-Bosch and Alegre 2000). Drought-like conditions created by 20% polyethylene glycol (PEG 6000) treatment to rice seedlings increased the levels of flavonoids and phenolics in the plant (Basu et al. 2010a). The level of these non-enzymatic antioxidants was found to be many times higher in the tolerant variety Pokkali, as compared to the sensitive cultivars IR-29 and Pusa Basmati (Basu et al. 2010a). The accumulation of an osmoprotectant proline was higher in drought-tolerant cultivars of chickpea as compared to its sensitive varieties under both normal and drought-stressed environments (Mafakheri et al. 2010). When rice seedlings were treated with salt concentrations (200 mM NaCl), the accumulation of anthocyanin and proline was found to be higher in the salt-tolerant cultivar Nona Bokra than the salt-sensitive cultivars like M-1-48 and Gobindobhog (Roychoudhury et al. 2008). The salt-tolerant cultivar of rice showed higher accumulation of flavonoids than the salt-sensitive cultivars (Chutipaijit et al. 2009).

2.7 Conclusion and Future Prospects

The ROS are inevitable byproducts of normal cell metabolic pathways associated with different cellular compartments or are produced by electronic flow in chloroplasts, mitochondria and plasma membrane. ROS generation in various cellular compartments is low under normal environmental conditions, but environmental stresses such as drought, salinity, chilling, heavy metal toxicity and UV-B radiations may disturb the cellular homeostasis and increase the ROS generation. The increased levels of ROS may cause extensive damage to the biomolecules like lipids, proteins and DNA, damaging the integral membrane properties and eventually leading to cell death. However, nature has gifted plants with a range of antioxidative defence mechanisms to detoxify the damaging effects of ROS and acclimatize them to various unfavourable environmental conditions. The antioxidative defence machinery is composed of the enzymatic antioxidants (e.g. superoxide dismutase, catalase, glutathione peroxidase and glutathione reductase) and non-enzymatic antioxidants (glutathione, proline, α-tocopherols, carotenoids and flavonoids). Although large development has been attained in recent years, there are still some uncertainties and lack of knowledge to understand the ROS formation and how they impact plants, due to their short half-life and extremely reactive nature. Although the compartmentalization of antioxidants is well established, the mechanism of ROS generation and their fate must be further explored. Transgenic approach for overexpression of antioxidant genes may be used to obtain plants with high tolerance under various stressful environmental conditions.

References

Abedi T, Pakniyat H (2010) Antioxidant enzyme changes in response to drought stress in ten cultivars of oilseed rape (*Brassica napus* L.) Czech J Genet Plant Breed 46:27–34

Agati G, Azzarello E, Pollastri S, Tattini M (2012) Flavonoids as antioxidants in plants: location and functional significance. Plant Sci 196:67–76

Aghaei K, Ehsanpour AA, Komatsu S (2009) Potato responds to salt stress by increased activity of antioxidant enzymes. J Integr Plant Biol 51:1095–1103

Alscher RG, Erturk N, Heatrh LS (2002) Role of superoxide dismutases (SODs) in controlling oxidative stress in plants. J Exp Bot 53:1331–1341

Apel K, Hirt H (2004) Reactive oxygen species: metabolism, oxidative stress, and signal transduction. Annu Rev Plant Biol 55:373–399

Asada K (1994) Production and action of active oxygen species in photosynthetic tissues. In: Foyer CH, Mullineaux PM (eds) Causes of photooxidative stress and amelioration of defense systems in plants. CRC Press, Boca Raton, pp 77–104

Asada K (1999) The water-water cycle in chloroplasts: scavenging of active oxygens and dissipation of excess photons. Annu Rev Plant Physiol Plant Mol Biol 50:601–639

Athar HUR, Khan A, Ashraf M (2008) Exogenously applied ascorbic acid alleviates salt-induced oxidative stress in wheat. Environ Exp Bot 63:224–231

Atkin OK, Macherel D (2009) The crucial role of plant mitochondria in orchestrating drought tolerance. Ann Bot 103:581–597

Azpilicueta CE, Benavides MP, Tomaro ML, Gallego SM (2007) Mechanism of CAT A3 induction by cadmium in sunflower leaves. Plant Physiol Biochem 45:589–595

Badawi GH, Kawano N, Yamauchi Y, Shimada E, Sasaki R, Kubo A et al (2004) Over–expression of ascorbate peroxidase in tobacco chloroplasts enhances the tolerance to salt stress and water deficit. Physiol Plant 121:231–238

Baker A, Graham AI (2002) Plant peroxisomes: biochemistry, cell biology and biotechnological applications. Kluwer Academic Publishers, Dordrecht

Barnes JD, Zheng Y, Lyons TM (2002) Plant resistance to ozone: the role of ascorbate. In: Omasa K, Saji H, Youssefian S, Kondo N (eds) Air pollution and plant biotechnology. Springer, Tokyo, pp 235–254

Bartoli CG, Gómez F, Martinez DE, Guiamet JJ (2004) Mitochondria are the main target for oxidative damage in leaves of wheat (*Triticum aestivum* L.) J Exp Bot 55:1663–1669

Basu S, Roychoudhury A, Saha PP, Sengupta DN (2010a) Differential antioxidative responses of indica rice cultivars to drought stress. Plant Growth Regul 60:51–59

Basu S, Roychoudhury A, Saha PP, Sengupta DN (2010b) Comparative analysis of some biochemical responses of three indica rice varieties during polyethylene glycol-mediated water stress exhibits distinct varietal differences. Acta Physiol Plant 32:551–563

Bhattacharjee S (2005) Reactive oxygen species and oxidative burst: roles in stress, senescence and signal. Curr Sci 89:1113–1121

Biehler K, Fock H (1996) Evidence for the contribution of the Mehler-peroxidase reaction in dissipating excess electrons in drought-stressed wheat. Plant Physiol 112:265–272

Bienert GP, Moller ALB, Kristiansen KA et al (2007) Specific aquaporins facilitate the diffusion of hydrogen peroxide across membranes. J Biol Chem 282:1183–1192

Blokhina O, Fagerstedt KV (2010) Reactive oxygen species and nitric oxide in plant mitochondria: origin and redundant regulatory systems. Physiol Plant 138:447–462

Boguszewska D, Grudkowska M, Zagdanska B (2010) Drought-responsive antioxidant enzymes in potato (*Solanum tuberosum* L.) Potato Res 53:373–382

Boo YC, Jung J (1999) Water deficit – induced oxidative stress and antioxidative defenses in rice plants. J Plant Physiol 155:255–261

Bowler C, Van Montagu M, Inze D (1992) Superoxide dismutase and stress tolerance. Annu Rev Plant Physiol Plant Mol Biol 43:83–116

Bright J, Desikan R, Hancock JT, Weir IS, Neill SJ (2006) ABA-induced NO generation and stomatal closure in Arabidopsis are dependent on H_2O_2 synthesis. Plant J 45:113–122
Chang-Quan W, Rui-Chang L (2008) Enhancement of superoxide dismutase activity in the leaves of white clover (*Trifolium repens* L.) in response to polyethylene glycol-induced water stress. Acta Physiol Plant 30:841–847
Chen Z, Gallie DR (2006) Dehydroascorbate reductase affects leaf growth, development, and function. Plant Physiol 142:775–787
Choudhury S, Panda P, Sahoo L, Panda SK (2013) Reactive oxygen species signaling in plants under abiotic stress. Plant Signal Behav 8:e23681
Chutipaijit S, Cha-Um S, Sompornpailin K (2009) Differential accumulations of proline and flavonoids in indica rice varieties against salinity. Pak J Bot 41:2497–2506
Cleland RE, Grace SC (1999) Voltammetric detection of superoxide production by photosystem II. FEBS Lett 457:348–352
Cona A, Rea G, Angelini R, Federico R, Tavladoraki P (2006) Functions of amine oxidases in plant development and defense. Trends Plant Sci 11:80–88
Corpas FJ, Barroso JB, del Río LA (2001) Peroxisomes as a source of reactive oxygen species and nitric oxide signal molecules in plant cells. Trends Plant Sci 6:145–150
Dalton DA, Baird LM, Langeberg L et al (1993) Subcellular localization of oxygen defense enzymes in soybean (*Glycine max* L. Merr.) root nodules. Plant Physiol 102:481–489
Dangl JL, Jones JDG (2001) Plant pathogens and integrated defense responses to infection. Nature 411:826–833
Dar MI, Khan FA, Rehman F (2015) Responses of antioxidative defense system and composition of photosynthetic pigments in *Brassica juncea* L. upon imidacloprid treatments. Abiotic Biotic Stress J. http://dx.doi.org/10.17582/journal.absjournal/2015/1.1.3.15
Dar MI, Naikoo MI, Rehman F, Naushin F, Khan FA (2016) Proline accumulation in plants: roles in stress tolerance and plant development. In: Iqbal N, Nazar R, Khan NA (eds) Osmolytes and plants acclimation to changing environment: emerging omics technologies. Springer, India, pp 155–166
Das K, Roychoudhury A (2014) Reactive oxygen species (ROS) and response of antioxidants as ROS-scavengers during environmental stress in plants. Front Environ Sci 2:53
Dat J, Vandenabeele S, Vranova E, Van Montagu M, Inze D, Van Breusegem F (2000) Dual action of the active oxygen species during plant stress responses. Cell Mol Life Sci 57:779–795
Davies KJA (2001) Oxidative stress, antioxidant defenses, and damage removal, repair, and replacement systems. IUBMB Life 50:279–289
del Río LA, Corpas FJ, Sandalio LM, Palma JM, Gómez M, Barroso JB (2002) Reactive oxygen species, antioxidant systems and nitric oxide in peroxisomes. J Exp Bot 53:1255–1272
del Rio LA, Sandalio LM, Corpas FJ, Palma JM, Barroso JB (2006) Reactive oxygen species and reactive nitrogen species in peroxisomes. Production, scavenging, and role in cell signaling. Plant Physiol 141:330–335
dePinto MC, DeGara L (2004) Changes in the ascorbate metabolism of apoplastic and symplastic spaces are associated with cell differentiation. J Exp Bot 55:2559–2569
Donaldson PA, Anderson T, Lane BG, Davidson AL, Simmonds DH (2001) Soybean plants expressing an active oligomeric oxalate oxidase from the wheat GF-2.8 (germin) gene are resistant to the oxalate-secreting pathogen *Sclerotina sclerotiorum*. Physiol Mol Plant Pathol 59:297–307
Ellouzi H, Hamed K, Cela J, Müller M, Abdelly C, Munné-Bosch S (2013) Increased sensitivity to salt stress in tocopherol-deficient *Arabidopsis* mutants growing in a hydroponic system. Plant Signal Behav 8:e23136
Elstner EF (1991) Mechanism of oxygen activation in different compartments. In: Pell EJ, Steffen KL (eds) Active oxygen/oxidative stress and plant metabolism. Am Soc Plant Physiol, Roseville, pp 13–25

Eltayeb AE, Kawano N, Badawi GH, Kaminaka H, Sanekata T, Shibahara T, Inanaga S, Tanaka K (2007) Overexpression of monodehydroascorbate reductase in transgenic tobacco confers enhanced tolerance to ozone, salt and polyethylene glycol stresses. Planta 225:1255–1264

Evans MD, Dizdaroglu M, Cooke MS (2004) Oxidative DNA damage and disease: induction, repair and significance. Mutat Res 567:1–61

Fam SS, Morrow JD (2003) The isoprostanes: unique products of arachidonic acid oxidation-a review. Curr Med Chem 10:1723–1740

Fini A, Brunetti C, DiFerdinando M, Ferrini F, Tattini M (2011) Stress- induced flavonoids biosynthesis and the antioxidant machinery of plants. Plant Signal Behav 6:709–711

Foyer CH, Noctor G (2005) Redox homeostasis and antioxidant signaling: a metabolic interface between stress perception and physiological responses. Plant Cell 17:1866–1875

Gapinska M, Skłodowska M, Gabara B (2008) Effect of short and long-term salinity on the activities of antioxidative enzymes and lipid peroxidation in tomato roots. Acta Physiol Plant 30:11–18

Garg N, Manchanda G (2009) ROS generation in plants: boon or bane? Plant Biosys 143:8–96

Gill SS, Tuteja N (2010) Reactive oxygen species and antioxidant machinery in abiotic stress tolerance in crop plants. Plant Physiol Biochem 48:909–930

Gill SS, Khan NA, Anjum NA, Tuteja N (2011) Amelioration of cadmium stress in crop plants by nutrients management: morphological, physiological and biochemical aspects. In: Anjum NA, Lopez-Lauri F (eds) Plant nutrition and abiotic stress tolerance III. Plant stress 5 (Special Issue 1). Global Science Books Ltd., Ikenobe, pp 1–23

Grant JJ, Loake GJ (2000) Role of reactive oxygen intermediates and cognate redox signaling in disease resistance. Plant Physiol 124:21–29

Gupta AS, Heinen JL, Holaday AS, Burke JJ, Allen RD (1993) Increased resistance to oxidative stress in transgenic plants that over express chloroplastic Cu/Zn superoxide dismutase. Proc Natl Acad Sci U S A 90:1629–1633

Halliwell B (2006) Reactive species and antioxidants. Redox biology is a fundamental theme of aerobic life. Plant Physiol 141:312–322

Halliwell B, Gutteridge JMC (1999) Free radicals in biology and medicine, 3rd edn. Oxford University Press, Oxford

Harinasut P, Poonsopa D, Roengmongkol K, Charoensataporn R (2003) Salinity effects on antioxidant enzymes in mulberry cultivar. Sci Asia 29:109–113

Hasanuzzaman M, Nahar K, Alam MM, Roychowdhury R, Fujita M (2013) Physiological, biochemical, and molecular mechanisms of heat stress tolerance in plants. Int J Mol Sci 14:9643–9684

Hatz S, Lambert JDC, Ogilby PR (2007) Measuring the lifetime of singlet oxygen in a single cell: addressing the issue of cell viability. Photochem Photobiol Sci 6:1106–1116

Hefny M, Abdel-Kader DZ (2009) Antioxidant-enzyme system as selection criteria for salt tolerance in forage sorghum genotypes (*Sorghum bicolor* L. Moench). In: Ashraf M, Ozturk M, Athar HR (eds) Salinity and water stress. Springer, The Netherlands, pp 25–36

Hernandez JA, Jimenez AP, Mullineaux SF (2000) Tolerance of pea (*Pisum sativum* L.) to long-term salt stress is associated with induction of antioxidant defenses. Plant Cell Environ 23:853–862

Higuchi ET (2006) Look back over the studies of lignin biochemistry. J Wood Sci 52:2–8

Hollander-Czytko H, Grabowski J, Sandorf I, Weckermann K, Weiler EW (2005) Tocopherol content and activities of tyrosine aminotransferase and cysteine lyase in *Arabidopsis* under stress conditions. J Plant Physiol 162:767–770

Hossain MA, Nakano Y, Asada K (1984) Monodehydroascorbate reductase in spinach chloroplasts and its participation in regeneration of ascorbate for scavenging hydrogen peroxide. Plant Cell Physiol 25:385–395

Hu X, Zhang A, Zhang J, Jiang M (2006) Abscisic acid is a key inducer of hydrogen peroxide production in leaves of maize plants exposed to water stress. Plant Cell Physiol 47:1484–1495

Igamberdiev AU, Seregelyes C, Manac N, Hill RD (2004) NADH- dependent metabolism of nitric oxide in alfalfa root cultures expressing barley hemoglobin. Planta 219:95–102

Jain K, Kataria S, Guruprasad KN (2003) Changes in antioxidant defenses of cucumber cotyledons in response to UV-B and to the free radicals generating compound AAPH. Plant Sci 165: 551–557

Jimenez A, Hernandez JA, Pastori G, delRio LA, Sevilla F (1998) Role of the ascorbate-glutathione cycle of mitochondria and peroxisomes in the senescence of pea leaves. Plant Physiol 118:1327–1335

Jubany-Mari T, Munné-Bosch S, López-Carbonell M, Alegre L (2009) Hydrogen peroxide is involved in the acclimation of the Mediterranean shrub, *Cistus albidus* L., to summer drought. J Exp Bot 60:107–120

Kandziora-Ciupa M, Ciepal R, Nadgorska-Socha A, Barczyk G (2013) A comparative study of heavy metal accumulation and antioxidant responses in *Vaccinium myrtillus* L. leaves in polluted and non-polluted areas. Environ Sci Pollut Res 20:4920–4932

Khan MIR, Khan NA (2014) Ethylene reverses photosynthetic inhibition by nickel and zinc in mustard through changes in PS II activity, photosynthetic-nitrogen use efficiency and antioxidant metabolism. Protoplasma 251:1007–1019

Khan MIR, Asgher M, Khan NA (2014) Alleviation of salt-induced photosynthesis and growth inhibition by salicylic acid involves glycinebetaine and ethylene in mungbean (*Vigna radiata* L.) Plant Physiol Biochem 80:67–74

Khan MIR, Nazir F, Asgher M, Per TS, Khan NA (2015) Selenium and sulfur influence ethylene formation and alleviate cadmium-induced oxidative stress by improving proline and glutathione production in wheat. J Plant Physiol 178:9–18

Khan MIR, Iqbal N, Masood A, Mobin M, Anjum NA, Khan NA (2016a) Modulation and significance of nitrogen and sulfur metabolism in cadmium challenged plants. Plant Growth Regul 78:1–11

Khan MIR, Khan NA, Masood A, Per TS, Asgher M (2016b) Hydrogen peroxide alleviates nickel-inhibited photosynthetic responses through increase in use-efficiency of nitrogen and sulfur, and glutathione production in mustard. Front Plant Sci 7:44

Kiffin R, Bandyopadhyay U, Cuervo AM (2006) Oxidative stress and autophagy. Antioxid Redox Signal 8:152–162

Kim MJ, Ciani S, Schachtman DP (2010) A peroxidase contributes to ROS production during *Arabidopsis* root response to potassium deficiency. Mol Plant 3:420–427

Krieger-Liszkay A, Fufezan C, Trebst A (2008) Singlet oxygen production in photosystem II and related protection mechanism. Photosynth Res 98:551–564

Kukreja S, Nandval AS, Kumar N, Sharma SK, Unvi V, Sharma PK (2005) Plant water status, H_2O_2 scavenging enzymes, ethylene evolution and membrane integrity of *Cicer arietinum* roots as affected by salinity. Biol Plant 49:305–308

Kwak JM, Mori IC, Pei ZM (2003) NADPH oxidase AtrbohD and AtrbohF genes function in ROS-dependent ABA signaling in *Arabidopsis*. EMBO J 22:2623–2633

Liang YC, Chen Q, Liu Q, Zhang WH, Ding RX (2003) Exogenous silicon (Si) increases antioxidant enzyme activity and reduces lipid peroxidation in roots of salt-stressed barley (*Hordeum vulgare* L.) Plant Physiol 160:1157–1164

Liu X, Hua X, Guo J, Qi D, Wang L, Liu Z et al (2008) Enhanced tolerance to drought stress in transgenic tobacco plants overexpressing *VTE1* for increased tocopherol production from *Arabidopsis thaliana*. Biotechnol Lett 30:1275–1280

Mafakheri A, Siosemardeh A, Bahramnejad B, Struik PC, Sohrabi Y (2010) Effect of drought stress on yield, proline and chlorophyll contents in three chick pea cultivars. Aust J Crop Sci 4:580–585

Maheshwari R, Dubey RS (2009) Nickel-induced oxidative stress and the role of antioxidant defense in rice seedlings. Plant Growth Regul 59:37–49

Malar S, Vikram SS, Favas PJC, Perumal V (2014) Lead heavy metal toxicity induced changes on growth and antioxidative enzymes level in water hyacinths [Eichhornia crassipes (Mart.)]. Bot Stud 55:54

Malik S, Ashraf M (2012) Exogenous application of ascorbic acid stimulates growth and photosynthesis of wheat (*Triticum aestivum* L.) under drought. Soil Environ 31:72–77

Martinez C, Montillet JL, Bresson E et al (1998) Apoplastic peroxidase generates superoxide anions in cells of cotton cotyledons undergoing the hypersensitive reaction to *Xanthomonas campestris* pv. *malvacearum* 18. Mol Plant-Microbe Interact 11:1038–1047

McDowell JM, Dangl JL (2000) Signal transduction in the plant immune response. Trends Biochem Sci 25:79–82

Mhamdi A, Queval G, Chaouch S, Vanderauwera S, Van Breusegem F, Noctor G (2010) Catalase function in plants: a focus on *Arabidopsis* mutants as stress-mimic models. J Exp Bot 61:4197–4220

Miller G, Suzuki N, Ciftci-Yilmaz S, Mittler R (2010) Reactive oxygen species homeostasis and signaling during drought and salinity stresses. Plant Cell Environ 33:453–467

Mishra V, Srivastava G, Prasad SM, Abraham G (2008) Growth, photosynthetic pigment and photosynthetic activity during seedling stage of cowpea (*Vigna unguiculata* L.) in response to UV-B and dimethoate. Pestic Biochem Physiol 92:30–37

Mishra S, Jha AB, Dubey RS (2011) Arsenite treatment induces oxidative stress, upregulates antioxidant system, and causes phytochelatin synthesis in rice seedlings. Protoplasma 248: 565–577

Mittler R (2002) Oxidative stress, antioxidants and stress tolerance. Trends Plant Sci 7:405–410

Mittler R, Vanderauwera S, Gollery M, Van Breusegem F (2004) Reactive oxygen gene network of plants. Trends Plant Sci 9:490–498

Moller IM, Jensen PE, Hansson A (2007) Oxidative modifications to cellular components in plants. Annu Rev Plant Biol 58:459–481

Montillet JL, Chamnongpol S, Rusterucci C, Dat J, van de Cotte B, Agnel JP, Battesti C, Inzé D, Van Breusegem F, Triantaphylides C (2005) Fatty acid hydroperoxides and H_2O_2 in the execution of hypersensitive cell death in tobacco leaves. Plant Physiol 138:1516–1526

Mullineaux PM, Rausch T (2005) Glutathione, photosynthesis and the redox regulation of stress-responsive gene expression. Photosynth Res 86:459–474

Munne-Bosch S, Alegre L (2000) Changes in carotenoids, tocopherols and diterpenes during drought and recovery, and the biological significance of chlorophyll loss in *Rosmarinus officinalis* plants. Planta 210:925–931

Murphy MP (2009) How mitochondria produce reactive oxygen species. Biochem J 417:1–13

Navrot N, Rouhier N, Gelhaye E, Jaquot JP (2007) Reactive oxygen species generation and antioxidant systems in plant mitochondria. Physiol Plant 129:185–195

Niyogi KK, Shih C, Chow WS, Pogson BJ, DellaPenna D, Bjorkman O (2001) Photoprotection in a zeaxanthin-and lutein-deficient double mutant of *Arabidopsis*. Photosynth Res 67:139–145

Noctor G, Foyer CH (1998) A re-evaluation of the ATP: NADPH budget during C3 photosynthesis. A contribution from nitrate assimilation and its associated respiratory activity? J Exp Bot 49:1895–1908

Noctor G, Gomez L, Vanacker H, Foyer CH (2002a) Interactions between biosynthesis, compartmentation and transport in the control of glutathione homeostasis and signaling. J Exp Bot 53:1283–1304

Noctor G, Veljovic-Jovanovic S, Driscoll S, Novitskaya L, Foyer CH (2002b) Drought and oxidative load in the leaves of C3 plants: a predominant role for photorespiration? Ann Bot 89:841–850

Noctor G, De Paepe R, Foyer CH (2007) Mitochondrial redox biology and homeostasis in plants. Trends Plant Sci 12:125–134

Palma JM, Corpas FJ, del Río LA (2009) Proteome of plant peroxisomes: new perspectives on the role of these organelles in cell biology. Proteomics 9:2301–2312

Parween T, Jan S, Mahmooduzzafar FT (2011) Alteration in nitrogen metabolism and plant growth during different developmental stages of green gram (*Vigna radiata* L.) in response to chlorpyrifos. Acta Physiol Plant 33:2321–2328

Pastore D, Trono D, Laus MN, Di Fonzo N, Flagella Z (2007) Possible plant mitochondria involvement in cell adaptation to drought stress. A case study: durum wheat mitochondria. J Exp Bot 58:195–210

Peng CL, Ou ZY, Liu N, Lin GZ (2005) Response to high temperature in flag leaves of super high-yielding rice Pei'ai 64S/E32 and Liangyoupeijiu. Rice Sci 12:179–186

Pfannschmidt T (2003) Chloroplast redox signals: how photosynthesis controls its own genes. Trends Plant Sci 8:33–41

Pinto E, Sigaud-Kutner TCS, Leitao MAS, Okamoto OK, Morse D, Colepicolo P (2003) Heavy metal induced oxidative stress in algae. J Phycol 39:1008–1018

Polidoros NA, Scandalias JG (1999) Role of hydrogen peroxide and different classes of antioxidants in the regulation of catalase and glutathione S-transferase gene expression in maize (*Zea mays* L.) Physiol Plant 106:112–120

Racchi ML, Bagnoli F, Balla I, Danti S (2001) Differential activity of catalase and superoxide dismutase in seedlings and in vitro micropropagated oak (*Quercus robur* L.) Plant Cell Rep 20:169–174

Rao MV, Paliyath G, Ormrod DP (1996) Ultraviolet-B- and ozone-induced biochemical changes in antioxidant enzymes of *Arabidopsis thaliana*. Plant Physiol 110:125–136

Rasmusson AG, Geisler DA, Moller IM (2008) The multiplicity of dehydrogenases in the electron transport chain of plant mitochondria. Mitochondrion 8:47–60

Rausch T, Wachter A (2005) Sulfur metabolism: a versatile platform for launching defense operations. Trends Plant Sci 10:503–509

Rhoads DM, Umbach AL, Subbaiah CC, Siedow JN (2006) Mitochondrial reactive oxygen species. Contribution to oxidative stress and interorganellar signaling. Plant Physiol 141:357–366

Roychoudhury A, Basu S (2012) Ascorbate-glutathione and plant tolerance to various abiotic stresses. In: Ajum NA, Umar SU, Ahmad A (eds) Oxidative stress in plants: causes, consequences and tolerance. IK International Publishing House, New Delhi, pp 177–258

Roychoudhury A, Ghosh S (2013) Physiological and biochemical responses of mung bean (*Vigna radiata* L. Wilczek) to varying concentrations of cadmium chloride or sodium chloride. Unique J Pharm Biol Sci 1:11–21

Roychoudhury A, Basu S, Sarkar SN, Sengupta DN (2008) Comparative physiological and molecular responses of a common aromatic indica rice cultivar to high salinity with non-aromatic indica rice cultivars. Plant Cell Rep 27:1395–1410

Sairam RK, Deshmukh PS, Saxena DC (1998) Role of antioxidant systems in wheat genotypes tolerance to water stress. Biol Plant 41:387–394

Scandalias JG (1990) Response of plant antioxidant defense genes to environmental stress. Adv Genet 28:1–41

Sekhar PN, Amrutha RN, Sangam S, Verma DP, Kishor PB (2007) Biochemical characterization, homology modelling and docking studies of ornithine delta aminotransferase- an important enzyme in proline biosynthesis of plants. J Mol Graph Model 26:709–719

Shalata A, Neumann PM (2001) Exogenous ascorbic acid (vitamin C) increases resistance to salt stress and reduces lipid peroxidation. J Exp Bot 52:2207–2211. doi:10.1093/jexbot/52.364.2207

Shao HB, Liang ZS, Shao MA, Sun Q (2005) Dynamic changes of anti- oxidative enzymes of 10 wheat genotypes at soil water deficits. Colloids Surf B: Biointerfaces 42:187–195

Sharma P, Dubey RS (2004) Ascorbate peroxidase from rice seedlings: properties of enzyme isoforms, effects of stresses and protective roles of osmolytes. Plant Sci 167:541–550

Sharma P, Dubey RS (2005) Modulation of nitrate reductase activity in rice seedlings under aluminium toxicity and water stress: role of osmolytes as enzyme protectant. J Plant Physiol 162:854–864

Sharma P, Dubey RS (2007) Involvement of oxidative stress and role of antioxidative defense system in growing rice seedlings exposed to toxic concentrations of aluminum. Plant Cell Rep 26:2027–2038

Sharma P, Jha AB, Dubey RS (2010) Oxidative stress and antioxidative defense system in plants growing under abiotic stresses. In: Pessarakli M (ed) Handbook of plant and crop stress, 3rd edn. CRC Press, Taylor and Francis Publishing Company, Boca Raton, pp 89–138

Sharma P Jha AB, Dubey RS, Pessarakli M (2012) Reactive oxygen species, oxidative damage, and antioxidative defense mechanism in plants under stressful conditions. J Bot 217037

Shringarpure R, Davies KJ (2002) Protein turnover by the proteasome in aging and disease. Free Radic Biol Med 32:1084–1089

Sieferman-Harms D (1987) The light harvesting function of carotenoids in photosynthetic membrane. Plant Physiol 69:561–568

Simova-Stoilova L, Vaseva I, Grigorova B, Demirevska K, Feller U (2010) Proteolytic activity and cysteine protease expression in wheat leaves under severe soil drought and recovery. Plant Physiol Biochem 48:200–206

Singh S, Khan NA, Nazar R, Anjum NA (2008) Photosynthetic traits and activities of antioxidant enzymes in black gram (*Vigna mungo* L. Hepper) under cadmium stress. Am J Plant Physiol 3:25–32

Smirnoff N (2005) Ascorbate, tocopherol and carotenoids: metabolism, pathway engineering and functions. In: Smirnoff N (ed) Antioxidants and reactive oxygen species in plants. Blackwell Publishing Ltd., Oxford, pp 53–86

Sreedevi S, Krishnan PN (2012) Detoxification of heavy metal induced oxidative stress in plants, enzymatic and non-enzymatic mechanisms- a mini review. In: Ajum NA, Umar SU, Ahmad A (eds) Oxidative stress in plants: causes, consequences and tolerance. IK International Publishing House, New Delhi, pp 291–307

Tanou G, Molassiotis A, Diamantidis G (2009) Hydrogen peroxide-and nitric oxide-induced systemic antioxidant prime-like activity under NaCl-stress and stress-free conditions in citrus plants. J Plant Physiol 166:1904–1913

Turrens JF (2003) Mitochondrial formation of reactive oxygen species. Plant Physiol 552:335–344

Tuteja N (2007) Mechanisms of high salinity tolerance in plants. Meth Enzymol Osmosens Osmosignal 428:419–438

Ushimaru T, Nakagawa T, Fujioka Y, Daicho K, Naito M, Yamauchi Y, Nonaka H, Amako K, Yamawaki K, Murata N (2006) Transgenic Arabidopsis plants expressing the rice dehydroascorbate reductase gene are resistant to salt stress. J Plant Physiol 163:1179–1184

Vaidyanathan H, Sivakumar P, Chakrabarty R, Thomas G (2003) Scavenging of reactive oxygen species in NaCl-stressed rice (*Oryza sativa* L.)-differential response in salt-tolerant and sensitive varieties. Plant Sci 165:1411–1418

Vanaja M, Charyawa NVN, Rao KVN (2000) Effect of cadmium on carbohydrate, nucleic acids and phenolic content in *Stigeoclonium tenue* Kutz. Indian J Plant Physiol 5:253–256

Verbruggen N, Hermans C (2008) Proline accumulation in plants: a review. Amino Acids 35:753–759

Wagner D, Przybyla D, Opden Camp R, Kim C, Landgraf F, Lee KP, Wursch M, Laloi C, Nater M, Hideg E, Apel (2004) The genetic basis of singlet oxygen-induced stress responses of *Arabidopsis thaliana*. Science 306:1183–1185

Wang Y, Ying Y, Chen J, Wang XC (2004) Transgenic *Arabidopsis* overexpressing Mn-SOD enhanced salt-tolerance. Plant Sci 167:671–677

Wang R, Liu S, Zhou F, Ding C, Hua C (2014) Exogenous ascorbic acid and glutathione alleviate oxidative stress induced by salt stress in the chloroplasts of *Oryza sativa* L. Z Naturforsch C 69:226–236

Wojtaszek P (1997) Oxidative burst: an early plant response to pathogen infection. Biochem J 322:681–692

Yang Z, Wu Y, Li Y, Ling HQ, Chu C (2009) OsMT1a, a type 1 metallothionein, plays the pivotal role in zinc homeostasis and drought tolerance in rice. Plant Mol Biol 70:219–229

Zaefyzadeh M, Quliyev RA, Babayeva SM, Abbasov MA (2009) The effect of the interaction between genotypes and drought stress on the superoxide dismutase and chlorophyll content in durum wheat landraces. Turk J Biol 33:1–7

Zlatev ZS, Lidon FC, Ramalho JC, Yordanov IT (2006) Comparison of resistance to drought of three bean cultivars. Biol Plant 50:389–394

Biotechnological Perspective of Reactive Oxygen Species (ROS)-Mediated Stress Tolerance in Plants

3

Thammineni Chakradhar, Srikrishna Mahanty, Ramesha A. Reddy, Kummari Divya, Palakolanu Sudhakar Reddy, and Malireddy K. Reddy

Abstract

All environmental cues lead to develop secondary stress conditions like osmotic and oxidative stress conditions that reduces average crop yields by more than 50% every year. The univalent reduction of molecular oxygen (O_2) in metabolic reactions consequently produces superoxide anions ($O_2^{\bullet-}$) and other reactive oxygen species (ROS) ubiquitously in all compartments of the cell that disturbs redox potential and causes threat to cellular organelles. The production of ROS further increases under stress conditions and especially in combination with high light intensity. Plants have evolved different strategies to minimize the accumulation of excess ROS like avoidance mechanisms such as physiological adaptation, efficient photosystems such as C4 or CAM metabolism and scavenging mechanisms through production of antioxidants and antioxidative enzymes. Ascorbate-glutathione pathway plays an important role in detoxifying excess ROS in plant cells, which includes superoxide dismutase (SOD) and ascorbate peroxidase (APX) in detoxifying $O_2^{\bullet-}$ radical and hydrogen peroxide (H_2O_2)

T. Chakradhar • S. Mahanty • R.A. Reddy • M.K. Reddy
Plant Molecular Biology Group, International Centre for Genetic Engineering and Biotechnology (ICGEB), Aruna Asaf Ali Marg, New Delhi 110 067, India

K. Divya
Cell, Molecular Biology & Genetic Engineering Group, International Crops Research Institute for the Semi-Arid Tropics (ICRISAT), Patancheru, Hyderabad 502 324, Telangana, India

P.S. Reddy (✉)
Plant Molecular Biology Group, International Centre for Genetic Engineering and Biotechnology (ICGEB), Aruna Asaf Ali Marg, New Delhi 110 067, India

Cell, Molecular Biology & Genetic Engineering Group, International Crops Research Institute for the Semi-Arid Tropics (ICRISAT), Patancheru, Hyderabad 502 324, Telangana, India
e-mail: p.sudhakarreddy@cgiar.org; palakolanusreddy@gmail.com

M.I.R. Khan, N.A. Khan (eds.), *Reactive Oxygen Species and Antioxidant Systems in Plants: Role and Regulation under Abiotic Stress*,
DOI 10.1007/978-981-10-5254-5_3

respectively, monodehydroascorbate reductase (MDHAR), dehydroascorbate reductase (DHAR) and glutathione reductase (GR) involved in recycling of reduced substrates such as ascorbate and glutathione. Efficient ROS management is one of the strategies used by tolerant plants to survive and perform cellular activities under stress conditions. The present chapter describes different sites of ROS generation and and their consequences under abiotic stress conditions and also described the approaches to overcome oxidative stress through genomics and genetic engineering.

Keywords
Ascorbate-glutathione cycle • Abiotic stress • Oxidative stress • Reactive oxygen species

Abbreviations

APX	Ascorbate peroxidase
ASA	Ascorbate
CAT	Catalase
DHAR	Dehydroascorbate reductase
GPX	Glutathione peroxidase
GR	Glutathione reductase
GSH	Reduced glutathione
GSSG	Oxidized glutathione
MDHAR	Monodehydroascorbate reductase
ROS	Reactive oxygen species
SOD	Superoxide dismutase

3.1 Introduction

In daily life, plants encounter different abiotic stresses such as water deficit, extreme temperatures, high salinity, high light intensity, heavy metals and more often combination of these stresses under field conditions. However, plants cannot escape from these harsh environmental stresses due to their sessile life. Although all plants are equipped with adaptive mechanisms, to encounter such environmental cues, difference in their allelic constituency has left few crop plants vulnerable. It is estimated that the average yields are reduced to 50%, due to such abiotic stress factors (Vij and Tyagi 2007). Therefore, to meet the food demand of increasing population and to cope up with ever-changing drastic climatic conditions due to the global warming effects, there is an urgent need to develop varieties/hybrids that yields better under abiotic stresses. All the primary stresses like drought, salinity and extreme temperatures leads to secondary stresses such as osmotic and oxidative stresses at cellular level. All living organisms utilize oxygen as reducing agent to generate chemical energy, i.e. ATP during active electron transport system. Paradoxically, the univalent reduction of O_2 in metabolic reactions consequently produces superoxide anions ($O_2^{\bullet -}$) ubiquitously in all compartments of the cell. In addition to the respiratory chain, the O_2 becomes an electron acceptor in photosynthetic electron transport chain in plants during light reaction and generates large amount of $O_2^{\bullet -}$, and their production further increases under various environmental stresses and especially in combination with high light intensity (Asada 2006). These $O_2^{\bullet -}$anions dismutate to H_2O_2, and in the presence of metal ions, and further $O_2^{\bullet -}$ anions react again with H_2O_2 to generate highly reactive hydroxyl radicals ($^{\bullet}OH$). These $^{\bullet}OH$ radical can react with DNA, proteins, lipids and other organic constituents of the cell and cause severe damage.

Plants have evolved different mechanisms to minimize the accumulation of excess ROS like avoidance mechanisms such as anatomical adaptation, suppression of photosynthesis and physiological adaptation such as C4 or CAM metabolism. These processes avoid production of excess ROS under environmental stress conditions. Scavenging mechanisms like production of antioxidants and antioxidative enzymes remove excess ROS produced during abiotic stresses. Efficient ROS detoxification is one of the strategies used by stress-tolerant plants for their survival. Therefore, in the present chapter, we focused on generation and maintenance of ROS in plant cell, and discussed approaches to minimize the damage caused by oxidative stress. Efforts made in transgenic technology using antioxidant genes, to overcome oxidative stress, have also been analysed. Finally, opportunities available in modern technologies like genomics and challenges to consider in overcoming oxidative stress tolerance are presented.

3.2 Antioxidants Involved in Scavenging Pathways by Regulating Molecular and Physiological Approaches

The balance between generation and detoxification of ROS is altered when plants are encountered with different abiotic stress conditions. ROS are also involved in signalling of plant growth, development and perception of responses in biotic and abiotic stresses. However, at higher concentration, they react and damage a number of biomolecules and eventually lead to programmed cell death (PCD) (Miller et al. 2008; Petrov and Van Breusegem 2012). Plants have evolved array of antioxidants and antioxidative enzymes along with other small molecules to detoxify ROS in different organelles. However, ROS generation and scavenging is imbalanced under stress conditions leading to oxidative stress (Tripathy and Oelmüller 2012). Enzymatic ROS scavenging of plants include SOD, catalase (CAT), glutathione peroxidise (GPX), glutathione S-transferase (GST), peroxiredoxins (PRX) and ascorbate-glutathione (AsA-GSH) pathway enzymes consisting of APX,

MDHAR, DHAR and GR. Non-enzymatic ROS detoxification system including ascorbate (AsA), glutathione (GSH), tocopherol, carotenoids and flavonoids also plays important role in ROS homeostasis (Khan and Khan 2014; Khan et al. 2014, 2015, 2016; Noctor et al. 2014; Gill and Tuteja 2010; You and Chan 2015).

3.2.1 Enzymatic ROS

The ROS detoxification in chloroplast, peroxisomes and cytosol is carried out by an important antioxidant mechanism in plants called ascorbate-glutathione pathway which comprises enzymes to detoxify the $O_2^{\bullet-}$ using ascorbate and glutathione as reducing substrates. SOD is the first candidate in line of defence in this pathway to reduce H_2O_2 to H_2O and O_2 using ascorbate as reducing agents. In this process, ascorbate is oxidized to monodehydroascorbate, which is spontaneously disproportionated into dehydroascorbate. Ascorbate is regenerated from monodehydroascorbate by MDHAR using reducing power of NADPH and by DHAR using GSH as reducing agent to convert dehydroascorbate into ascorbate. GSH is recycled back from oxidized GSH through GR using NADPH (Pandey et al. 2015).

3.2.1.1 Superoxide Dismutase (SOD)

SODs are metalloenzymes that are designated as initial mode of defence against toxic effects of ROS in all the cellular compartments generating $O_2^{\bullet-}$ in nearly all aerobic organisms (Touati 1997). SODs generate less toxic molecule H_2O_2 by alternately adding or removing electron from $O_2^{\bullet-}$. Two superoxide molecules react with each other and dismutate to O_2 and H_2O_2 and this follows second-order reaction, whereas reaction catalysed by SOD is first order and is necessary to stop damage caused by superoxide radical to different biomolecules. The presence of SOD at the site of radical formation is crucial as phospholipid membrane is impermeable to charged $O_2^{\bullet-}$ radical (Takahashi and Asada 1983). SOD dismutates $O_2^{\bullet-}$ into less toxic H_2O_2, thereby preventing formation of more harmful $^{\bullet}OH$ via Haber-Weiss reaction (Gill and Tuteja 2010). SODs are classified into iron (Fe-SOD), manganese (Mn-SOD), copper-zinc (Cu-Zn SOD) and nickel (Ni-SOD) based on cofactor requirement. Fe-SOD localizes to chloroplast and is sensitive to H_2O_2, while Mn-SOD located both in mitochondria and peroxisomes and resistant to H_2O_2 and Cu-Zn SOD localizes to chloroplast and cytoplasm and is sensitive to H_2O_2 (Gill and Tuteja 2010). Using *in silico* approach, Nath et al. (2014) identified one *SOD* copper chaperone gene and other seven *SOD* isoforms in rice including two cytosolic Cu-Zn SODs, one putative Cu-Zn SOD-like, one plastidic SOD, two Fe-SODs and one Mn-SOD. These isoforms display differential expression during different developmental stages implicating specific role played by each of the gene. Seven *Arabidopsis* SODs were profiled for expression levels in response to oxidative stresses, and that increase in Fe-SOD levels in response to UV irradiation and high light stress was reported, but not to ozone exposure (Kliebenstein et al. 1998). When pea plants were exposed to high salinity,

Fe-SOD levels increased with corresponding decrease in Cu-Zn SOD, but when plants were treated with less severity salt stress, enhanced Cu-Zn SOD and Mn-SOD activities were observed (Gomez et al. 1999; Hernandez et al. 1995).

3.2.1.2 Ascorbate Peroxidise (APX)

APX belongs to class I family of peroxidases, which contains heme as a prosthetic group. APX catalyses transfer of electrons from specific substrate, ascorbate, to H_2O_2, thereby reducing peroxide to water as a part of ascorbate-glutathione pathway. APX activity is highly sensitive to ascorbate concentration, which is reflected in decline of its activity under limited ascorbate concentration. Studies have revealed that APX plays an important role in scavenging excess ROS, therefore maintaining ROS homeostasis and oxidative protection in plants under drought, heat, methyl viologen and high light stress (Caverzan et al. 2012; Zhang et al. 2013). *Arabidopsis thaliana* genome codes for eight isoforms of APXs, including cytosolic (APX1, APX2 and APX6), microsome membrane-bound (APX3, APX4 and APX5) and chloroplastic (stromal APX and thylakoid APX) (Panchuk et al. 2002; Dabrowska et al. 2007). Rice genome also contains eight *APX* genes including two cytosolic (*OsAPX1* and *OsAPX2*), two peroxisomal (*OsAPX3* and *OsAPX4*) and three chloroplastic (*OsAPX5*, *OsAPX7* and *OsAPX8*), and *OsAPX6* localizes to mitochondria (Teixeira et al. 2006). APX isoforms are developmental and stress modulated in response to drought, salinity, extreme temperatures, pathogen attack, UV radiation and heavy metal stress (Shigeoka et al. 2002; Dabrowska et al. 2007). Expression profiles of APX isoforms were studied by exposing rice seedlings to salt stress and found that *OsAPX1*, *OsAPX3*, *OsAPX4*, *OsAPX5* and *OsAPX6* transcripts were not altered, but *OsAPX2*, *OsAPX7* and *OsAPx8* transcripts were differently modulated under salinity stress (Teixeira et al. 2006). The role of APX in cold stress tolerance was established by Sato et al. (2011) who found that overexpression of cytosolic *OsAPX1* has significantly improved male sterility at continuous cold stress of 12 °C than WT plants that exhibited complete male sterility. This study has lot of impact on improving hybrid rice production. However, *OsAPX2* overexpression enhances salinity tolerance in transgenic *Arabidopsis* and *Medicago sativa* (Guan et al. 2012; Lu et al. 2007). The role of *OsAPX2* in abiotic stress was dissected through knockout mutant and overexpression studies in rice lines. In a study, Zhang et al. (2013) created knockout mutants for cytosolic *APX2* gene in rice, using random T-DNA insertion method, and found that mutants are hypersensitive to all abiotic stresses. Contrasting to this phenotype, rice plants overexpressing *OsAPX2* exhibited increased tolerance to drought, salinity and cold. The transcript levels of pea cytoplasmic APX increases 4-fold in response to drought and significantly increased to 15-fold after the stress withdrawal (Mittler and Zilinskas 1994). The results from above-mentioned experiments suggested that APX transcript and protein are induced in response to various abiotic stress conditions and therefore play important role in oxidative stress tolerance.

3.2.1.3 Monodehydroascorbate Reductase (MDHAR)

Monodehydroascorbate is a primary oxidation product of ascorbate which is reduced back to ascorbate by MDHAR using NADPH as specific electron donor. MDHAR is a FAD enzyme, and the crystal structure of the enzyme from *Oryza sativa* along with its cofactors was elucidated (Park et al. 2016). The overall protein structure is similar to iron-sulphur reductase except the active site forming long loop. Based on point mutation, it was found that Arg320 plays an important role in substrate binding, and Tyr349 is involved reduction reaction in the transfer of electron from NADPH to substrate via FAD (Park et al. 2016). Since ascorbate acts as antioxidant and also as reducing substrate for the APX activity, its levels are to be maintained high during stress conditions. Recycling of monodehydroascorbate to ascorbate can be enhanced by higher expression and activity of MDHAR. Overexpression of MDHAR from mangrove plant confers salinity stress and showed better yield through increased tiller and grain weight in rice (Sultana et al. 2012). In another experiment, transgenic overexpression of LeMDHAR in *Arabidopsis* chloroplast has conferred methyl viologen (MV)-induced oxidative damage through increased levels of reduced ascorbate (Li et al. 2010). Transgenic tobacco plants overexpressing halophytic mangrove chloroplastic MDHAR confer salinity stress tolerance by enhancing redox state of ascorbate and APX activity (Kavitha et al. 2010).

3.2.1.4 Dehydroascorbate Reductase (DHAR)

Oxidation of ascorbate produces short-lived radical monodehydroascorbate, which is converted to ascorbate by MDHAR, or disproportionates non-enzymatically to ascorbate and dehydroascorbate (DHA). DHAR regenerate reduced ascorbate from dehydroascorbate using glutathione as reducing substrate, thus contributing to cellular redox homeostasis (Chen and Gallie 2006). The molecular basis of enzyme DHAR was elucidated from crystal structure of the enzyme isolated from *Oryza sativa* by Do et al. (2016). The structure of the protein in native confirmation along with ascorbate and glutathione-bound form provides information regarding binding sites and interacting residues. The cys20 was involved in electron transfer at the active site to reduce dehydroascorbate to ascorbate by ping-pong kinetic mechanism. DHAR has been isolated, characterized and validated for its role in stress tolerance in different genetic backgrounds and organelles. (Chew et al. 2003; Kataya and Reumann 2010; Shimaoka et al. 2000; Mittova et al. 2000; Teixeira et al. 2005). DHAR plays a vital role in abiotic stress tolerance and this is supported by abundancy of its transcripts under different abiotic stress conditions (Fan et al. 2014; Lu et al. 2008). Mutants with low or no expression of DHAR revealed loss of Rubisco activity, low CO_2 assimilation and increased leaf ageing (Chen and Gallie 2006). In contrast, plants with high DHAR expression exhibited delayed leaf senescence, higher levels of chlorophylls and improved photosynthetic efficiency. Chen and Gallie (2004) demonstrated that high DHAR activity in guard cells interferes ABA-mediated stomatal closure, allowing increased transpiration under drought. This condition might pose negative impact on plant survival of transgenic plants developed with DHAR overexpression. However, this effect could be minimized through tissue-specific expression of DHAR.

3.2.1.5 Glutathione Reductase (GR)

GR is an important flavoprotein oxidoreductase antioxidative enzyme involved in regeneration of reduced form of glutathione using NADPH as reducing substrate, thereby contributing to redox homeostasis. GR contains FAD and highly conserved GXGXXA motif in the NADPH-binding site, and arginine present therein facilitates binding of NADPH to this domain. Reduction of GSSG to GSH by NADPH at the active site is catalysed by cysteine residues (Trivedi et al. 2013). GR activity has been detected in chloroplast, mitochondria, peroxisomes and cytosol.

3.2.2 Non-enzymatic ROS Scavengers

Non-enzymatic antioxidants essentially include the redox buffers ascorbate and glutathione, as well as tocopherol, flavonoids, alkaloids and carotenoids. Ascorbate and glutathione pools existing at high concentrations (5–20 mM AsA and 1–5 mM GSH) during stress conditions indicate their role in oxidative stress-mediated abiotic defence. This is further evident in transgenic mutants of ascorbate and glutathione, where the plants were found hypersensitive to abiotic stress conditions (Creissen et al. 1999). Maintaining high ratio of reduced/oxidized levels of cellular ascorbate and glutathione is necessary for smooth functioning of ROS-scavenging activity. Little is known about flavonoids and carotenoids in ROS detoxification in plants. However, flavonoids and carotenoids are supposedly acting as auxiliary or secondary or accessory pigments in leaves that absorb the excess light energy not taken by chlorophyll and help in the prevention of photo-oxidation of the same.

3.2.2.1 Ascorbate

Ascorbate is a low molecular weight, water-soluble and highly abundant antioxidant that mitigates the bad effects of enhanced ROS. Ascorbate is generally present in all the tissues and cellular compartments in millimolar concentration with chloroplast stroma reported to contain highest concentration up to 50 mM (Gill and Tuteja 2010; Smirnoff 2000). Ascorbate presents mostly in reduced form under optimal growth conditions. Ascorbate is a multifunctional molecule involved in normal growth and development and in response to stresses by directly detoxifying several ROS and also by donating electrons for detoxification of ROS in enzymatic catalysed reactions (Smirnoff 2000). Ascorbate is also involved in regeneration of vitamin E. The levels of ascorbate are dependent on synthesis, oxidation, recycling and transport. Ascorbate is oxidized to monodehydroascorbate when it reacts with superoxide, singlet oxygen and H_2O_2. Recycling of ascorbate takes place from disproportion of monodehydroascorbate and also through enzymatic reactions catalysed by MDHAR and DHAR. The elucidation of biosynthesis of ascorbate came from experiments of Conklin et al. (1996), where they have isolated and tested ascorbate-deficient mutants of *Arabidopsis*. Analysis of *vtc*1 mutant which contains only 25% of ascorbate compared to wild type by genetics, molecular and biochemical methods showed that this gene encodes for GDP mannose pyrophosphorylase. The ascorbate levels are closely linked to light and photosynthesis, as *Arabidopsis* plants grown under light contain increased ascorbate than dark-grown plants and also correlate with ascorbate biosynthesis enzyme transcript levels (Yabuta et al. 2007). Foliar application of ascorbic acid in *Zea mays* alleviates ill effects of water stress by reducing antioxidative enzyme activities (Dolatabadian et al. 2009). In response to ozone treatment in winter wheat, apoplastic ascorbate in leaves correlates with sensitive phenotypes of varieties (Feng et al. 2010).

3.2.2.2 Glutathione

Glutathione is a sulphur-contacting tripeptide (GSH, γ-L-glutamyl-L-cysteinylglycine) present in millimolar concentrations in reduced form in different cellular compartments. Glutathione plays a vital role in plant growth and development, evident by its role in phytochrome signalling and cellular redox homeostasis. Its role in stress condition is prominent and discussed largely (Noctor et al. 2011). GSH biosynthesis takes place from cystine, glutamine and glycine in two ATP-dependent reactions. The first reaction is rate-limiting step that is catalysed by glutamate-cysteine ligase to form γ-glutamylcysteine. In the second reaction, glutathione synthetase adds glycine to γ-glutamylcysteine to form glutathione (GSH). The reduced form of GSH is utilized as a substrate for multiple cellular processes to form oxidized GSSG. The ratio of GSH to GSSG is critical for maintaining cellular redox balance. The GSH scavenges $O_2^{\bullet -}$, H_2O_2 and $^{\bullet}OH$ radical and also regenerates powerful antioxidant ascorbate through ascorbate-glutathione cycle. In addition, GSH is also used as a precursor in synthesis of phytochelatins, in response to heavy metal stress and sulphur assimilation (Noctor et al. 2011). Several studies have strongly demonstrated that GSH concentration correlates with abiotic stress-induced oxidative stress tolerance in plants. Two cultivars of *Vigna radiata* L. were analysed for salt stress tolerance and found that variety CO 4 showed greater stress susceptibility and associated with higher GSSG and lower GSH concentration than Pusa Bold (Sumithra et al. 2006). Involvement of GSH in light signalling came from studies of rax1 mutant whose glutathione content is 50% less than wild type that display constitutive expression of high light-induced gene, *APX*2 (Ball et al. 2004).

In addition to antioxidative scavenging mechanism, reduction of ROS generation mechanisms during abiotic stresses also plays an important role in ROS homeostasis. For example, alternative oxidases (AOX) can divert the electrons from electron transport chain in mitochondria, instead of oxygen to produce $O_2^{\bullet -}$, and thereby prevent excess generation of ROS (Maxwell et al. 1999; You and Chan 2015). Other mechanisms include anatomical and morphological changes such as leaf movement and folding and rearrangement of photosynthetic apparatus depending on the amount of light energy to be harvested based on availability CO_2 (Mittler 2002).

3.3 ROS Generation Sites in a Plant Cell

ROS production is an inevitable part of aerobic metabolism of a living organism due to the partial reduction nature of molecular oxygen. ROS are produced continuously at low concentration (below threshold levels) in normal plant cells, at sites that are actively engaged in electron transportation reactions (Choudhury et al. 2013). According to an estimation, about 1% of total O_2 consumed by plants is being utilized to generate ROS in various cellular organelles like chloroplasts, mitochondria and peroxisomes (Bhattacharjee 2005). ROS is also known as reactive oxygen intermediates (ROI) or active oxygen species (AOS). ROS with potent damaging effect includes $O_2^{\bullet-}$, singlet oxygen ($^1O_2$), $^{\bullet}OH$, perhydroxyl radical ($HO_2^{\bullet}$), H_2O_2, alkoxy radical ($RO^{\bullet}$), peroxy radical (ROO) and organic hydroperoxide (ROOH) (Konig et al. 2012; Mignolet-Spruyt et al. 2016).

3.3.1 Chloroplast

The thylakoid membrane system present in chloroplast harbours all components of the light-capturing photosynthetic apparatus. In chloroplast, the presence of triplet chlorophyll and electron transport chain (ETC) in PSI and PSII are the major sites of ROS ($O_2^{\bullet-}$, $^1O_2$ and H_2O_2) production. Under unstressed condition, the electrons flows from the excited photosystem centres to reduce $NADP^+$ to NADPH. Then it enters the Calvin cycle and reduces the final electron acceptor, CO_2. Under water-stressed condition with the combination of limited CO_2 and high light intensity, the deprivation of $NADP^+$ causes overloading in electron transport chain (ETC) system, causing leakage of electrons from ferredoxin to O_2, reducing it to $O_2^{\bullet-}$ (Karupandian et al. 2011; Das and Roychoudhury 2014). This reaction is called Mehler reaction. Transfer of excess electrons or electron leakage to molecular oxygen (O_2) is also observed from 2Fe-2S and 4Fe-4S clusters in PSI and QA-QB complex of PS II leading to the production of $O_2^{\bullet-}$ radical (Sharma et al. 2012). The $O_2^{\bullet-}$continuously dismutated to H_2O_2 enzymatically by SOD found on the external 'stromal' membrane surface or may be protonated to $HO_2^{\bullet-}$ on the internal, 'lumen' membrane surface (Miller et al. 2008). The singlet oxygen ($^1O_2$) is a natural by-product of photosynthesis, mainly formed at PS II even under low-light conditions. At Fe-S centres where Fe2+ is available, H_2O_2 may be transformed to much more toxic $^{\bullet}OH$ ion through the Fenton reaction. So oxygen generated in the chloroplasts during photosynthesis accepts electrons passing through the photosystems, resulting in the formation of ROS (Zolla and Rinalducci 2002; Gill and Tuteja 2010).

3.3.2 Mitochondria

The mitochondrial ETC system and various enzymes present in its matrix produce ROS. Unlike animal cell, plant mitochondria have oxygen-rich environment and contain photorespiration, a condition that favours ROS generation. Respiratory complex I, II and III are designated major sites for $O_2^{\bullet-}$ and H_2O_2 production in mitochondrial ETC (Huang et al. 2016). Plant mitochondria are also involved in pathogen-mediated defence reaction and PCD (Zhu et al. 2014; Wu et al. 2015). When the concentration of NAD^+-linked substrates is limited at complex I, it leads to reverse electron flow from complex III to complex I. As a result, the production of ROS is enhanced at complex I. The hydrolysis of ATP regulates the reverse electron flow (Turrens 2003; Møller et al. 2007; Noctor et al. 2007; Paterson et al. 2015). In a plant cell under normal aerobic conditions, ETC and ATP synthesis are tightly coupled, but under water stress, the inhibition and modification of ETC components cause the over-reduction of electron carrier like ubiquinone pool which enhances the ROS production (Rhoads et al. 2006; Blokhina and Fagerstedt 2010). During water stress, the lower rate of chloroplast ATP synthesis is compensated by increasing the synthesis rate of ATP in mitochondria; hence, the mitochondrial ROS production is also increased (Atkin and Macherel 2009). The enzymes present in the mitochondria matrix like aconitase directly produce the ROS, and enzyme-like 1-galactono-γ-lactone dehydrogenase (GAL) indirectly produces ROS by supplying electrons to the ETC (Rasmusson et al. 2008). The primary ROS, superoxide $O_2^{\bullet-}$, formed by monovalent reduction in the ETC dismutates quickly into H_2O_2 by the MnSOD (Moller 2001). It has been estimated that 1–5% of total O_2 consumed by mitochondria converts to H_2O_2. This H_2O_2 can be further converted to extremely active hydroxyl radical ($^{\bullet}OH$) by reacting with reduced Fe^{2+} and Cu^+ in the Fenton reaction (Moller 2001; Rhodas et al. 2006). The mitochondria have two major enzymes, alternative oxidase (AOX) and Mn-SOD, to counteract the oxidative stress. The AOX lower down the ROS by catalysing the O_2-dependent oxidation of ubiquinone pool (Jezek and Hlavata 2005). It is found that membrane-bound uncoupling proteins reduce production of ROS by diverting electrons to complex I, III and IV. Transgenic expression of uncoupling protein I in *Arabidopsis* exhibited reduced ROS generation clubbed with tolerance to multiple abiotic stresses (Barreto et al. 2014).

3.3.3 Peroxisome

Peroxisomes are minute subcellular organelle with single lipid bilayer membrane and harbours important oxidative reactions of the cell (fatty acid β-oxidation). H_2O_2 is the major ROS generated in peroxisomes (Luis and López-Huertas 2006). Peroxisome also produces $O^{\bullet-}$, like chloroplasts and mitochondria, as a consequence of the normal metabolic process. The major metabolic chemistries that produce H_2O_2 in peroxisomes include the photorespiratory glycolate oxidase reaction, the fatty acid β-oxidation, the enzymatic reaction of flavin oxidases and the disproportionation of O2•− radical. Under water stress conditions, the stomata remain closed (Sandalio and Rumero puretas 2015). So, there is a considerable reduction in the ratio of CO_2 to O_2, which causes increased photorespiration, leading to glycolate formation. This glycolate is oxidized by the glycolate oxidase in peroxisome and produces H_2O_2, making it the leading producer of H_2O_2 (Noctor et al. 2002; Mittler et al. 2004; Rodríguez-Serrano et al. 2016). As mentioned earlier, fatty acid β-oxidation is another source of H_2O_2 formation in peroxisomes through action of acyl-CoA oxidase (Palma et al. 2009). It has been established that the $O_2^{\bullet -}$ is generated at two major sites of peroxisome: one is the peroxisomal matrix containing the xanthine oxidase (XOD) that catalyses both xanthine and hypoxanthine into uric acid and generate $O_2^{\bullet -}$ as a by-product and second is the peroxisomal membranes where a small ETC composed of a flavoprotein NADH and Cyt b is involved. The three polypeptides (PMPs) with molecular masses of 18, 29 and 32 kDa integrated in the peroxisomal membrane were found to be involved in $O_2^{\bullet -}$ production. It is found that the 18- and 32-kDa PMPs use NADH as electron donor for $O_2^{\bullet -}$ production. The 29-kDa PMP was clearly dependent on NADPH and was able to reduce cytochrome c with NADPH as electron donor (Lopez-Huertas et al. 2000). Out of the three polypeptides, the 18 kDa is main producer of $O_2^{\bullet -}$, which was believed to be a cytochrome possibly belonging to the b-type group. The PMP32 is very closely related to the MDHAR, and the PMP29 is related to the peroxisomal NADPH, cytochrome P450 reductase. Catalase and SOD are the major ROS-scavenging enzymes found in peroxisomes (Hayashi and Nishimura 2003; Palma et al. 2009). It is noticed that oxidative stress defence mechanism of peroxisomes plays a role in heavy metal stress and xenobiotic stress (2,4-D) (Luis and López-Huertas 2006).

3.3.4 Plasma Membrane

The ubiquitous electron transporting oxidoreductases present in the plasma membrane produces ROS. One such is the NADPH oxidase integrated in the plasma membrane found producing ROS under normal and stressed condition. The plant NADPH oxidase is also known as respiratory burst oxidase homolog (Sagi and Fluhr 2006; Sharma et al. 2012). The core domain of NADPH oxidase is identified by six transmembrane helices supported by two heme groups containing conserved histidines. The NADPH oxidase contains hydrophilic domains for NADPH and FAD at C-terminal and two EF hand motifs at N-terminal for calcium binding. NADPH oxidase catalyses transfer of electrons from cytoplasmic NADPH to exterior O_2 to form $O_2^{\bullet -}$. $O_2^{\bullet -}$ is dismutated to H_2O_2 either spontaneously or by SOD activity (Miller et al. 2009). ROS generated by membrane-bound NADPH oxidase activity acts as signalling molecules under various biotic and abiotic stress stimuli and elicit cellular defence response (Torres et al. 2002; Foreman et al. 2003; Kwak et al. 2003; Jones et al. 2007).

3.3.5 Cell Wall

ROS generates continuously at the outer surface of plasma membrane and cell wall membranes of the epidermis and vascular tissue, where active electron transfer is involved. H_2O_2 is produced by four possible reactions, importantly being located at the external portion of plasma membrane, and is catalysed by NOX. Enzymes of oxidoreductase peroxidases, laccases, poly(di)amine oxidases (PDAOs), oxalate oxidases (OXOs) and SODs play a role in H_2O_2 generation in different reactions. Unlike PDAOs and OXOs that directly generate H_2O_2, NOX and peroxidases favour the formation of superoxide molecule, which later dismutes to H_2O_2 (Schopfer 2001; Spiteller 2003; Higuchi 2006). The di- and polyamine oxidases present in cell wall membranes of meristematic tissue generate significant quantity of radicle oxygen molecules (H_2O_2), while catalysing polyamines (putrescine, spermidine, cadaverine, etc.) to peroxides. Other cell wall-bound enzyme lipoxygenase (LOX) catalyses hydroperoxidation of polyunsaturated fatty acids (PUFA) making it active source for ROS like $^{\bullet}OH$, $O_2^{\bullet-}$, H_2O_2 and $^1O_2$ (Spiteller 2003; Higuchi 2006; Kim et al. 2010). The lipid hydroperoxides (PUFA-OOH) generated can undergo reductive cleavage catalysed by reduced metals, such as $Fe2^+$. The lipid $RO^{\bullet}$ produced (PUFA-O) initiates sequential reactions that lead to generation of toxic ROS (Liszkay et al. 2004; Karuppanapandian et al. 2006a, b, c, 2009; Karuppanapandian et al. 2008).

3.3.6 Apoplast

The cell wall-located enzymes are responsible for ROS production in apoplast and have been proven (Apel and Hirt 2004; Heyno et al. 2011). CO_2 that enters the apoplastic space will be converted to soluble, diffusible form and enters the cytoplasm to further participate in photosynthesis. The cell wall-associated oxalate oxidase, also known as germin, produces H_2O_2 and CO_2 from oxalic acid (Lane 2002; Cona et al. 2006). Other apoplastic ROS-forming enzymes include cell wall-bound oxidases, peroxidases (POXs), NADPH oxidase and polyamines (Mittler 2002; Kwak et al. 2003). Amine oxidases catalyse the oxidative deamination of polyamines (i.e., putrescine, spermine and spermidine) using FAD as a cofactor. Under extreme environmental conditions, stress signals combined with abscisic acid (ABA) make the apoplast a significant site for H_2O_2 production (Hernandez et al. 2001; Zhu 2001; Hu et al. 2006).

3.3.7 Endoplasmic Reticulum

The endoplasmic reticulum is the active site for ROS generation, as it harbours site for PSII that allows fast electron transportation through CytP450 (Mittler 2002). Organic substrate, RH, interacts with the CytP450 followed by reduction by a flavoprotein to form a free radical intermediate (CytP450 R^-) that can readily react with $3O_2$ to give rise to an oxygenated complex (CytP450-ROO^-); now both the intermediate and $3O_2$ bear one unpaired electron. This oxygenated complex may be reduced by Cyt *b*, or, occasionally, the complex may automatically form the P450-RH by generating $O_2^{\bullet-}$ (Watanabe and Lam 2006; Shi et al. 2010).

3.4 ROS Chemistry and Effect Under Abiotic Stress

The molecular oxygen in its ground state in the cellular environment is stable and it has no deleterious effect. The ground state triplet molecular oxygen is a biradical having two unpaired electrons with parallel spin which makes it paramagnetic (Tripathy and Oelmüller 2012). To participate in oxidation, triplet oxygen molecule would require a partner having pair of electrons with parallel spinning in its outermost orbit. But the ground state O_2 can easily be converted to reactive oxygen species either by absorbing energy or through electron transfer reaction (Apel and Hirt 2004). The former helps in the formation of singlet oxygen, whereas the latter results in the sequential reduction to superoxide ($O_2^{\bullet -}$), hydrogen peroxide (H_2O_2) and hydroxyl radical ($^{\bullet}OH$). However, if the triplet oxygen absorbs sufficient energy, the spin restriction is removed and the spin of one of its unpaired electrons is reversed, favouring the generation of singlet oxygen ($^1O_2$) (Apel and Hirt 2004; Halliwell 2006). In plants, ROS are continuously produced as by-product of the metabolic pathways localized in various cellular compartments and also produced due to the spillage of electron from the electron transport chain system to the cellular environment which reduces the available molecular oxygen (Moller et al. 2007; Miller et al. 2008).

3.4.1 Singlet Oxygen ($^1O_2$)

In plants, light-harvesting complex (LHC) and the photosynthetic reaction centre II are the primary centres for $^1O_2$ production. Inefficient transfer of energy results in the generation of triplet state of chlorophyll that reacts with triplet oxygen to produce the highly reactive $^1O_2$ (Carmody et al. 2016; Zhang et al. 2014):

$$\text{Chl} \rightarrow 3\text{Chl}\ 3\text{Chl} + 3\text{O}_2 \rightarrow \text{Chl}+{}^1\text{O}_2$$

Under conditions of excess light energy, the QA and QB plastoquinone electron acceptors of PSII in the electron transport chain become over-reduced leading to unfinished charge separation between P680 and pheophytin (Dietz et al. 2016). This results in the formation of triplet state of the reaction centre Chl P680 (3P680) thus leading to the formation of $^1O_2$. The release of $^1O_2$ is also detected in isolated PS II particles (Macpherson et al. 1993), thylakoids (Hideg et al. 1998; Fryer et al. 2002) and cytochrome *b6*/*f* complex (Suh et al. 2000). Alternatively, the nanoradicle $^1O_2$ is formed when superoxide anion (O_2*-) reacts with hydroxyl radicle (*OH) in places where electron transfer is low at pace. Plants under abiotic stress conditions close their stomata partially as adaptive mechanism. This leads to the accumulation of intracellular CO_2, a condition that favours the formation of $^1O_2$ in chloroplast, because of accumulation of electrons at PSII. $^1O_2$ is a highly reactive radical oxygen molecule that is attributed for ROS-induced cell death under high light intensity. Though $1O_2$ has very less half-time period, of about 3 μs (Hatz et al. 2007), its high diffusion capacity (100 nm) can cause the cellular damage under adverse conditions. Plants have evolved several antioxidant mechanisms to quench off $^1O_2$, and most of them involve non-enzymatic methods. Few of efficient scavengers of $^1O_2$ involve carotenoids, tocopherols and plastoquinones that exist in thylakoid membranes of chloroplast. Apart from the negative effect, $^1O_2$ act as signalling molecule along with chloroplast protein Executar 1 (Ex1) that triggers stress acclimation process under pathogen attack and abiotic stress (Krieger-Liszkay et al. 2008).

3.4.2 Superoxide Radical ($O_2^{\bullet-}$)

Superoxides ($O_2^{\bullet-}$) are produced due to the transfer of electrons to O_2 instead of reducing NADP during excess electron transfer reaction in PSI and PSII. $O_2^{\bullet-}$ first formed ROS in plant cell that further catalysed or dismuted to form more toxic ROS molecules. Superoxide is a moderately reactive ROS with approximately 2–4 μs of half-life and does not cause extensive damage by itself. In plant tissues, about 1–2% of O_2 consumption leads to the generation of $O_2^{\bullet-}$ (Moller 2001; Sharma et al. 2012). Superoxide radical ($O_2^{\bullet-}$) can also undergo further reactions to generate more reactive ROS like $^{\bullet}OH$ and more possibly $^1O_2$ (Bielski et al. 1983; Elstner 1987). $O_2^{\bullet-}$ radical has both oxidizing and reducing property (neutrophilic). The O_2 $^{\bullet-}$ undergoes protonation to give up $HO_2^{\bullet}$ a strong oxidizing agent, which directly attacks the PUFA (Bielski et al. 1983). The $O_2^{\bullet-}$ can also donate an electron to iron (Fe_3^{+}) to yield a reduced form of iron (Fe_2^{+}) and $^1O_2$. The reduced Fe_2^{+} catalysed by SOD produces H_2O_2 that is further reduced by accepting electron to form highly toxic hydroxyl radicle ($^{\bullet}OH$), and this is known as Fenton reaction. Alternatively, $O_2^{\bullet-}$ dismutated to H_2O_2 and led to formation of $^{\bullet}OH$ radicle, a process called Haber-Weiss reaction (Halliwell 2006):

$$O_2{}^{\bullet-} + Fe^{3+} \rightarrow {}^1O_2 + Fe^{2+}$$

$$2O_2{}^{\bullet-} + 2H^+ \rightarrow O_2 + H_2O_2Fe_3{}^+$$

$$Fe_2{}^+ + H_2O_2Fe_3{}^+ \rightarrow Fe_3{}^+ + OH^- + {}^{\bullet}OH(\text{Fenton Reaction})$$

3.4.3 Hydrogen Peroxide (H_2O_2)

H_2O_2 is a potent reactive oxygen species that interacts with organic molecules that contain Fe^{2+}, leading to formation of highly reactive OH radical. The major source of H_2O_2 includes chloroplasts and mitochondria, where imbalance in electron transfer favours its generation. Type III peroxidases, NAD(P)H oxidases, Mehler reaction are few other sites of H_2O_2 generation (Gill and Tuteja 2010). It can occur both non-enzymatically through dismutation of H_2O_2 under low pH conditions and enzymatically, mostly catalysed by SOD. Under water stress conditions, the closure of stomata limits CO_2 fixation, thus favouring oxygenation of ribulose 1, -5-bisphosphate (RuBisCO) which leads to the formation of H_2O_2. As discussed earlier, H_2O_2 has a dual role in plant growth and development. At low concentrations, it acts as signalling molecule for many physiological processes like senescence (Peng et al. 2005), photorespiration and photosynthesis (Noctor et al. 2004), stomatal movement (Bright et al. 2006) and cell cycle (Tanou et al. 2009). In comparison to other ROS, H_2O_2 has a longer half-life of 1 ms, and hence, it can diffuse longer distances in cell and even cross plant cell membranes through aquaporins (Li et al. 2014) and induce oxidative damage. Generally, H_2O_2 is less toxic than other ROS, but its high intracellular concentration can oxidize cysteine (-SH) and methionine (-SCH3) residues of Calvin cycle enzymes (Dat et al. 2000; Halliwell 2006).

3.5 Approaches for Overcoming the Oxidative Stress Tolerance in Plants

As discussed above, all other stress conditions act through a common mechanism, i.e. oxidative stress. Plants have evolved with different enzymatic, non-enzymatic and antioxidant mechanisms to combat and adapt to the damage caused by oxidative stress. Transferring of genes/genomic regions coding for antioxidants from tolerant genotypes to susceptible species either through transgenic or breeding methods is of prime importance. But, besides the negative consequences, ROS act as sensory molecules in biotic and abiotic stress conditions and also participate in crosstalk with other signalling pathways (Mitler et al. 2004; Suzuki et al. 2012; Baxter et al. 2014; Noetor et al. 2014). Hence, it is important to bear in cognizance, about the threshold levels of ROS molecules, while designing strategies/approaches to develop crops with improved stress tolerance.

3.5.1 Molecular Breeding

Plant breeding techniques are indispensable tools in developing high-yielding varieties/hybrids. The ever-changing global climate, clubbed with continuous evolution of new pests, has thrown challenges to plant breeders in crop improvement. Breeders have exploited genetic variation at all levels, in developing improved varieties. Linkage drag and time involved in transferring of useful alleles and selection of right recombinant are the potential problems associated with traditional breeding methods. Molecular marker technology has greatly facilitated plant breeding in identification of genomic regions (QTLs), linked with trait of interest (Salvi and Tuberosa 2005; Walia et al. 2007; Marino et al. 2009; Pandit et al. 2010). But the recent developments in next-generation sequence-based (NGS) technologies offer SNP-based assays that have revolutionized plant breeding. Genome-wide analysis using NGS methods helps in identification of superior alleles and also helps in identification of evolutionary conserved domains. Genome-wide analysis of genes of GR in rice has shed light on evolutionary conserved domains, which helps in better understanding of its role under normal and stressed conditions (Trivedi et al. 2013). Genome-wide SNP analysis of antioxidative genes in rice has revealed functional polymorphism between resistant and susceptible genotypes that can be used to identify superior alleles in rice population studies (Prakash et al. 2016). Integration of conventional plant breeding methods, together with advanced genomic tools and phenotyping methods, will rapidly deliver superior breeding material with improved genetic gains.

Parallel to the developments of NGS technologies, research in genetics has discovered novel genetic resources that can utilize the benefits of genomics to gain the maximum genetic gains. Multi-parental crossing designs developed recently combine the alleles across the locations and also increase the diversity and resolution of quantitative trait loci (QTL) mapping studies (Giraud et al. 2014). One such population developed in maize is NAM population by crossing 25 diverse inbred lines with common parent (Elshire et al. 2011) that combines the advantages of both linkage- and association-based mappings. Unfortunately, very limited work has been done in oxidative stress tolerance with reference to molecular breeding. Melonaldehyde is the by-product of lipid peroxidation during extreme oxidative stress damage in crop plants. MDA content is often considered as an index to measure the extent of damage caused by oxidative stress. Jing et al. (2009) identified two QTLs associated with MDA content in rice-inbred population. Markers associated with these QTLs could be used to evaluate the genotypes for oxidative stress tolerance. During drought studies on pearl millet, Kholová et al. (2011) opinioned the possible presence of *APX5* gene in drought-tolerant QTL, as the APX5 activity was high in NIL (near-isogenic line) population introgressed with drought-tolerant QTL. Further studies require confirming this hypothesis, and designing of marker assay will help in selecting the stress-tolerant breeding material in pearl millet. Applying dynamic QTL analysis method, QTLs for antioxidant enzymes SOD, APX and MDA were detected in wheat-inbred population that explains more than 10% phenotypic variation (Jiang et al. 2013). Dynamic QTL analysis, used in this experiment, is a novel and reliable method to detect genomic regions based on both static and dynamic expression of genes, which improves the accuracy and sensitivity of QTL detection. Hence, genomic information generated could be effectively deployed in wheat improvement programmes through marker-assisted selection (MAS). In another study using three different tomato populations, ascorbate-related QTLs were mapped on chromosome numbers 2, 8, 9, 10 and 12. Further characterization of above QTLs revealed the presence of ascorbate regeneration genes *MDHAR* and *GME* in QTL region, indicating the reliability of the experiment (Stevens et al. 2007). The allelic polymorphism in the above QTLs can be exploited for tomato breeding programs on abiotic stress tolerance. In another study on tomato, 28 QTLs were detected governing non-enzymatic antioxidants ascorbate, vitamin C, total phenols and flavonoids, using back cross population (BC_2F_2) (Okmen et al. 2011). Markers associated with these QTLs could be used in marker-assisted breeding (MAB) for improving tomato breeding material.

3.5.2 Genetic Engineering

Limitations in conventional plant breeding in improving crop productivity can be overcome by genetic engineering approach. As discussed above, all abiotic stress like drought, salt and heat conditions leads to induction of secondary stresses like oxidative stress through generation of ROS at the cellular level. Hence, it is most appropriate to target antioxidant genes to overcome abiotic stresses, through gene transfer technology. Though all antioxidant enzymes described participate in scavenging activity, it is proved that CAT, APX and SOD upregulate more abundantly during drought stress, whereas transcripts of GR, POD, MDHAR and DHAR express under cold stress. CAT induces preferentially under salt stress compared to other genes (Zhang et al. 2015). Transgenic plants developed with improved stress tolerance using various enzymatic and non-enzymatic methods have been listed in Table 3.1. As indicated earlier in this chapter, ROS also invokes defence response in plants; hence, expression levels of transgenes are crucial to keep in pace of both metabolic pathways, i.e. signalling activity and scavenging pathway.

3.5.2.1 Non-enzymatic

Plants evolved with an efficient non-enzymatic mechanisms (ascorbate, AsA; glutathione, GSH; a-tocopherol; phenolic compounds; alkaloids; flavonoids; and carotenoids) that could be exploited to control ROS. Many such antioxidants have been isolated, characterized from various sources and tested their efficiency in different backgrounds through transgenic approach. Few of such examples are discussed here. Hemavathi et al. (2010) found that transgenic potato (*Solanum tuberosum* L. cv. Taedong Valley) plants overexpressing the *l-gulono-c-lactone oxidase* (*GLOase*) gene exhibited increased ascorbic acid (AsA) (141%) compared to the control plants, and transgenics exhibited better performance under simulated stress conditions induced by methyl viologen, NaCl and mannitol. Direct correlation between enhanced AsA and abiotic stress tolerance in transgenics was observed in this experiment. Alteration in GSH expression made through transgenic method has also conferred stress tolerance to great extent in plants. In another interesting study, transgenic mustard (*B. juncea*), overexpressing antioxidant enzymes *glutamine synthetase (GS)* and *gamma-glutamylcysteine synthetase (g-ECS)* genes, has displayed tolerance to abiotic stress caused by different heavy metals (Cd, Zn, As and Pb). This tolerance has been attributed to higher accumulation of reduced glutamine (GSH) necessary for phytochelatin synthesis (PC) that actively participates in phytoremediation (Reisinger et al. 2008). Liu et al. (2008) generated tobacco transgenic plants by overexpressing *VTE1* gene and encoding tocopherol cyclase (VTE1), an important enzyme involved in tocopherol biosynthesis. This study revealed that plants overexpressing *VTE1* gene have increased tolerance. Zhang et al. (2012) identified that overexpression of a transcription factor 'AtERF98' in *Arabidopsis* has resulted in higher accumulation of antioxidant ascorbic acid that confers drought and salinity tolerance in transgenics. Further, this study indicated the direct role of ethylene responsive factor (ERF) in AsA biosynthesis. It has also been found that overexpression of proline biosynthetic

Table 3.1 List of transgenic plants developed for ROS-scavenging enzymes in different plants through overexpression show improved tolerance to various abiotic stresses including oxidative stress

Gene	Gene source	Transgenic plant	Function	References
Cu-Zn SOD	*Oryza sativa*	*Nicotiana tabacum*	Salinity and drought	Badawi et al. (2004)
Cu-Zn SOD	*Avicennia marina*	*Oryza sativa*	Salinity, drought and oxidative	Prashanth et al. (2008)
MnSOD	*Nicotiana plumbaginifolia*	*Triticum aestivum*	Oxidative and photo-oxidative	Melchiorre et al. (2009)
MnSOD	*Tamarix androssowii*	*Populus davidiana X Populus bolleana*	Salinity	Wang et al. (2005b)
CAT3	*Brassica juncea*	*Nicotiana tabacum*	Heavy metal	Gichner et al. (2004)
katE	*Escherichia coli*	*Nicotiana tabacum*	Salinity	Al-Taweel et al. (2007)
MDAR1	*Arabidopsis thaliana*	*Nicotiana tabacum*	Salinity, ozone and drought	Etrayeb et al. (2007)
cAPX	*Pisum sativum*	*Lycopersicon esculentum*	Drought, heat, cold and UV light	Wang et al. (2005a)
swpa4	*Ipomoea batatas*	*Nicotiana tabacum*	Salinity, osmotic and oxidative	Kim et al. (2008)
APX1	*Hordeum vulgare*	*Arabidopsis thaliana*	Salinity	Xu et al. (2008)
StAPX	*Solanum lycopersicum*	*Nicotiana tabacum*	Salinity and drought	Sun et al. (2010)
OsAPXa	*Oryza sativa*	*Oryza sativa*	Cold	Sato et al. (2011)
DHAR	*Oryza sativa*	*Arabidopsis thaliana*	Salinity	Chen and Gallie (2005)
DHAR	*Arabidopsis thaliana*	*Nicotiana tabacum*	Drought and ozone	Ushimaru et al. (2006)
DHAR	*Arabidopsis thaliana*	*Nicotiana tabacum*	Salinity and drought	Etrayeb et al. (2007)
GR	*Arabidopsis thaliana*	*Gossypium hirsutum*	Cold and photo-oxidative	Kornyeyev et al. (2003)
GPX	*Chlamydomonas*	*Nicotiana tabacum*	Salinity cold and oxidative	Yoshimura et al. (2004)
GPX-2	*Synechocystis*	*Arabidopsis thaliana*	Salinity, drought, cold, heavy metal, oxidative and MV	Gaber et al. (2006)
VTE1	*Arabidopsis thaliana*	*Nicotiana tabacum*	Drought	Liu et al. (2008)
P5CS	*Arabidopsis thaliana, Oryza sativa*	*Petunia hybrida*	Drought	Yamada (2005)

(continued)

Table 3.1 (continued)

Gene	Gene source	Transgenic plant	Function	References
P5CS	*Vigna aconitifolia*	*Triticum aestivum*	Drought	Vendruscolo et al. (2007)
GLOase	Strawberry	*Solanum tuberosum*	Accumulation of vitamin C with enhanced abiotic stress	Hemavathi et al. (2010)
GS	*Escherichia coli*	*Brassica juncea*	Heavy metal tolerance	Reisinger et al. (2008)
AtERF98	*Arabidopsis thaliana*	*Arabidopsis thaliana*	Activation of ascorbic acid synthesis	Zhang et al. (2012)
GLOase	Rat cells	*Solanum tuberosum*	L-ascorbic acid accumulation and tolerance to salinity and MV	Hemavathi et al. (2010)
P5CS	*Arabidopsis thaliana*	*Solanum tuberosum*	Accumulation of proline in response to salinity	Hmida-Sayari et al. (2005)
P5CS	*Vigna aconitifolia*	*Oryza sativa*	Drought and salinity	Su and Wu (2004)
EsSPDS1	*Eutrema salsugineum*	*Nicotiana tabacum*	Drought	Zhou et al. (2015)
MnSOD	Yeast	*Oryza sativa*	Salinity	Tanak et al. (1999)
APX2 and 3	*Arachis hypogaea*	*Arabidopsis thaliana*	Heat tolerance	Chiang et al. (2015)
Cytosolic GR	*Brassica campestris*	*Oryza sativa*	Photo-oxidative stress	Kouril et al. (2003)
GR	*Pennisetum glaucum*	*E. coli*	Heat and MV	Achary et al. (2015)
MDHAR	*Acanthus ebracteatus*	*Oryza sativa*	Salinity	Sultana et al. (2012)
DHAR	*Arabidopsis thaliana*	*Nicotiana tabacum*	Aluminium	Yin et al. (2010)
AtMDHAR1	*Arabidopsis thaliana*	*Nicotiana tabacum*	Salinity, ozone and PEG	Eltayeb et al. (2007)
DHAR	*Eutrema salsugineum*	*Nicotiana tabacum*	Ascorbic acid biosynthesis and drought	Zhou et al. (2015a)
DREB1A/CBF3	*Arabidopsis thaliana*	*Solanum lycopersicum*	Drought stress	Rai et al. (2013)
APX and Cu-Zn SOD	*Nicotiana tabacum*	*Nicotiana tabacum*	Methyl viologen and oxidative damage	Kwon et al. (2002)
APX and Cu-Zn SOD	*Manihot esculenta*	*Solanum tuberosum*	Heat, oxidative stress and MV	Tang et al. (2006)
APX, Cu-Zn SOD and DHAR	*Manihot esculenta*	*Nicotiana tabacum*	Salinity and paraquat	Xu et al. (2014)

(continued)

Table 3.1 (continued)

Gene	Gene source	Transgenic plant	Function	References
APX and Cu-Zn SOD	*Manihot esculenta*	*Manihot esculenta*	MV and cold	Xu et al. (2014)
APX and Cu-Zn SOD	*Arachis hypogaea*	*Nicotiana tabacum*	Salinity	Negi et al. (2015)
Cu-Zn SOD	*Pisum sativum*	*Nicotiana tabacum*	MV and cold	Gupta et al. (1993)
Cu-Zn SOD	*Kandelia candel*	*Nicotiana tabacum*	Oxidative and salinity	Jing et al. (2015)
Cu-Zn SOD	*Arachis hypogaea*	*Nicotiana tabacum*	Drought and salinity	Negi et al. (2015)
Mn SOD +APX	*Nicotiana tabacum*	*Festuca arundinacea*	Multiple abiotic stresses	Lee et al. (2007)
Cu-Zn SOD + CAT	*Zea mays*	*Brassica campestris*	Salinity and SO_2	Tseng et al. (2007)
SOD +APX	*Spinacia oleracea/Pisum sativum*	*Prunus domestica cv. Claudia Verde*	Salinity	Diaz-Vivancos et al. (2013)
MDHAR + DHAR	*Brassica rapa*	*Arabidopsis thaliana*	Freezing oxidative	Shin et al. (2013)

pathway genes has resulted in enhancing the abiotic stress tolerance in transgenic plants. The potato transgenic plants overexpressing pyrroline-5-carboxylate synthetase (P5CS) cDNA from *A. thaliana* have exhibited increased proline levels under salt stress and showed less altered tuber yield and weight in comparison to control plants (Hmida-Sayari et al. 2005). Su and Wu (2004) reported that both constitutive expression and stress-inducible expression of the P5CS cDNA in transgenic *O. sativa* have led to the accumulation of P5CS mRNA and Pro which resulted in higher tolerance to salt and water deficiency. In a similar study, Vendruscolo et al. (2007) developed wheat transgenics expressing VaP5CS cDNA under the control of stress-inducible promoter AIPC and transgenics, which performed well under water deficit stress. This study also revealed that P5CS induced proline acts in oxidative stress management than osmatic adjustment under water stress. Other than as osmolyte proline also act as scavenging molecule by quenching OH^- and $^1O_2$ radicals (Trovato et al. 2008). Tissue-specific expression of EsSPDS1 encoding a novel cellular polyamine has showed resistance to multiple abiotic stresses through strict regulation of ROS genes (Zhou et al. 2015a).

3.5.2.2 Enzymatic

A number of transgenic plants including *Arabidopsis*, tobacco, rice, tomato, maize, sweet potato and potato have been developed with manipulated expression of antioxidant enzymes (SOD, APX, DHAR, MDHAR, GR and CAT) that showed increased tolerance to drought, low or high temperatures and salinity stress (Prashanth et al. 2008; Al-Taweel et al. 2007; Kim et al. 2008; Etrayeb et al. 2007; Ushimaru et al. 2006; Table 3.1). Ascorbate (AsA)-glutathione (GSH) pathway is a key metabolic pathway that harbours enzymes like SOD, APX, DHAR, MDHAR and GR, with efficient ROS detoxification capacities. Hence majority of transgenic work is focused on genes involved in this pathway. Majority of the researchers focused on transgenics with overexpression of SOD for improving abiotic stress tolerance (Caverzan et al. 2016). It was noteworthy that overexpression of a single gene could make the transgenic rice tolerant to different stresses. For instance, overexpression of a yeast *MnSOD* gene in transgenic rice resulted in increased salt tolerance (Tanaka et al. 1999), whereas overexpression of the same gene from pea could make the transgenic rice drought tolerant (Wang et al. 2005b). Overexpression of the cytosolic *Cu-Zn SOD* gene from mangrove in transgenic rice could make the plant tolerant to both salinity and drought (Prashanth et al. 2008). This could be due to efficiency of SOD, to detoxify superoxide radical, the first generated ROS in plants that prevents the formation of subsequent ROS molecules, which are more toxic. An *OsAPX* gene overexpressed in transgenic rice could enhance tolerance to chilling at the booting stage (Sato et al. 2011). Overexpression of peanut *APX2* and *3* genes in *Arabidopsis* has improved seed germination rate, and transgenic displayed tremendous heat tolerance compared to WT, through efficient elimination of cellular H_2O_2 (Chiang et al. 2015). Transgenic rice overexpressing a cytosolic GR of *Brassica campestris* showed tolerance towards intensive photo-oxidative stress (Kouril et al. 2003). Overexpression of cDNA clone of GR from *Pennisetum* has conferred heat and methyl viologen (MV) tolerance in *E. coli*. This is due to fat scavenging activity of reduced glutathione pools, generated by overexpression (Achary et al. 2015). Sultana et al. (2012) have developed salt-tolerant rice transgenic by overexpressing a *MDHAR* gene isolated from the mangrove plant (*Acanthus ebracteatus*). Nevertheless, there are exceptions to the general assumptions that transgenic plant would always enhance stress tolerance. For instance, transgenic tobacco and tomato plants overexpressing petunia *Cu-Zn SOD* failed to exhibit any increased tolerance to oxidative or cold stress (Tepperman and Dunsmuir 1990). Likewise, transgenic cotton plant overexpressing an AtGR did not confer any protection from photoinhibition under chilling stress (Logan et al. 2003). Yin et al. (2010) demonstrated that transgenic tobacco plants overexpressing DHAR exhibited better root growth, low levels of H_2O_2 and less lipid peroxidation when compared with their wild counterparts under heavy metal stress (Al). The elevated levels of AsA and APX activity could have performed better scavenging activity in the above experiment. The overexpression of AtMDHAR1 in tobacco increases net photosynthesis rates under salt, ozone and PEG stresses (Eltayeb et al. 2007). Overexpression of dehydroascorbate reductase (DHAR) cDNA in tobacco has improved oxidative damage through fast regeneration of ascorbate (Zhou et al. 2015a). In a significant study, Herbette et al. (2011) made an interesting observation that, along with tolerance to abiotic stresses, transgenic tomato plants overexpressing GPX become susceptible to pathogen attack. This study has opinioned that the antioxidant role of GPX in both biotic and abiotic pathways and crosstalk mechanism might be responsible for this activity. Controlled expression of GPX may solve this problem.

Apart from various enzymatic and non-enzymatic antioxidants, transgenic expression of signalling molecules and regulatory elements involved in oxidative defence pathways also plays a significant role in developing stress-tolerant crop plants. Regulatory genes or transcription factors (TFs) that control large set of downstream genes involved in ROS-scavenging genes are ideal choice to achieve stress tolerance in crop plants. In *Arabidopsis*, overexpression of mitogen-activated kinase kinase 1 (MKK1) leads to increased tolerance to abiotic stresses through upregulation of genes involved in ascorbate-glutathione pathway (Wrzaczek et al. 2013; Xing et al. 2008). In other example, overexpression of transcription factors *Zat12* or *JERF3*, *Zat10* resulted in upregulation of various transcripts involved in ROS-scavenging pathways leading to higher tolerance to salt, drought or osmotic stresses (Sakamoto et al. 2004; Davletova et al. 2005). Rai et al. (2013) have reported that overexpression of AtDREB1A/CBF3 of *Arabidopsis* under the control of stress-inducible promoter (rd29A) in tomato (cv. Kashi Vishesh) showed enhanced levels of ROS-scavenging enzymes and antioxidants with increased tolerance to drought-induced oxidative stress.

3.5.2.3 Pyramiding of Antioxidant Genes

It is well known that expression of single foreign gene in plants will make them stress tolerant to a certain level in different backgrounds. As science advances, it is realized that simultaneous co-expression of genes involved in metabolic pathway could increase stress tolerance to a great extent (Halpin 2005; Vemanna et al. 2013). Various strategies have been used for multigene transfer including iterative, co-transformation and multigene linking (Halpin et al. 2001). Gene transformation through iterative strategies is more conventional and includes cross-fertilization (Zhao et al. 2003) and retransformation approaches (Singhla-Pareek et al. 2003). In the multigene linking strategy, multiple transgenes are introduced simultaneously into a plant by linking multiple transgene expression cassettes onto a single T-DNA that co-expresses in host plant (Chen et al. 2006, 2010). Among all methods, multigene linking strategy becomes most successful and convenient.

Pyramiding of APX and Cu-Zn SOD has been achieved in tobacco, potato, sweet potato and tall fescue plants that overexpressed in cytosol and chloroplast (Kwon et al. 2002; Tang et al. 2006; Lee et al. 2007; Lim et al. 2007; Xu et al. 2014; Negi et al. 2015). Pyramiding of SOD, APX and DHAR has been done in transgenic tobacco plant (Lee et al. 2007). Transgenic tobacco having overexpressed *Cu-Zn SOD* showing tolerance to oxidative stress (Gupta et al. 1993), upon retransformation with the chloroplastic *APX*, showed enhanced tolerance to paraquat (Kwon et al. 2002). Later into the above transgenic tobacco, a chloroplastic *DHAR* gene was transferred that showed further enhancement of tolerance to oxidative stress (Lee et al. 2007). Likewise, transgenic tobacco through *in vitro* pyramiding of cytosolic Cu-Zn SOD and APX was developed that showed enhanced tolerance to drought stress as a result of overexpression of both the enzymes (Faize et al. 2011). The pyramiding of Cu-Zn SOD and chloroplastic APX in potato (Tang et al. 2006) and in sweet potato (Lim et al. 2007) enhanced the plant tolerance to chilling, high temperature and paraquat stresses. Transgenic tobacco overexpressing cotton glutathione S-transferase (GST) and *Chlamydomonas* glutathione peroxidase (GPX) showed enhanced resistance to paraquat as well as chilling (Yu et al. 2003; Yoshimura et al. 2004). Also, gene pyramiding with double transgenic plants overexpressing both GST and GPX enhances the seedling growth during chilling and salt stress (Roxas et al. 1997). Co-expression of *Cu* Zn SOD + APX from peanut has greatly improved salt and drought tolerance in tobacco, through minimizing oxidative damage (Negi et al. 2015). It also enhanced germination percentage, indicating the role of oxidative pathway enzymes in seed germination. Similarly co-expression of maize ZmCu-ZmSOD and ZmCAT showed significant increase in photosynthetic performance along with NaCl-induced salt tolerance in transgenic cabbage (*Brassica campestris* L.) better than the independent performance of either ZmCu-Zn SOD or ZmCAT in similar background (Tseng et al. 2007). Xu et al. (2014) have overexpressed native *Cu-Zn SOD* and APX2 in cassava plants that has resulted in higher tolerance to oxidative stress induced by MV and H_2O_2 along with cold tolerance (4 °C for 2 days) (Xu et al. 2014). Lu et al. (2010) have demonstrated that overexpression of SOD and APX in sweet potato have resulted in increased expression of antioxidant enzymes, i.e. SOD, APX and CAT, thus protecting the plant from oxidative stress damage under stress conditions. Zhao and Zhang (2006) explained that co-expression of the *GST* and *CAT1* genes in rice has greatly enhanced their tolerance to salt (200 mM NaCl) and paraquat-induced stresses. While SOD and catalase activity played a crucial role in conferring salt tolerance, GST activity was found only during paraquat treatment. The generation of H_2O_2 and MDA decreased in the transgenics than in non-transgenics under the same conditions. Martret et al. (2011) observed that tobacco chloroplast transformants expressing genes encoding DHAR, GR, and GST exhibit altered antioxidant metabolism that has improved tolerance to salinity and chilling. This improved protection could be explained by synergistic effects of DHAR with GR or GST with GR. The expression of these combinations of transgenes also increased the regeneration of AsA (1.6-fold) and GSH (2.4-fold) and participated in a more rapid scavenging of $O2^{\bullet-}$ and H_2O_2 prior to their interaction with target molecules. Simultaneous expression of Brassica MDHAR and DHAR cDNA clones under the control of oxidative induced promoter SWPA2 has tremendous increased oxidative stress causing freezing tolerance in *Arabidopsis* compared to individual overexpression (Shin et al. 2013). This study combines the benefit of controlled expression and gene pyramiding.

There are many genes other than oxidative pathways that control abiotic stress, through activation of ROS-scavenging pathway. *Expansins* are one such gene belonging to cell wall proteins, inducing cell wall loosening, and participate in plant growth and development processes. *TaEXPB23*, a wheat *expansin* gene, was proved for oxidative stress tolerance, when overexpressed in tobacco plants (Han et al. 2015). The resultant transgenic tobacco plants revealed increased peroxidase activity particularly in their cell walls conferring oxidative stress tolerance (Chen et al. 2016). This study revealed that *expansin* genes are one of the best probable candidates to produce crop plants tolerant to abiotic stress through transgenic technology.

3.6 Conclusion and Future Prospects

ROS generation is a common mechanism happening in almost all abiotic stresses that creates secondary stress condition, i.e. oxidative stress. A recent study dealing with meta-analysis of publicly available transcriptomes in rice revealed that ROS detoxifiers (scavengers) are the major differentially expressed genes during abiotic stress (de Abreu Neto and Frei 2015). This explains the significance of ROS scavengers in controlling all abiotic stress in crop plants. But progress in science has stressed that maintaining basal level of ROS is essential for plant cell, as it involves various physiological functions including signal transduction and programmed cell death (Mittler 2017). Hence, focus has to be made on controlled expression of transgenes which further depends on the choice of promoters, while designing transgenic approach. T_7 RNA polymerase-based expressing system with induced promoter that has proved successful in our laboratory (unpublished data) could be more judicious for heterologous expression of multigene constructs. Research efforts also have to be made on determining the threshold levels of ROS that makes the cell to function normally. Though overexpression of single genes has proved successful in stress tolerance for long time, improved effect could be achieved by transferring multigenes involved in pathways. Technologies have to be standardized for multigene transfer that express successfully in host plant using a single vector. Site-specific recombination cloning (gateway cloning) holds promising to transfer many genes in single T-DNA. Care must be taken while choosing the genes used for oxidative stress tolerance as antioxidant enzymes perform differentially under different abiotic stress conditions.

Due to the regulatory issues associated with GMOs, transgenic approach may be considered as lost approach, when the gene of interest is not available within the germplasm of host plant. Marker-assisted breeding (molecular breeding) may be considered as the right choice to identify and transfer the QTLs associated with trait of interest. Due to complexity of the trait, and functional sharing with salt, drought and heat stress, oxidative stress was given less focus in QTL identification and subsequent marker development programmes. Not many specific QTLs were identified associated with oxidative stress, but many QTL identified and markers associated were developed for heat tolerance that shares common genes for detoxifying ROS generated through oxidative stress. Bita and Gerats (2013) have described various QTLs associated with heat tolerance in different crops. Characterization of these QTLs revealed ROS-scavenging genes explaining commonality between the pathways. Recent innovations in genomics have resulted in designing of SNP-based assays that are more closely linked to the trait of interest which are more reliable. Along with regular research in transgenic and molecular breeding efforts to overcome oxidative stress tolerance, focus also has to put on exploring the interacting molecules of *ROS* genes during the combined attack of biotic and abiotic stress conditions.

Acknowledgement PSR acknowledges the Department of Science and Technology, Govt. of India, for the fellowship and research grant through the INSPIRE Faculty Award and Young Scientist Scheme.

References

Achary VM, Reddy CS, Pandey P, Islam T, Kaul T, Reddy MK (2015) Glutathione reductase a unique enzyme: molecular cloning, expression and biochemical characterization from the stress adapted C4 plant, *Pennisetum glaucum* (L.) R. Br. Mol Biol Rep 42:947–962

Al-Taweel K, Iwaki T, Yabuta Y, Shigeoka S, Murata N, Wadano A (2007) A bacterial transgene for catalase protects translation of d1 protein during exposure of salt-stressed tobacco leaves to strong light. Plant Physiol 145:258–265

Apel K, Hirt H (2004) Reactive oxygen species: metabolism, oxidative stress, and signal transduction. Annu Rev Plant Biol 55:373–399

Asada K (2006) Production and scavenging of reactive oxygen species in chloroplasts and their functions. Plant Physiol 141:391–396

Atkin OK, Macherel D (2009) The crucial role of plant mitochondria in orchestrating drought tolerance. Ann Bot 103:581–597

Badawi GH, Yamauchi Y, Shimada E, Sasaki R, Kawano N, Tanaka K, Tanaka K (2004) Enhanced tolerance to salt stress and water deficit by overexpressing superoxide dismutase in tobacco (Nicotiana tabacum) chloroplasts. Plant Sci 166(4):919–928

Ball L, Accotto GP, Bechtold U, Creissen G, Funck D, Jimenez A, Kular B, Leyland N, Mejia-Carranza J, Reynolds H, Karpinski S, Mullineaux PM (2004) Evidence for a direct link between glutathione biosynthesis and stress defense gene expression in *Arabidopsis*. Plant Cell 16:2448–2462

Barreto P, Okura VK, Neshich IAP, Maia IG, Arruda P (2014) Overexpression of UCP1 in tobacco induces mitochondrial biogenesis and amplifies a broad stress response. BMC Plant Biol 14:144

Baxter A, Mittler R, Suzuki N (2014) ROS as key players in plant stress signalling. J Exp Bot 65:1229–1240

Bhattacharjee S (2005) Reactive oxygen species and oxidative burst: roles in stress, senescence and signal transduction in plants. Curr Sci India 89:1113–1121

Bielski BHJ, Arudi RL, Sutherland MW (1983) A study of the reactivity of HO2/O2 with unsaturated fatty-acids. J Biol Chem 258:4759–4761

Bita CE, Gerats T (2013) Plant tolerance to high temperature in a changing environment: scientific fundamentals and production of heat stress-tolerant crops. Front Plant Sci 4:273

Blokhina O, Fagerstedt KV (2010) Reactive oxygen species and nitric oxide in plant mitochondria: origin and redundant regulatory systems. Physiol Plant 138:447–462

Bright J, Desikan R, Hancock JT, Weir IS, Neill SJ (2006) ABA-induced NO generation and stomatal closure in *Arabidopsis* are dependent on H_2O_2 synthesis. Plant J 45:113–122

Carmody M, Crisp PA, d'Alessandro S, Ganguly D, Gordon M, Havaux M, Albrecht-Borth V, Pogson BJ (2016) Uncoupling high light responses from singlet oxygen retrograde signaling and spatial-temporal systemic acquired acclimation. Plant Physiol 171:1734–1749

Caverzan A, Passaia G, Rosa SB, Ribeiro CW, Lazzarotto F, Margis-Pinheiro M (2012) Plant responses to stresses: role of ascorbate peroxidase in the antioxidant protection. Genet Mol Biol 35:1011–1019

Caverzan A, Casassola A, Brammer SP (2016) Antioxidant responses of wheat plants under stress. Genet Mol Biol 39:1–6
Chen Z, Gallie DR (2004) The ascorbic acid redox state controls guard cell signaling and stomatal movement. Plant Cell 16:1143–1116
Chen Z, Gallie DR (2005) Increasing tolerance to ozone by elevating folia ascorbic acid confers greater protection against ozone than increasing avoidance. Plant Physiol 138:1673–1689
Chen Z, Gallie DR (2006) Dehydroascorbate reductase affects leaf growth, development, and function. Plant Physiol 142:775–787
Chen Q-J, Zhou H-M, Chen J, Wang X-C (2006) A Gateway-based platform for multigene plant transformation. Plant Mol Biol 62(6):927–936
Chen QJ, Xie M, Ma XX, Dong L, Chen J, Wang XC (2010) Missa is a highly efficient in vivo DNA assembly method for plant multiple- gene transformation. Plant Physiol 153(1):41–51
Chen YH, Han YY, Zhang M, Zhou S, Kong XZ, Wang W (2016) Overexpression of the wheat expansin gene TaEXPA2 improved seed production and drought tolerance in transgenic tobacco plants. PLoS One 11(4)
Chew O, Whelan J, Millar AH (2003) Molecular definition of the ascorbate-glutathione cycle in *Arabidopsis* mitochondria reveals dual targeting of antioxidant defenses in plants. J Biol Chem 278:46869–46877
Chiang CM, Chien HL, Chen LFO, Hsiung TC, Chiang MC, Chen SP, Lin KH (2015) Overexpression of the genes coding ascorbate peroxidase from *Brassica campestris* enhances heat tolerance in transgenic *Arabidopsis thaliana*. Biol Plant 59:305–315
Choudhury S, Panda P, Sahoo L, Panda SK (2013) Reactive oxygen species signalling in plants under abiotic stress. Plant Signal Behav 8:e23681
Cona A, Rea G, Angelini R, Federico R, Tavladoraki P (2006) Functions of amine oxidases in plant development and defence. Trends Plant Sci 11:80–88
Conklin PL, Williams EH, Last RL (1996) Environmental stress sensitivity of an ascorbic acid-deficient *Arabidopsis* mutant. Proc Natl Acad Sci U S A 93:9970–9974
Creissen G Elevated glutathione biosynthetic capacity in the chloroplasts of transgenic tobacco plants paradoxically causes increased oxidative stress. Plant Cell Online 11(7):1277–1292
Dabrowska G, Katai A, Goc A, Szechynska-Hebda M, Skrzypek E (2007) Characteristics of the plant ascorbate peroxidase family. Acta Biol Cracov Ser Bot 49:7–17
Das K, Roychoudhury A (2014) Reactive oxygen species (ROS) and response of antioxidants as ROS-scavengers during environmental stress in plants. Front Environ Sci 2:53
Dat J, Vandenabeele S, Vranova E, Van Montagu M, Inze D, Van Breusegem F (2000) Dual action of the active oxygen species during plant stress responses. Cell Mol Life Sci 57:779–795
Davletova S, Schlauch K, Coutu J, Mittler R (2005) The zinc-finger protein Zat12 plays a central role in reactive oxygen and abiotic stress signaling in *Arabidopsis*. Plant Physiol 139:847–856
de Abreu Neto JB, Frei M (2015) Microarray meta-analysis focused on the response of genes involved in redox homeostasis to diverse abiotic stresses in rice. Front Plant Sci 6:1260
Diaz-Vivancos P, Faize M, Barba-Espin G, Faize L, Petri C, Antonio Hernández J, Burgos L (2013) Ectopic expression of cytosolic superoxide dismutase and ascorbate peroxidase leads to salt stress tolerance in transgenic plums. Plant Biotechnol J 11(8):976–985
Dietz KJ, Turkan I, Krieger-Liszkay A (2016) Redox- and reactive oxygen species-dependent signaling into and out of the photosynthesizing chloroplast. Plant Physiol 171:1541–1550
Do H, Kim IS, Jeon BW, Lee CW, Park AK, Wi AR, Shin SC, Park H, Kim YS, Yoon HS, Kim HW, Lee JH (2016) Structural understanding of the recycling of oxidized ascorbate by dehydroascorbate reductase (OsDHAR) from *Oryza sativa L. japonica*. Sci Rep 6:19498
Dolatabadian A, Sanavy SAMM, Sharifi M (2009) Alleviation of water deficit stress effects by foliar application of ascorbic acid on *Zea mays* L. J Agron Crop Sci 195:347–355
Elshire RJ, Glaubitz JC, Sun Q, Poland JA, Kawamoto K, Buckler ES, Mitchell SE (2011) A robust, simple genotyping-by-sequencing (GBS) approach for high diversity species. PLoS One 6:e19379
Elstner EF (1987) Metabolism of activated oxygen species. In: Davies DD (ed) Biochemistry of plants. Academic, London, p 253e315

Eltayeb AE, Kawano N, Badawi GH, Kaminaka H, Sanekata T, Shibahara T, Inanaga S, Tanaka K (2007) Overexpression of monodehydroascorbate reductase in transgenic tobacco confers enhanced tolerance to ozone, salt and polyethylene glycol stresses. Planta 225:1255–1264

Faize M, Burgos L, Faize L, Piqueras A, Nicolas E, Barba-Espin G, Clemente-Moreno MJ, Alcobendas R, Artlip T, Hernandez JA (2011) Involvement of cytosolic ascorbate peroxidase and Cu/Zn-superoxide dismutase for improved tolerance against drought stress. J Exp Bot 62:2599–2613

Fan HF, Ding L, Du CX, Wu X (2014) Effect of short-term water deficit stress on antioxidative systems in cucumber seedling roots. Bot Stud 55:46

Feng ZZ, Pang J, Nouchi I, Kobayashi K, Yamakawa T, Zhu JG (2010) Apoplastic ascorbate contributes to the differential ozone sensitivity in two varieties of winter wheat under fully open-air field conditions. Environ Pollut 158:3539–3545

Foreman J, Demidchik V, Bothwell JHF, Mylona P, Miedema H, Torres MA, Linstead P, Costa S, Brownlee C, Jones JDG, Davies JM, Dolan L (2003) Reactive oxygen species produced by NADPH oxidase regulate plant cell growth. Nature 422:442–446

Fryer MJ, Oxborough K, Mullineaux PM, Baker NR (2002) Imaging of photo-oxidative stress responses in leaves. J Exp Bot 53:1249–1254

Gaber A, Yoshimura K, Yamamoto T, Yabuta Y, Takeda T, Miyasaka H, Nakano Y, Shigeoka S Glutathione peroxidase-like protein of Synechocystis PCC 6803 confers tolerance to oxidative and environmental stresses in transgenic Arabidopsis. Physiol Plant 128(2):251–262

Gichner T, Patková Z, Száková J, Demnerová K (2004) Cadmium induces DNA damage in tobacco roots, but no DNA damage, somatic mutations or homologous recombination in tobacco leaves. Mutat Res Genet Toxicol Environ Mutagen 559(1-2):49–57

Gill SS, Tuteja N (2010) Reactive oxygen species and antioxidant machinery in abiotic stress tolerance in crop plants. Plant Physiol Biochem 48:909–930

Giraud H, Lehermeier C, Bauer E, Falque M, Segura V, Bauland C, Camisan C, Campo L, Meyer N, Ranc N, Schipprack W, Flament P, Melchinger AE, Menz M, Moreno-Gonzalez J, Ouzunova M, Charcosset A, Schon CC, Moreau L (2014) Linkage disequilibrium with linkage analysis of multiline crosses reveals different multiallelic QTL for hybrid performance in the flint and dent heterotic groups of maize. Genetics 198:1717–1734

Gomez JM, Hernandez JA, Jimenez A, del Rio LA, Sevilla F (1999) Differential response of antioxidative enzymes of chloroplasts and mitochondria to long-term NaCl stress of pea plants. Free Radic Res 31:S11–S18

Guan QJ, Takano T, Liu SK (2012) Genetic transformation and analysis of rice OsAPx2 gene in *Medicago sativa*. PLoS One 7:e41233

Gupta AS, Heinen JL, Holaday AS, Burke JJ, Allen RD (1993) Increased resistance to oxidative stress in transgenic plants that overexpress chloroplastic Cu/Zn superoxide-dismutase. Proc Natl Acad Sci USA 90:1629–1633

Halliwell B (2006) Reactive species and antioxidants. Redox biology is a fundamental theme of aerobic life. Plant Physiol 141:312–322

Halpin C (2005) Gene stacking in transgenic plants – the challenge for 21st century plant biotechnology. Plant Biotechnol J 3:141–155

Halpin C, Barakate A, Askari BM, Abbott JC, Ryan MD (2001) Enabling technologies for manipulating multiple genes on complex pathways. Plant Mol Biol 47:295–310

Han YY, Chen YH, Yin SH, Zhang M, Wang W (2015) Over-expression of TaEXPB23, a wheat expansin gene, improves oxidative stress tolerance in transgenic tobacco plants. J Plant Physiol 173:62–71

Hatz S, Lambert JDC, Ogilby PR (2007) Measuring the lifetime of singlet oxygen in a single cell: addressing the issue of cell viability. Photochem Photobiol Sci 6:1106–1116

Hayashi M, Nishimura M (2003) Entering a new era of research on plant peroxisomes. Curr Opin Plant Biol 6:577–582

Hemavathi UCP, Akula N, Young KE, Chun SC, Kim DH, Park SW (2010) Enhanced ascorbic acid accumulation in transgenic potato confers tolerance to various abiotic stresses. Biotechnol Lett 32:321–330

Herbette S, de Labrouhe DT, Drevet JR, Roeckel-Drevet P (2011) Transgenic tomatoes showing higher glutathione peroxidase antioxidant activity are more resistant to an abiotic stress but more susceptible to biotic stresses. Plant Sci 180:548–553

Hernandez JA, Olmos E, Corpas FJ, Sevilla F, Delrio LA (1995) Salt-induced oxidative stress in chloroplasts of pea-plants. Plant Sci 105:151–167

Hernandez JA, Ferrer MA, Jimenez A, Barcelo AR, Sevilla F (2001) Antioxidant systems and O (2)(.-)/H(2)O(2) production in the apoplast of pea leaves. Its relation with salt-induced necrotic lesions in minor veins. Plant Physiol 127:817–831

Heyno E, Mary V, Schopfer P, Krieger-Liszkay A (2011) Oxygen activation at the plasma membrane: relation between superoxide and hydroxyl radical production by isolated membranes. Planta 234:35–45

Hideg E, Kalai T, Hideg K, Vass I (1998) Photoinhibition of photosynthesis in vivo results in singlet oxygen production detection via nitroxide-induced fluorescence quenching in broad bean leaves. Biochemistry 37:11405–11411

Higuchi T (2006) Look back over the studies of lignin biochemistry. J Wood Sci 52:2–8

Hmida-Sayari A, Gargouri-Bouzid R, Bidani A, Jaoua L, Savoure A, Jaoua S (2005) Overexpression of Delta (1)-pyrroline-5-carboxylate synthetase increases proline production and confers salt tolerance in transgenic potato plants. Plant Sci 169:746–752

Hu XL, Zhang AY, Zhang JH, Jiang MY (2006) Abscisic acid is a key inducer of hydrogen peroxide production in leaves of maize plants exposed to water stress. Plant Cell Physiol 47:1484–1495

Huang S, Van Aken O, Schwarzlander M, Belt K, Millar AH (2016) The roles of mitochondrial reactive oxygen species in cellular signaling and stress response in plants. Plant Physiol 171:1551–1559

Jezek P, Hlavata L (2005) Mitochondria in homeostasis of reactive oxygen species in cell, tissues, and organism. Int J Biochem Cell B 37:2478–2503

Jiang P, Wan ZY, Wang ZX, Li SS, Sun QQ (2013) Dynamic QTL analysis for activity of antioxidant enzymes and malondialdehyde content in wheat seed during germination. Euphytica 190:75–85

Jing J, Jie-yun Z, Ye-yang F, Bo S (2009) Mapping of QTLs for leaf malondialdehyde content associated with stress tolerance in rice. Rice Sci 16:72–74

Jing X, Hou P, Lu Y, Deng S, Li N, Zhao R, Sun J, Yang W, Han Y, Lang T, Ding M, Shen X, Chen S (2015) Overexpression of copper/zinc superoxide dismutase from mangrove Kandelia candel in tobacco enhances salinity tolerance by the reduction of reactive oxygen species in chloroplast. Front Plant Sci 6

Jones MA, Raymond MJ, Yang ZB, Smirnoff N (2007) NADPH oxidase-dependent reactive oxygen species formation required for root hair growth depends on ROP GTPase. J Exp Bot 58:1261–1270

Karuppanapandian T, Sinha PB, Haniya AMK, Manoharan K (2006a) Differential antioxidative responses of ascorbate-glutathione cycle enzymes and metabolites to chromium stress in green gram (*Vigna radiata* L. Wilczek) leaves. J Plant Biol 49:440–447

Karuppanapandian T, Sinha PB, Kamarul Haniya A, Premkumar G, Manoharan K (2006b) Aluminium-induced changes in antioxidative enzyme activities, hydrogen peroxide content and cell wall peroxidase activity in green gram (*Vigna radiata* L. cv. Wilczek) roots. J Plant Biol 33:241–246

Karuppanapandian T, Sinha PB, Premkumar G, Manoharan K (2006c) Chromium toxicity: correlated with increased in degradation of photosynthetic pigments and total soluble protein and increased peroxidase activity in green gram (*Vigna radiata* L.) seedlings. J Swamy Bot Cl 23:117–122

Karuppanapandian T, Saranyadevi AR, Jeyalakshmi K, Manoharan K (2008) Mechanism, control and regulation of leaf senescence in plants. J Plant Biol 35:141–155

Karuppanapandian T, Sinha PB, Haniya AK, Manoharan K (2009) Chromium-induced accumulation of peroxide content, stimulation of antioxidative enzymes and lipid peroxidation in green gram (*Vigna radiata* L. cv. Wilczek) leaves. Afr J Biotechnol 8:475–479

Karuppanapandian T, Wang HW, Prabakaran N, Jeyalakshmi K, Kwon M, Manoharan K, Kim W (2011) 2,4-dichlorophenoxyacetic acid-induced leaf senescence in mung bean (*Vigna radiata* L. Wilczek) and senescence inhibition by co-treatment with silver nanoparticles. Plant Physiol Biochem 49:168–177

Kataya AR, Reumann S (2010) Arabidopsis glutathione reductase 1 is dually targeted to peroxisomes and the cytosol. Plant Signal Behav 5:171–175

Kavitha K, George S, Venkataraman G, Parida A (2010) A salt-inducible chloroplastic monodehydroascorbate reductase from halophyte *Avicennia marina* confers salt stress tolerance on transgenic plants. Biochimie 92:1321–1329

Khan MIR, Khan NA (2014) Ethylene reverses photosynthetic inhibition by nickel and zinc in mustard through changes in PS II activity, photosynthetic-nitrogen use efficiency and antioxidant metabolism. Protoplasma 251:1007–1019

Khan MIR, Asgher M, Khan NA (2014) Alleviation of salt-induced photosynthesis and growth inhibition by salicylic acid involves glycinebetaine and ethylene in mungbean (*Vigna radiata* L.) Plant Physiol Biochem 80:67–74

Khan MIR, Nazir F, Asgher M, Per TS, Khan NA (2015) Selenium and sulfur influence ethylene formation and alleviate cadmium-induced oxidative stress by improving proline and glutathione production in wheat. J Plant Physiol 178:9–18

Khan MIR, Iqbal N, Masood A, Mobin M, Anjum NA, Khan NA (2016) Modulation and significance of nitrogen and sulfur metabolism in cadmium challenged plants. Plant Growth Regul 78:1–11

Kholová J, Tom Hash C, Kocová M, Vadez V (2011) Does a terminal drought tolerance QTL contribute to differences in ROS scavenging enzymes and photosynthetic pigments in pearl millet exposed to drought? Environ Exp Bot 71(1):99–106

Kim YH, Kim CY, Song WK, Park DS, Kwon SY, Lee HS, Bang JW, Kwak SS (2008) Overexpression of sweet potato swpa4 peroxidase results in increased hydrogen peroxide production and enhances stress tolerance in tobacco. Planta 227:867–881

Kim MJ, Ciani S, Schachtman DP (2010) A peroxidase contributes to ROS production during *Arabidopsis* root response to potassium deficiency. Mol Plant 3:420–427

Kliebenstein DJ, Monde RA, Last RL (1998) Superoxide dismutase in *Arabidopsis*: an eclectic enzyme family with disparate regulation and protein localization. Plant Physiol 118:637–650

Konig J, Muthuramalingam M, Dietz KJ (2012) Mechanisms and dynamics in the thiol/disulfide redox regulatory network: transmitters, sensors and targets. Curr Opin Plant Biol 15:261–268

Kornyeyev D, Logan BA, Payton PR, Allen RD, Scott Holaday A (2003) Elevated chloroplastic glutathione reductase activities decrease chilling-induced photoinhibition by increasing rates of photochemistry, but not thermal energy dissipation, in transgenic cotton. Funct Plant Biol 30 (1):101

Kouril R, Lazar D, Lee H, Jo J, Naus J (2003) Moderately elevated temperature eliminates resistance of rice plants with enhanced expression of glutathione reductase to intensive photooxidative stress. Photosynthetica 41:571–578

Krieger-Liszkay A, Fufezan C, Trebst A (2008) Singlet oxygen production in photosystem II and related protection mechanism. Photosynth Res 98:551–564

Kwak JM, Mori IC, Pei ZM, Leonhardt N, Torres MA, Dangl JL, Bloom RE, Bodde S, Jones JD, Schroeder JI (2003) NADPH oxidase AtrbohD and AtrbohF genes function in ROS-dependent ABA signaling in *Arabidopsis*. EMBO J 22:2623–2633

Kwon SY, Jeong YJ, Lee HS, Kim JS, Cho KY, Allen RD, Kwak SS (2002) Enhanced tolerances of transgenic tobacco plants expressing both superoxide dismutase and ascorbate peroxidase in chloroplasts against methyl viologen-mediated oxidative stress. Plant Cell Environ 25:873–882

Lane BG (2002) Oxalate, germins, and higher-plant pathogens. IUBMB Life 53:67–75

Lee SH, Ahsan N, Lee KW, Kim DH, Lee DG, Kwak SS, Kwon SY, Kim TH, Lee BH (2007) Simultaneous overexpression of both CuZn superoxide dismutase and ascorbate peroxidase in transgenic tall fescue plants confers increased tolerance to a wide range of abiotic stresses. J Plant Physiol 164:1626–1638

Li F, Wu QY, Sun YL, Wang LY, Yang XH, Meng QW (2010) Overexpression of chloroplastic monodehydroascorbate reductase enhanced tolerance to temperature and methyl viologen-mediated oxidative stresses. Physiol Plant 139:421–434

Li G, Santoni V, Maurel C (2014) Plant aquaporins: roles in plant physiology. Biochim Biophys Acta 1840:1574–1582

Lim S, Kim YH, Kim SH, Kwon SY, Lee HS, Kim JS, Cho KY, Paek KY, Kwak SS (2007) Enhanced tolerance of transgenic sweet potato plants that express both CuZnSOD and APX in chloroplasts to methyl viologen-mediated oxidative stress and chilling. Mol Breed 19:227–239

Liszkay A, van der Zalm E, Schopfer P (2004) Production of reactive oxygen intermediates (O(2)(.-), H(2)O(2), and (.)OH) by maize roots and their role in wall loosening and elongation growth. Plant Physiol 136:3114–3123

Liu X, Hua X, Guo J, Qi D, Wang L, Liu Z, Jin Z, Chen S, Liu G (2008) Enhanced tolerance to drought stress in transgenic tobacco plants overexpressing VTE1 for increased tocopherol production from *Arabidopsis thaliana*. Biotechnol Lett 30:1275–1280

Logan BA, Monteiro G, Kornyeyev D, Payton P, Allen RD, Holaday AS (2003) Transgenic overproduction of glutathione reductase does not protect cotton, *Gossypium hirsutum* (Malvaceae), from photoinhibition during growth under chilling conditions. Am J Bot 90:1400–1403

Lopez-Huertas E, Charlton WL, Johnson B, Graham IA, Baker A (2000) Stress induces peroxisome biogenesis genes. EMBO J 19:6770–6777

Lu ZQ, Liu DL, Liu SK (2007) Two rice cytosolic ascorbate peroxidases differentially improve salt tolerance in transgenic *Arabidopsis*. Plant Cell Rep 26:1909–1917

Lu P, Sang WG, Ma KP (2008) Differential responses of the activities of antioxidant enzymes to thermal stresses between two invasive eupatorium species in China. J Integr Plant Biol 50:393–401

Lu YY, Deng XP, Kwak SS (2010) Overexpression of CuZn superoxide dismutase (CuZn SOD) and ascorbate peroxidase (APX) in transgenic sweet potato enhances tolerance and recovery from drought stress. Afr J Biotechnol 9:8378–8391

Luis AR, López-Huertas E (2006) ROS generation in peroxisomes and its role in cell signaling. Plant Cell Physiol 57:1364–1376

Macpherson AN, Telfer A, Barber J, Truscott TG (1993) Direct-detection of singlet oxygen from isolated photosystem-Ii reaction centers. Biochim Biophys Acta 1143:301–309

Marino R, Ponnaiah M, Krajewski P, Frova C, Gianfranceschi L, Pe ME, Sari-Gorla M (2009) Addressing drought tolerance in maize by transcriptional profiling and mapping. Mol Gen Genomics 281:163–179

Martret LB, Poage M, Shiel K, Nugent GD, Dix PJ (2011) Tobacco chloroplast transformants expressing genes encoding dehydroascorbate reductase, glutathione reductase, and glutathione-S-transferase, exhibit altered anti-oxidant metabolism and improved abiotic stress tolerance. Plant Biotechnol J 9:661–673

Maxwell DP, Wang Y, McIntosh L (1999) The alternative oxidase lowers mitochondrial reactive oxygen production in plant cells. Proc Natl Acad Sci USA 96:8271–8276

Melchiorre M, Robert G, Trippi V, Roberto R, Ramiro Lascano H (2009) Superoxide dismutase and glutathione reductase overexpression in wheat protoplast: photooxidative stress tolerance and changes in cellular redox state. Plant Growth Regul 57(1):57–68

Mignolet-Spruyt L, Xu E, Idanheimo N, Hoeberichts FA, Muhlenbock P, Brosche M, Van Breusegem F, Kangasjarvi J (2016) Spreading the news: subcellular and organellar reactive oxygen species production and signalling. J Exp Bot 67:3831–3844

Miller G, Shulaev V, Mittler R (2008) Reactive oxygen signaling and abiotic stress. Physiol Plant 133:481–489

Miller G, Schlauch K, Tam R, Cortes D, Torres MA, Shulaev V, Dangl JL, Mittler R (2009) The plant NADPH oxidase RBOHD mediates rapid systemic signaling in response to diverse stimuli. Sci Signal 2:ra45

Mittler R (2002) Oxidative stress, antioxidants and stress tolerance. Trends Plant Sci 7:405–410

Mittler R (2017) ROS are good. Trends Plant Sci 22(1):11–19.

Mittler R, Zilinskas BA (1994) Regulation of pea cytosolic ascorbate peroxidase and other antioxidant enzymes during the progression of drought stress and following recovery from drought. Plant J 5:397–405

Mittler R, Vanderauwera S, Gollery M, Van Breusegem F (2004) Reactive oxygen gene network of plants. Trends Plant Sci 9:490–498

Mittova V, Volokita M, Guy M, Tal M (2000) Activities of SOD and the ascorbate-glutathione cycle enzymes in subcellular compartments in leaves and roots of the cultivated tomato and its wild salt-tolerant relative Lycopersicon pennellii. Physiol Plant 110:42–51

Moller IM (2001) Plant mitochondria and oxidative stress: electron transport, NADPH turnover, and metabolism of reactive oxygen species. Annu Rev Plant Physiol 52:561–591

Moller IM, Jensen PE, Hansson A (2007) Oxidative modifications to cellular components in plants. Annu Rev Plant Biol 58:459–481

Nath K, Kumar S, Poudyal RS, Yang YN, Timilsina R, Park YS, Nath J, Chauhan PS, Pant B, Lee CH (2014) Developmental stage-dependent differential gene expression of superoxide dismutase isoenzymes and their localization and physical interaction network in rice (*Oryza sativa* L.) Genes Genom 36:45–55

Negi NP, Shrivastava DC, Sharma V, Sarin NB (2015) Overexpression of CuZnSOD from Arachis hypogaea alleviates salinity and drought stress in tobacco. Plant Cell Rep 34(7):1109–1126

Noctor G, Veljovic-Jovanovic S, Driscoll S, Novitskaya L, Foyer CH (2002) Drought and oxidative load in the leaves of C3 plants: a predominant role for photorespiration? Ann Bot 89:841–850

Noctor G, Dutilleul C, De Paepe R, Foyer CH (2004) Use of mitochondrial electron transport mutants to evaluate the effects of redox state on photosynthesis, stress tolerance and the integration of carbon/nitrogen metabolism. J Exp Bot 55:49–57

Noctor G, De Paepe R, Foyer CH (2007) Mitochondrial redox biology and homeostasis in plants. Trends Plant Sci 12:125–134

Noctor G, Queval G, Mhamdi A, Chaouch S, Foyer CH (2011) Glutathione. Arabidopsis Book 9:1–32

Noctor G, Mhamdi A, Foyer CH (2014) The roles of reactive oxygen metabolism in drought: not so cut and dried. Plant Physiol 164:1636–1648

Okmen B, Sigva HO, Gurbuz N, Ulger M, Frary A, Doganlar S (2011) Quantitative trait loci (QTL) analysis for antioxidant and agronomically important traits in tomato (*Lycopersicon esculentum*). Turk J Agric For 35:501–514

Palma JM, Corpas FJ, del Rio LA (2009) Proteome of plant peroxisomes: new perspectives on the role of these organelles in cell biology. Proteomics 9:2301–2312

Panchuk II, Volkov RA, Schoffl F (2002) Heat stress- and heat shock transcription factor-dependent expression and activity of ascorbate peroxidase in *Arabidopsis*. Plant Physiol 129:838–853

Pandey P, Singh J, Achary VMM, Reddy MK (2015) Redox homeostasis via gene families of ascorbate-glutathione pathway. Front Environ Sci 3:25

Pandit A, Rai V, Bal S, Sinha S, Kumar V, Chauhan M, Gautam RK, Singh R, Sharma PC, Singh AK, Gaikwad K, Sharma TR, Mohapatra T, Singh NK (2010) Combining QTL mapping and transcriptome profiling of bulked RILs for identification of functional polymorphism for salt tolerance genes in rice (*Oryza sativa* L.) Mol Gen Genomics 284:121–136

Park AK, Kim IS, Do H, Jeon BW, Lee CW, Roh SJ, Shin SC, Park H, Kim YS, Kim YH, Yoon HS, Lee JH, Kim HW (2016) Structure and catalytic mechanism of monodehydroascorbate reductase, MDHAR, from *Oryza sativa L. japonica*. Sci Rep 6:33903

Patterson HC, Gerbeth C, Thiru P, Vogtle NF, Knoll M, Shahsafaei A, Samocha KE, Huang CX, Harden MM, Song R, Chen C, Kao J, Shi J, Salmon W, Shaul YD, Stokes MP, Silva JC, Bell GW, MacArthur DG, Ruland J, Meisinger C, Lodish HF (2015) A respiratory chain controlled signal transduction cascade in the mitochondrial intermembrane space mediates hydrogen peroxide signaling. Proc Natl Acad Sci U S A 112:5679–5688

Peng CL, Ou ZY, Liu N, Lin GZ (2005) Response to high temperature in flag leaves of super high-yielding rice Pei'ai64S/E32andLiangyoupeijiu. Rice Sci 12:179–186

Petrov VD, Van Breusegem F (2012) Hydrogen peroxide-a central hub for information flow in plant cells. Aob Plants 2012:pls 014

Prakash C, Mithra SV, Singh PK, Mohapatra T, Singh NK (2016) Unraveling the molecular basis of oxidative stress management in a drought tolerant rice genotype Nagina 22. BMC Genomics 17:774

Prashanth SR, Sadhasivam V, Parida A (2008) Over expression of cytosolic copper/zinc superoxide dismutase from a mangrove plant *Avicennia marina* in indica rice var Pusa Basmati-1 confers abiotic stress tolerance. Transgenic Res 17:281–291

Rai GK, Rai NP, Rathaur S, Kumar S, Singh M (2013) Expression of rd29A::AtDREB1A/CBF3 in tomato alleviates drought-induced oxidative stress by regulating key enzymatic and non-enzymatic antioxidants. Plant Physiol Biochem 69:90–100

Rasmusson AG, Geisler DA, Moller IM (2008) The multiplicity of dehydrogenases in the electron transport chain of plant mitochondria. Mitochondrion 8:47–60

Reisinger S, Schiavon M, Terry N, Pilon-Smits EAH (2008) Heavy metal tolerance and accumulation in Indian mustard (*Brassica juncea* L.) expressing bacterial gamma-glutamylcysteine synthetase or glutathione synthetase. Int J Phytoremediation 10:440–454

Rhoads DM, Umbach AL, Subbaiah CC, Siedow JN (2006) Mitochondrial reactive oxygen species. Contribution to oxidative stress and interorganellar signaling. Plant Physiol 141:357–366

Rodriguez-Serrano M, Romero-Puertas MC, Sanz-Fernandez M, Hu J, Sandalio LM (2016) Peroxisomes extend peroxules in a fast response to stress via a reactive oxygen species-mediated induction of the peroxin PEX11a. Plant Physiol 171:1665–1674

Roxas VP, Smith RK, Allen ER, Allen RD (1997) Overexpression of glutathione S-transferase glutathione peroxidase enhances the growth of transgenic tobacco seedlings during stress. Nat Biotechnol 15:988–991

Sagi M, Fluhr R (2006) Production of reactive oxygen species by plant NADPH oxidases. Plant Physiol 141:336–340

Sakamoto H, Maruyama K, Sakuma Y, Meshi T, Iwabuchi M, Shinozaki K, Yamaguchi-Shinozaki K (2004) Arabidopsis Cys2/His2-type zinc-finger proteins unction as transcription repressors under drought, cold, and high-salinity stress conditions. Plant Physiol 136:2734–2746

Salvi S, Tuberosa R (2005) To clone or not to clone plant QTLs: present and future challenges. Trends Plant Sci 10:297–304

Sandalio LM, Romero-Puertas MC (2015) Peroxisomes sense and respond to environmental cues by regulating ROS and RNS signalling networks. Ann Bot 116:475–485

Sato Y, Masuta Y, Saito K, Murayama S, Ozawa K (2011) Enhanced chilling tolerance at the booting stage in rice by transgenic overexpression of the ascorbate peroxidase gene, OsAPXa. Plant Cell Rep 30:399–406

Schopfer P (2001) Hydroxyl radical-induced cell-wall loosening in vitro and in vivo: implications for the control of elongation growth. Plant J 28:679–688

Sharma P, Jha AB, Dubey RS, Pessarakli M (2012) Reactive oxygen species, oxidative damage, and antioxidative defense mechanism in plants under stressful conditions. J Bot 2012:217037

Shi R, Sun YH, Li QZ, Heber S, Sederoff R, Chiang VL (2010) Towards a systems approach for lignin biosynthesis in *Populus trichocarpa*: transcript abundance and specificity of the monolignol biosynthetic genes. Plant Cell Physiol 51:144–163

Shigeoka S, Ishikawa T, Tamoi M, Miyagawa Y, Takeda T, Yabuta Y, Yoshimura K (2002) Regulation and function of ascorbate peroxidase isoenzymes. J Exp Bot 53:1305–1319

Shimaoka T, Yokota A, Miyake C (2000) Purification and characterization of chloroplast dehydroascorbate reductase from spinach leaves. Plant Cell Physiol 41:1110–1118

Shin SY, Kim MH, Kim YH, Park HM, Yoon HS (2013) Co-expression of monodehydroascorbate reductase and dehydroascorbate reductase from *Brassica rapa* effectively confers tolerance to freezing-induced oxidative stress. Mol Cell 36:304–315

Singla-Pareek SL, Reddy MK, Sopory SK (2003) Genetic engineering of the glyoxalase pathway in tobacco leads to enhanced salinity tolerance. Proc Natl Acad Sci U S A 100:14672–14677

Smirnoff N (2000) Ascorbate biosynthesis and function in photoprotection. Philos Trans R Soc B 355:1455–1464

Spiteller G (2003) The relationship between changes in the cell wall, lipid peroxidation, proliferation, senescence and cell death. Physiol Plant 119:5–18

Stevens R, Buret M, Duffe P, Garchery C, Baldet P, Rothan C, Causse M (2007) Candidate genes and quantitative trait loci affecting fruit ascorbic acid content in three tomato populations. Plant Physiol 143:1943–1953

Su J, Wu R (2004) Stress-inducible synthesis of proline in transgenic rice confers faster growth under stress conditions than that with constitutive synthesis. Plant Sci 166:941–948

Suh HJ, Kim CS, Jung J (2000) Cytochrome b6/f complex as an indigenous photodynamic generator of singlet oxygen in thylakoid membranes. Photochem Photobiol 71:103–109

Sultana S, Khew CY, Morshed MM, Namasivayam P, Napis S, Ho CL (2012) Overexpression of monodehydroascorbate reductase from a mangrove plant (AeMDHAR) confers salt tolerance on rice. J Plant Physiol 169:311–318

Sumithra K, Jutur PP, Dalton Carmel B, Reddy AR (2006) Salinity-induced changes in two cultivars of Vigna radiata: responses of antioxidative and proline metabolism. Plant Growth Regul 50(1):11–22

Sun W-H, Duan M, Shu D-F, Yang S, Meng Q-W (2010) Over-expression of StAPX in tobacco improves seed germination and increases early seedling tolerance to salinity and osmotic stresses. Plant Cell Rep 29(8):917–926

Suzuki N, Koussevitzky S, Mittler R, Miller G (2012) ROS and redox signalling in the response of plants to abiotic stress. Plant Cell Environ 35:259–270

Takahashi MA, Asada K (1983) Superoxide anion permeability of phospholipid-membranes and chloroplast thylakoids. Arch Biochem Biophys 226:558–566

Tanaka Y, Hibino T, Hayashi Y, Tanaka A, Kishitani S, Takabe T, Yokota S, Takabe T (1999) Salt tolerance of transgenic rice overexpressing yeast mitochondrial Mn-SOD in chloroplasts. Plant Sci 148:131–138

Tang L, Kwon SY, Kim SH, Kim JS, Choi JS, Cho KY, Sung CK, Kwak SS, Lee HS (2006) Enhanced tolerance of transgenic potato plants expressing both superoxide dismutase and ascorbate peroxidase in chloroplasts against oxidative stress and high temperature. Plant Cell Rep 25:1380–1386

Tanou G, Molassiotis A, Diamantidis G (2009) Hydrogen peroxide- and nitric oxide-induced systemic antioxidant prime-like activity under NaCl-stress and stress-free conditions in citrus plants. J Plant Physiol 166:1904–1913

Teixeira FK, Menezes-Benavente L, Galvao VC, Margis-Pinheiro M (2005) Multigene families encode the major enzymes of antioxidant metabolism in *Eucalyptus grandis* L. Genet Mol Biol 28:529–538

Teixeira FK, Menezes-Benavente L, Galvao VC, Margis R, Margis-Pinheiro M (2006) Rice ascorbate peroxidase gene family encodes functionally diverse isoforms localized in different subcellular compartments. Planta 224:300–314

Tepperman JM, Dunsmuir P (1990) Transformed plants with elevated levels of chloroplastic SOD are not more resistant to superoxide toxicity. Plant Mol Biol 14:501–511

Torres MA, Dangl JL, Jones JD (2002) Arabidopsis gp91phox homologues AtrbohD and AtrbohF are required for accumulation of reactive oxygen intermediates in the plant defense response. Proc Natl Acad Sci U S A 99:517–522

Touati D (1997) Superoxide dismutases in bacteria and pathogen protists. Cold Spring Harb Monogr Arch 34:447–493

Tripathy BC, Oelmüller R (2012) Reactive oxygen species generation and signaling in plants. Plant Signal Behav 7:1621–1633

Trivedi DK, Gill SS, Yadav S, Tuteja N (2013) Genome-wide analysis of glutathione reductase (GR) genes from rice and Arabidopsis. Plant Signal Behav 8:e23021

Trovato M, Mattioli R, Costantino P (2008) Multiple roles of proline in plant stress tolerance and development. Rendiconti Lincei 19(4):325–346

Tseng MJ, Liu CW, Yiu JC (2007) Enhanced tolerance to sulfur dioxide and salt stress of transgenic Chinese cabbage plants expressing both superoxide dismutase and catalase in chloroplasts. Plant Physiol Biochem 45:822–833

Turrens JF (2003) Mitochondrial formation of reactive oxygen species. J Physiol 552:335–344

Ushimaru T, Nakagawa T, Fujioka Y, Daicho K, Naito M, Yamauchi Y, Nonaka H, Amako K, Yamawaki K, Murata N (2006) Transgenic *Arabidopsis* plants expressing the rice dehydroascorbate reductase gene are resistant to salt stress. J Plant Physiol 163:1179–1184

Vemanna RS, Chandrashekar BK, Rao HMH, Sathyanarayanagupta SK, Sarangi KS, Nataraja KN, Udayakumar M (2013) A modified multisite gateway cloning strategy for consolidation of genes in plants. Mol Biotechnol 53:129–138

Vendruscolo ECG, Schuster I, Pileggi M, Scapim CA, Molinari HBC, Marur CJ, Vieira LGE (2007) Stress-induced synthesis of proline confers tolerance to water deficit in transgenic wheat. J Plant Physiol 164(10):1367–1376

Vij S, Tyagi AK (2007) Emerging trends in the functional genomics of the abiotic stress response in crop plants. Plant Biotechnol J 5:361–380

Walia H, Wilson C, Zeng L, Ismail AM, Condamine P, Close TJ (2007) Genome-wide transcriptional analysis of salinity stressed japonica and indica rice genotypes during panicle initiation stage. Plant Mol Biol 63:609–623

Wang YJ, Wisniewski M, Meilan R, Cui MG, Webb R, Fuchigami L (2005a) Overexpression of cytosolic ascorbate peroxidase in tomato confers tolerance to chilling and salt stress. J Am Soc Hortic Sci 130:167–173

Wang FZ, Wang QB, Kwon SY, Kwak SS, Su WA (2005b) Enhanced drought tolerance of transgenic rice plants expressing a pea manganese superoxide dismutase. J Plant Physiol 162:465–472

Watanabe N, Lam E (2006) Arabidopsis Bax inhibitor-1 functions as an attenuator of biotic and abiotic types of cell death. Plant J 45:884–894

Wrzaczek M, Brosche M, Kangasjarvi J (2013) ROS signaling loops – production, perception, regulation. Curr Opin Plant Biol 16(5):575–582

Wu J, Sun Y, Zhao Y, Zhang J, Luo L, Li M, Wang J, Yu H, Liu G, Yang L et al (2015) Deficient plastidic fatty acid synthesis triggers cell death by modulating mitochondrial reactive oxygen species. Cell Res 25:621–633

Xing Y, Jia WS, Zhangl JH (2008) AtMKK1 mediates ABA-induced CAT1 expression and H2O2 production via AtMPK6-coupled signaling in *Arabidopsis*. Plant J 54:440–451

Xu WF, Shi WM, Ueda A, Takabe T (2008) Mechanisms of salt tolerance in transgenic Arabidopsis thaliana carrying a peroxisomal ascorbate peroxidase gene from barley. Pedosphere 18(4):486–495

Xu J, Yang J, Duan XG, Jiang YM, Zhang P (2014) Increased expression of native cytosolic Cu/Zn superoxide dismutase and ascorbate peroxidase improves tolerance to oxidative and chilling stresses in cassava (*Manihot esculenta* Crantz). BMC Plant Biol 14:208

Yabuta Y, Mieda T, Rapolu M, Nakamura A, Motoki T, Maruta T, Yoshimura K, Ishikawa T, Shigeoka S (2007) Light regulation of ascorbate biosynthesis is dependent on the photosynthetic electron transport chain but independent of sugars in *Arabidopsis*. J Exp Bot 58:2661–2671

Yamada M (2005) Effects of free proline accumulation in petunias under drought stress. J Exp Bot 56(417):1975–1981

Yin LN, Wang SW, Eltayeb AE, Uddin MI, Yamamoto Y, Tsuji W, Takeuchi Y, Tanaka K (2010) Overexpression of dehydroascorbate reductase, but not monodehydroascorbate reductase, confers tolerance to aluminum stress in transgenic tobacco. Planta 231:609–621

Yoshimura K, Miyao K, Gaber A, Takeda T, Kanaboshi H, Miyasaka H, Shigeoka S (2004) Enhancement of stress tolerance in transgenic tobacco plants overexpressing Chlamydomonas glutathione peroxidase in chloroplasts or cytosol. Plant J 37:21–33

You J, Chan ZL (2015) ROS regulation during abiotic stress responses in crop plants. Front Plant Sci 6:1092

Yu T, Li YS, Chen XF, Hu J, Chang X, Zhu YG (2003) Transgenic tobacco plants overexpressing cotton glutathione S-transferase (GST) show enhanced resistance to methyl viologen. J Plant Physiol 160:1305–1311

Zhang ZJ, Wang J, Zhang RX, Huang RF (2012) The ethylene response factor AtERF98 enhances tolerance to salt through the transcriptional activation of ascorbic acid synthesis in *Arabidopsis*. Plant J 71:273–287

Zhang ZG, Zhang Q, Wu JX, Zheng X, Zheng S, Sun XH, Qiu QS, Lu TG (2013) Gene knockout study reveals that cytosolic ascorbate peroxidase 2(OsAPX2) plays a critical role in growth and reproduction in Rice under drought, salt and cold stresses. PLoS One 8:e57472

Zhang S, Apel K, Kim C (2014) Singlet oxygen-mediated and EXECUTERdependent signalling and acclimation of *Arabidopsis thaliana* exposed to light stress. Philos Trans R Soc Lond Ser B Biol Sci 369:20130227

Zhang Y, Li Z, Peng Y, Wang X, Peng D, Li Y, He X, Zhang X, Ma X, Huang L, Yan Y (2015) Clones of FeSOD, MDHAR, DHAR genes from white clover and gene expression analysis of ROS-scavenging enzymes during abiotic stress and hormone treatments. Molecules 20:20939–20954

Zhao FY, Zhang H (2006) Salt and paraquat stress tolerance results from co-expression of the *Suaeda salsa* glutathione S-transferase and catalase in transgenic rice. Plant Cell Tissue Org 86:349–358

Zhao J-Z, Cao J, Li Y, Collins HL, Roush RT, Earle ED, Shelton AM (2003) Transgenic plants expressing two Bacillus thuringiensis toxins delay insect resistance evolution. Nat Biotechnol 21(12):1493–1497

Zhou C, Sun Y, Ma Z, Wang J (2015a) Heterologous expression of EsSPDS1 in tobacco plants improves drought tolerance with efficient reactive oxygen species scavenging systems. S Afr J Bot 96:19–28

Zhou C, Sun YJ, Ma ZY, Wang JF (2015b) Overexpression of EsDHAR1 improved tolerance in transgenic tobacco with increased ascorbic acid levels. Oxid Commun 38:677–688

Zhu JK (2001) Plant salt tolerance. Trends Plant Sci 6:66–71

Zhu Q, Dugardeyn J, Zhang C, Mühlenbock P, Eastmond PJ, Valcke R, De Coninck B, Öden S, Karampelias M, Cammue BPA et al (2014) The *Arabidopsis thaliana* RNA editing factor SLO2, which affects the mitochondrial electron transport chain, participates in multiple stress and hormone responses. Mol Plant 7:290–310

Zolla L, Rinalducci S (2002) Involvement of active oxygen species in degradation of light-harvesting proteins under light stresses. Biochemistry 41:14391–14402

ROS Compartmentalization in Plant Cells Under Abiotic Stress Condition

4

Vandana Gautam, Ravdeep Kaur, Sukhmeen Kaur Kohli, Vinod Verma, Parminder Kaur, Ravinder Singh, Poonam Saini, Saroj Arora, Ashwani Kumar Thukral, Yuriy Victorovich Karpets, Yuriy Evgenovich Kolupaev, and Renu Bhardwaj

Abstract

Reactive oxygen species (ROS) are generated in various plant organelles under normal conditions and play an important role in different physiological progressions. But under abiotic stress, excessive ROS generation takes place which causes damage to normal functioning of plants. ROS play a dual role as they cause cellular damage and are also involved in abiotic stress signaling. Therefore, it is important to investigate the features of appearance of physiological effects of ROS depending on their cellular localization under the abiotic stress. Plants possess certain antioxidative mechanisms to deal with excess ROS in the cells, which involves enzymatic and nonenzymatic antioxidants. In the review, the mechanisms of ROS formation in different cellular compartments like mitochondria, peroxisomes, chloroplasts, nucleus, vacuole, cell wall, and plasma membranes are considered and summarized.

Keywords

Abiotic stress • Reactive oxygen species • ROS in mitochondria • Compartmentalization

V. Gautam • R. Kaur • S.K. Kohli • V. Verma • P. Kaur • R. Singh • P. Saini • S. Arora
A.K. Thukral • R. Bhardwaj (✉)
Department of Botanical and Environment Sciences, Guru Nanak Dev University, Amritsar, Punjab 143005, India
e-mail: renubhardwaj82@gmail.com

Y.V. Karpets • Y.E. Kolupaev
Dokuchaev Kharkiv National Agrarian University, Kharkiv 62483, Ukraine

M.I.R. Khan, N.A. Khan (eds.), *Reactive Oxygen Species and Antioxidant Systems in Plants: Role and Regulation under Abiotic Stress*,
DOI 10.1007/978-981-10-5254-5_4

Abbreviations

$^1O_2$	Singlet oxygen
ABA	Abscisic acid
ADP	Adenosine diphosphate
AOs	Amine oxidases
AOX	Alternative oxidase
APX	Ascorbate peroxidases
Ca^{2+}	Calcium cation
Cad^{2+}	Cadaverine
CAT	Catalase
CO_2	Carbon dioxide
COX	Cyclooxygenase
Cu	Copper
Cys	Cystine
Cyt	Cytosol
DAO	Diamine oxidase
DCFH	1, 2′-7′-dichlorodihydrofluorescein
DHE	Dihydroethidium
DNA	Deoxyribose nucleic acid
ETC	Electron transport chain
FAD	Flavin adenine dinucleotide
g	Gram
GFP	Green fluorescent proteins
GSH	Glutathione
GSSG	Glutathione disulfide
H_2O_2	Hydrogen peroxide
HE	Hydroethidine
HOCl	Hypochlorous acid
HPLC	High-performance liquid chromatography
kD	Kilodaltons
MAOs	Monoamine oxidases
MAP	Mitogen-activated protein
mg	Milligram
min	Minute
MitoAR	4-aminophenyl aryl ether
MitoHR	4-hydroxy aryl ether group
MitoSOX	Mitochondrial-targeted dihydroethidium
MS	Mass spectroscopy
NADP	Nicotinamide adenine dinucleotide phosphate
NADPH	Reduced nicotinamide adenine dinucleotide phosphate
nmol	Nanomole
NO	Nitric oxide
O_2	Oxygen

$O_2^{\bullet-}$	Superoxide anion
$OH^{\bullet}$	Hydroxyl radical
$ONOO^-$	Peroxynitrite
PAOs	Polyamine oxidases
PCD	Programmed cell death
PET	Photoinduced electron transfer
PMA	Phorbol-12-myristate-13-acetate
PML	Promyelocytic leukemia
Prx	Peroxiredoxins
PS	Photosystem
PS I	Photosystem I
PS II	Photosystem II
Put^{2+}	Putrescine
RBOH	Respiratory burst oxidase homologs
RBOHD	Respiratory burst oxidase homolog protein D
RBOHF	Respiratory burst oxidase homolog protein F
roGFPs	Hyper, redox-sensitive green fluorescent proteins
ROS	Reactive oxygen species
rxYFP	Redox-sensitive yellow fluorescent protein
SOD	Superoxide dismutase
SOSG	Singlet oxygen sensor green
Spd^{3+}	Spermidine
Spm^{4+}	Thermospermine
TCA	Quinine-triazine-based excitation probe
TCAO	Oxidized quinine-triazine-based excitation probe
Trxs	Thioredoxins
UPLC	Ultra-performance liquid chromatography
UV	Ultraviolet
Zn	Zinc

4.1 Introduction

Abiotic stresses are a condition, put forth by the environment, which hinders the normal growth, functioning, productivity, and survival of plants (Wang et al. 2004). These stresses can be physical (drought, flooding, wind, radiation, cold, and heat) or chemical (heavy metals, high salinity, etc.) in nature. In modern age of industrialization, the abiotic stresses are major threats to the agriculture system (Thakur et al. 2010; Conde et al. 2011). Abiotic stresses significantly induce protein dysfunction and modify carbon assimilation and photosynthetic interference in plants (Alam et al. 2013; Hasanuzzaman et al. 2014; Khan and Khan 2014; Khan et al. 2014, 2015, 2016a, b). Common outcome of those consequences is reactive oxygen species (ROS) generation. ROS are the partially reduced forms of atmospheric oxygen (O_2), comprising of non-radical molecules, e.g., $^1O_2$ (singlet oxygen) and

H_2O_2 (hydrogen peroxide), in addition to the free radicals like $OH^{\bullet}$ (hydroxyl radical) and $O_2^{\bullet-}$ (superoxide anion). ROS are generated in different cell organelles (mitochondria, chloroplasts, cell wall, vacuoles, etc.) as a normal product of plant cellular metabolism (Table 4.1; Laloi et al. 2007). The overproduction of ROS is directly linked to the increased generation of oxidizing species and the subsequent decline of antioxidant responses as well (Lee et al. 2012). The role of free radicals and ROS has been investigated by many workers since early days of emergence of interest in the damage caused due to oxidative stress in biological systems (Maheshwari and Dubey 2009; Sharma et al. 2012). ROS have long been considered as lethal bodies with the intention of causing damaging effects. They react in the body with proteins, lipids, and nucleic acids resulting in the disproportion among oxidants and antioxidants (Mishra et al. 2011). Despite their negative activity, they are well-known signaling molecules playing roles in normal physiological responses in cells as well as in tolerance to different environmental stresses depending upon the concentration of ROS (Golldack et al. 2014; Choudhary et al. 2013).

Disturbance of balance between the production and removal of free radicals and reactive oxygen species leads to oxidative stress (Suzuki et al. 2012). There exist antioxidant mechanisms that provide defense against oxidative stress caused by ROS overproduction. These mechanisms include nonenzymatic and enzymatic defense. Nonenzymatic defense system is comprised of antioxidants like ascorbate, carotenoids, tocopherols, glutathione, etc., whereas enzymatic defense system includes catalase, superoxide dismutase, ascorbate peroxidase, guaiacol peroxidase, monodehydroascorbate reductase, glutathione reductase, and dehydroascorbate reductase (Hideg et al. 2011; Heyno et al. 2011). The spatial and chronological position of ROS is important to understand signaling behavior (Gupta and Igamberdiev 2015). An overview of data on the features of generation and appearance of ROS functions in various compartments of plant cells was the purpose of present review.

Table 4.1 Sites and processes of ROS production

S. no.	Localization	ROS	Process
1	Mitochondria	$O_2^{\bullet-}$, $OH^{\bullet}$, H_2O_2	Respiration electron transport, NADPH oxidase
2	Chloroplast	$O_2^{\bullet-}$, $^1O_2$	Photosynthesis electron transport, photosystems I and II
3	Peroxisomes	$O_2^{\bullet-}$, $OH^{\bullet}$, H_2O_2	Fatty acid β-oxidation, glycolate oxidase reaction
4	Plasma membrane	$O_2^{\bullet-}$, H_2O_2	NADPH oxidase
5	Cell wall	$O_2^{\bullet-}$, H_2O_2, $OH^{\bullet}$	Electron transport chain, peroxidase, Mn^{2+}, and NADH
6	Nucleus	H_2O_2	D-amino acid oxidase, transcription factors

Based on Baker and Graham (2002) and Mittler et al. (2004)
$^1O_2$ singlet oxygen, $O_2^{\bullet-}$ superoxide anion, H_2O_2 hydrogen peroxide, $OH^{\bullet}$ hydroxyl radical

4.2 Compartmentalization of ROS

In plants, ROS are produced in cell organelles such as plasma membrane, cell walls, mitochondria, peroxisomes, endoplasmic reticulum, apoplast, chloroplast, nuclei, etc. (Sharma et al. 2012). These ROS are generated in different organelles by outflow of electrons onto O_2 in chloroplast, mitochondria, and plasma membrane by activity of electron transport chain (Gill and Tuteja 2010). In dark conditions, mitochondria are the main source of generation of ROS, and in light conditions, peroxisomes and chloroplast are the main source (Foyer and Noctor 2003).

4.2.1 Mitochondrial Compartmentalization

Mitochondria are considered to be the prime subcellular location for production of ROS since long time and also one of the organelles where discrete compartmentalization of the produced ROS takes place and the signaling cascade gets activated (Maron and Michel 2012). Production of ROS occurs when rate of relocation of ROS exceeds the capacity of AOX and COX pathways (Rhoads et al. 2006a).

Mitochondria contain a large number of electron carriers. Oxidation-reduction potential of those from them, which form the initial and middle sites of chains, is often more negative than −0.3 V (electrochemical potential of pair $O_2/O_2^{\bullet-}$). It means that the casual interaction of these carriers with the molecular oxygen can lead to one-electron reduction of O_2 to $O_2^{\bullet-}$. In plants, as well as in animals, the electron transport complexes I and III are considered as the main sites of electron leakages (Cvetkovska and Vanlerberghe 2013). In the isolated mitochondria, the generation of ROS (H_2O_2 or $O_2^{\bullet-}$) can vary in wide range (from 0.2 to 30 nmol/(min. • protein mg)) (Rhoads et al. 2006b). The speed of ROS generation in mitochondria of animal cells depends first of all on the reduction rate of electron transport chain and membrane potential. The dissipation of membrane potential happens at oxidative phosphorylation of ADP. Therefore if in mitochondria there is enough of ADP and it is phosphorylated actively, then the dissipation of proton gradient reduces the membrane potential and probability of $O_2^{\bullet-}$ generation (Cvetkovska and Vanlerberghe 2013). It is shown that in the stress conditions, the electron transport chain in mitochondria can become reduced because of oppression of energy-intensive processes, owing to what the generation of ROS by these organelles amplifies (Purvis 1997). There are proofs that mitochondria can be the main "suppliers" of ROS during the aging and dying out of leaves (Jimenez et al. 1998). The data which confirms the significant contribution of mitochondria to the processes of oxidative damages of plants in the drought conditions are obtained. So, in the stress conditions in wheat leaves, the content of carbonized proteins in mitochondria was ten times more, than in chloroplasts (Bartoli et al. 2004).

Hypoxic conditions lead to production of ROS in the intramitochondrial spaces other than its matrix region, which might be due to redox sensitivity of intramitochondrial region (Pani et al. 2001). Microscopic study was carried out to administer the behavior of mitochondria when ROS was produced in response to

UV-C radiation stress (Gao et al. 2008). Another report of production of ROS was given by Piantadosi (2008) who observed that carbon monoxide and cyanide cause suppression of complex IV which leads to increased production of ROS in mitochondria. Subcellular localization of intramitochondrial compartmentalization in response to high ROS productions under hypoxic conditions was demonstrated by using protein sensors (fluorescent in nature; Waypa et al. 2010).

In plants, the crosstalk between the transport of electrons, oxidative phosphorylation, and ROS generation is more sophisticated, than in animals, in connection with the availability of alternative oxidase, catalyzing the oxidation of ubiquinone and reduction of molecular oxygen to water. At the same time, the probability of $O_2^{\bullet-}$ formation is prevented owing to the leakage of electron from the complex III. Along with it, the transport of electrons, bypassing over of complex III, cytochrome *c*, and complex IV, decreases the re-reduction of mitochondria, their membrane potential, and, as a result, probability of ROS formation (Moller 2001a). Thus, the alternative oxidase of mitochondria can be considered as the component of antioxidative system.

It has been noted that complexes I and III are considered as the main sites of ROS formation in mitochondria (Cvetkovska and Vanlerberghe 2013). However, *Escherichia coli* cells and various body organs of mammals have shown that the complex II (succinate dehydrogenase, succinate:ubiquinone oxidoreductase) can also generate the superoxide anion radical and hydrogen peroxide with the high speed (Grivennikova and Vinogradov 2013). These enzymes catalyze one of the reactions in the cycle of tricarboxylic acids and deliver the electrons to the respiratory chain through a ubiquinone. It is established that in *Arabidopsis* mutants with the reduced activity of succinate dehydrogenase, the ROS formation in mitochondria decreased, including at the influence of agents of oxidative stress (Gleason et al. 2011). Superoxide radical is primary ROS generated in mitochondria. The electron transport chain of mitochondria has substantial energy to directly reduce O_2 (Rhodes et al. 2006). This O_2 can be readily reduced to H_2O_2 in the presence of superoxide dismutase enzyme (Quan et al. 2008). It has been estimated that approximately 15% of O_2 is converted to H_2O_2. There are reports of involvement of various proteins such as p66shc and monoamine oxidases (MAOs) in the production of ROS (Kaludercic et al. 2014a, b; Fig. 4.1).

Due to production and compartmentalization of ROS in mitochondria, its DNA is damaged, which consequently results in modification in processes such as replication and repair (Gupta and Igamberdiev 2015). Mitochondrial assimilation of ROS leads to peroxidation of the membrane lipids (polyunsaturated fatty acids). Lipid peroxidation leads to formation of toxic compounds such as cytosolic lipid alkenes, aldehydes, and hydroxyl alkenes (Moller 2001b).

4.2.2 Chloroplast Compartmentalization

Different forms of ROS are produced from various locations in the chloroplast (Fig. 4.2). Electron transport chains (ETC) in photosystem I (PSI) and photosystem

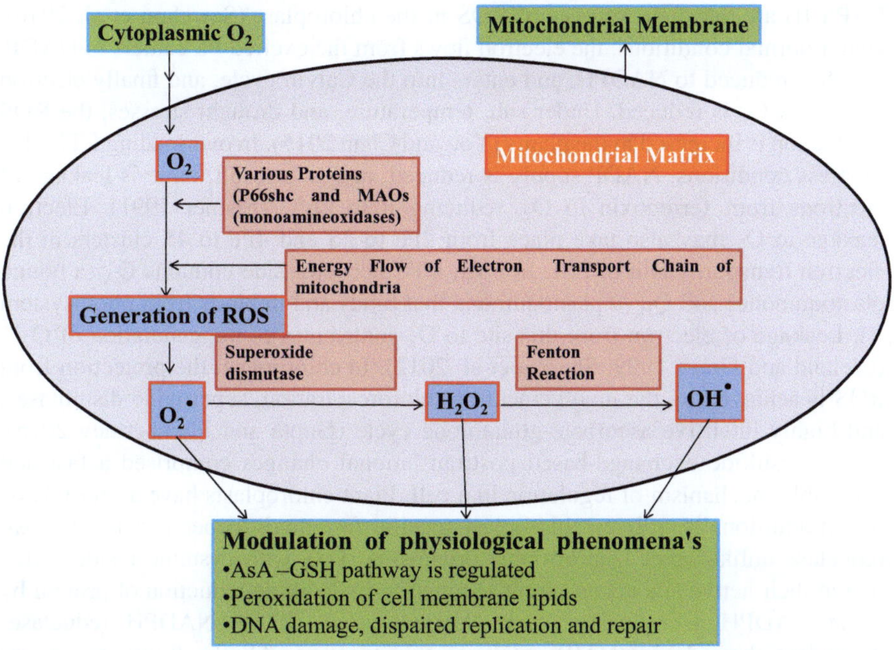

Fig. 4.1 ROS mitochondrial compartmentalization in plants (Adapted from Jajic et al. 2015)

Fig. 4.2 Production and scavenging of ROS in a chloroplast (Adapted from Jajic et al. 2015)

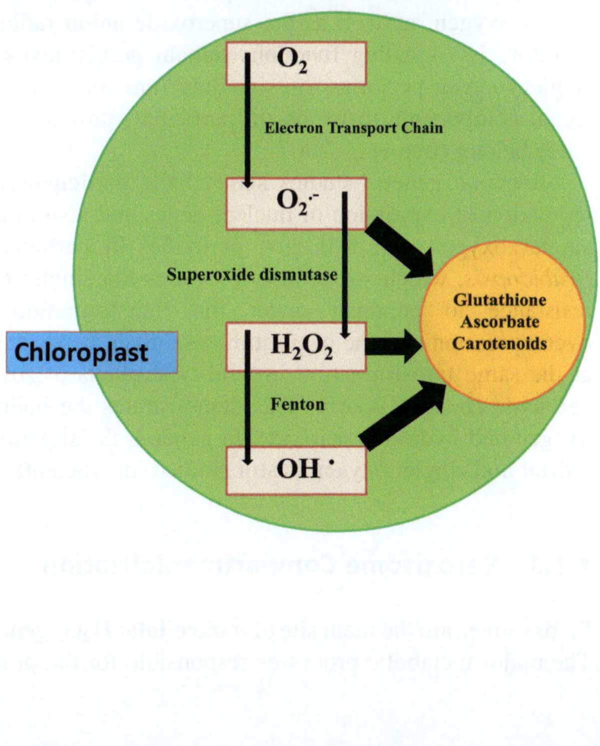

II (PS II) are the main sources of ROS in the chloroplast (Sewelam et al. 2016). Under normal conditions, the electron flows from the excited PS centers to NADP, which is reduced to NADPH, and enters into the Calvin cycle, and finally electron acceptor CO_2 is reduced. Under salt, temperature, and drought stresses, the ROS production is increased in the plants (You and Chan 2015). In overloading ETC, due to stress conditions, NADP supply is reduced, and as a result, there is leakage of electrons from ferredoxin to O_2, reducing it to $O_2^{\bullet-}$ (Elstner 1991). Electron leakage to O_2 may also take place from 2Fe to 2S and 4Fe to 4S clusters in the electron transport chain of PSI. In PSII, ETC acceptor side contains Q_A (a bound plastoquinone) and Q_B (a plastoquinone that binds and unbinds from photosystem II). Leakage of electron from this site to O_2 contributes to the generation of $O_2^{\bullet-}$ (Cleland and Grace 1999; Sharma et al. 2012). In chloroplast, the protection from ROS is achieved by the involvement of chlororespiration, superoxide dismutases, and highly intensive ascorbate-glutathione cycle (Gupta and Igamberdiev 2015). Dithiol-disulfide exchange-based posttranslational changes comprised a fast and reversible mechanism of regulation in a cell. Plant chloroplasts have a set of Trxs, which additionally utilizes chloroplast-specific ferredoxin-dependent thioredoxin reductase unlike other heterotrophic organisms. Trxs with cysteine residues that pair at their active site act as essential function in disulfide reduction of protein by using NADPH as reducing agent (Jacquot et al. 2009). NADPH reductase-dependent thioredoxin (NTR) catalyzes this reaction. All the living organisms such as plants, animals, and bacteria possess two-component NTR/Trx redox systems (Meyer et al. 2005). Along with superoxide anion radical, the singlet oxygen can be formed in chloroplasts when photosystem II is functioning. The singlet oxygen, as well as the superoxide anion radical, differs a short lifetime; therefore its signaling functions remain poorly investigated. It is supposed that singlet oxygen performs the signaling functions due to the involvement of other signal components to which, in particular, proteins EXECUTER 1 and 2 (EX1, EX2) belong (Lee et al. 2007).

Molecular-genetic studies showed the participation of singlet oxygen in the regulation of expression of nuclear genes and also in the interaction of signals of singlet oxygen and hydrogen peroxide. In particular, the mutant line *flu* of *Arabidopsis*, which strenuously produces the singlet oxygen, completely lost the resistance to paraquat under the transformation, which had caused the overexpression of gene of plastidic ascorbate peroxidase (Laloi et al. 2007). Thus at the same transformation in wild-type plants of *Arabidopsis*, the resistance to herbicide changed poorly. This demonstrates the interaction of signals of singlet oxygen and hydrogen peroxide. In general, the signaling role of superoxide anion radical and singlet oxygen is still studied insufficiently.

4.2.3 Peroxisome Compartmentalization

Peroxisomes are the main site of intracellular H_2O_2 generation (Del Rio et al. 2006). The major metabolic processes responsible for the production of H_2O_2 in various types of peroxisomes are the β-oxidation of fatty acids, disproportionation of $O_2^{\bullet-}$ radicals, glycolate oxidase reaction, and flavin oxidase enzymatic reaction (Baker and Graham 2002). During the process of photorespiration in peroxisomes, glycolate oxidation by glycolate oxidase accounts for the major generation of H_2O_2 (Noctor et al. 2002). In leaf peroxisomes, xanthine oxidase produces superoxide radicals via xanthine or hypoxanthine oxidation to uric acid. SOD dismutates superoxide anion to H_2O_2, and catalase degrades H_2O_2 (Mittler et al. 2004; Fig. 4.3).

ROS-generating enzymes are urate oxidase, sarcosine oxidase, acyl-CoA oxidase, xanthine oxidase, glycolate oxidase, D-amino acid oxidase, polyamine oxidase, and hydroxyl acid oxidases 2 and 3 (Schrader and Fahimi 2006; Wanders and Waterham 2006). Eukaryotic peroxisomes contain glutathione peroxidase, manganese superoxide dismutase, epoxide hydrolase, peroxiredoxin-1, peroxiredoxin-5, and copper/zinc superoxide dismutase (Wander and Waterham 2006). Peroxisomes contribute mainly to maintain the cellular redox homeostasis by involving key enzymes such as catalase inside the peroxisomes. Catalase is the main enzyme for the H_2O_2 metabolism in peroxisomes, and it appears that under physiological conditions, H_2O_2 diffusion is stopped through its fast transformation to O_2. Peroxisomes are more oxidizing organelle as compared to mitochondria and nuclei (Kaludercic et al. 2014b). It has been observed that cat2 mutant that grows under low light possesses enhanced diaminobenzidine staining (Bueso et al. 2007). It has been also studied that cat2 and cat2:cat3 knockout plants contain twofold enhancement in extractable H_2O_2 (Hu et al. 2010; Das et al. 2015).

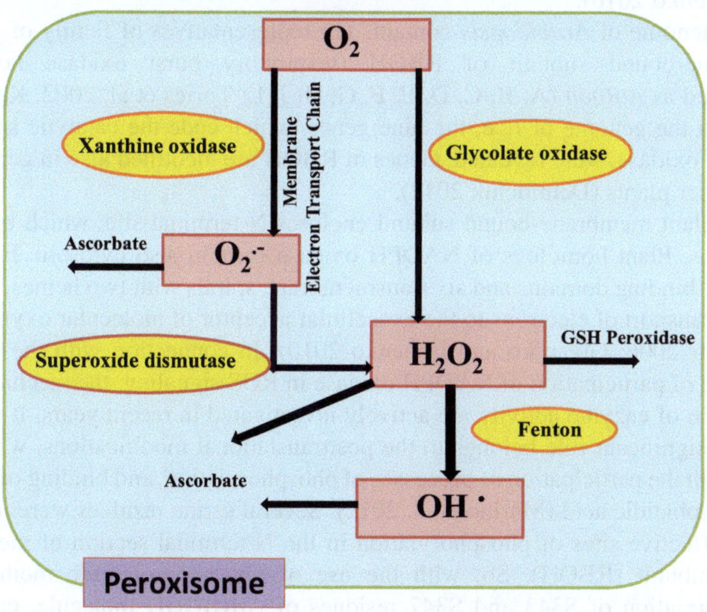

Fig. 4.3 Production and scavenging of ROS in peroxisome (Adapted from Jajic et al. 2015)

4.2.4 Plasma Membrane Compartmentalization

Plasma membrane is ROS production center due to the presence of membrane-bound NADPH oxidase (EC 1.6.3.1; Apel and Hirt 2004). NADPH oxidase is considered as the main enzymatic producer of ROS in animal and plant cells (Sagi and Fluhr 2006; Glyan'ko and Ischenko 2010). This enzymatic complex restores the molecular oxygen with the formation of superoxide anion radical:

$$\mathrm{NADPH} + 2\mathrm{O}_2 \rightarrow \mathrm{NADP}^{+} + 2\mathrm{O}_2^{\bullet -} + \mathrm{H}^{+}.$$

In the reaction, it is used as the cytoplasmic NADPH, from which one of the electrons with the participation of FAD and heme is transferred through the membrane to its external side and attached to the molecular oxygen with the formation of superoxide anion radical (Glyan'ko and Ischenko 2010).

NADPH oxidase of plants differs from that one in animals. So, enzyme of phagocytes represents the difficult system, which consists of six heterogeneous subunits, including two membrane bound ($gp91^{phox}$ – glycoprotein of 91 kDa phagocyte oxidase-specific – and $p22^{phox}$) and four cytosolic ($p47^{phox}$, $p40^{phox}$, $p67^{phox}$, and Rac1 or Rac2 (GTPases)), which at the activation of enzyme under the influence of wide spectrum of effectors combine in the complex, which generates $O_2^{\bullet -}$ (Sagi and Fluhr 2006). In plants, it has been found only as the homologs of membrane-bound subunit $gp91^{phox}$ (Nox2-NADPH oxidase) and cytosolic subunit of family of small GTPases (Rac proteins, which in relation to plants is more often called as Rop (Rho-related GTPases from plants)) (Glyan'ko and Ischenko 2010).

The genome of *Arabidopsis* contains ten representatives of family of genes of membrane-bound subunit of RBOH (respiratory burst oxidase homologs) designated as *AtRboh* (A, B, C, D, E, F, G, H, J, L; Torres et al. 2002; Kaye et al. 2011). In the genome of rice, the nine genes, which code the catalytic subunit of NADPH oxidase, were revealed. Genes of RBOH are identified also in genomes of some other plants (Demidchik 2012).

The plant membrane-bound subunit encloses N-terminal site, which binds calcium ions. Plant homologs of NADPH oxidase contain also cytosolic FAD- and NADPH-binding domains and six transmembrane spirals with two hemes, required for the transport of electrons to the exocellular acceptor of molecular oxygen (Sagi and Fluhr 2006; Glyan'ko and Ischenko 2010). In connection with the obtained evidence of participation of NADPH oxidase in ROS signaling, the mechanisms of regulation of enzyme activity are actively investigated in recent years. It is shown that the significant role belongs to the posttranslational modifications, which take place with the participation of processes of phosphorylation, and binding of calcium and phosphatidic acid (Marino et al. 2012). Several serine residues were identified as the effective sites of phosphorylation in the N-terminal section of membrane-bound subunit (RBOH). So, with the use of phosphoproteomic methods, the phosphorylation of S343 and S347 residues of AtRBOHD molecule, caused by the influence of pathogenic elicitors on *Arabidopsis* plants, was shown (Nühse et al.

2007). At the same time, it is established that the phosphorylation of specified sites is necessary, but it is not enough for the full activation of RBOHD. In molecules of other form of NADPH oxidase – AtRBOHC – it is revealed that the residues of S318 and S322, which are phosphorylated, and the gene mutation on them cause the changes in ROS generation (Takeda et al. 2008).

In regulation of activity of NADPH oxidase, the calcium ions play a special role. According to one of the models given by Wong et al. (2007), calcium activates the Ca^{2+}-dependent protein kinase, which phosphorylates the N-terminal site of membrane-bound subunit (RBOH) of NADPH oxidase and causes its conformational changes, which are serving to binding with it of cytosolic component – Rop protein (GTPase). As a result, the formation of active dimer takes place that leads to the intensifying of ROS generation (Wong et al. 2007). Later on, in the example of one of the forms of membrane-bound subunit of rice NADPH oxidase (OsRBOHB), the existence of EF-hands was shown in N-terminal area that indicates about the formation of dimer with the participation of calcium (Oda et al. 2010). Thus, the increase of activity of NADPH oxidase with the participation of calcium, apparently, is bound not only to the activation of protein kinase by it but also to the direct interaction of Ca^{2+} with catalytic subunit. The effect of synergism of Ca^{2+}-ionophore ionomycin and inhibitor of protein phosphatases calyculin A in the activating influence on NADPH oxidase of *Arabidopsis* is established (Ogasawara et al. 2008). It is supposed that the phosphorylated catalytic subunit of AtRBOHD, attaching calcium in the location of EF-hands, receives more profound conformational changes, than non-phosphorylated one. Thereof the more essential activation of enzyme occurs under the simultaneous influence of calcium ionophore and inhibitor of phosphatases. Recently, it was shown that the inhibitor of protein kinases K-252a suppressed the activation of AtRBOHD and AtRBOHF, which had occurred under the influence of Ca^{2+} (Kimura et al. 2012). Authors offered the model, according to which the phosphorylation of RBOHD and RBOHF in *Arabidopsis* occurs before the process of calcium binding. At the same time, the phosphorylation of RBOH causes in itself the enzyme activation that leads to the intensifying of generation of ROS, which are involved in the opening of calcium channels. The influx of calcium in cytosol causes the additional activation of NADPH oxidase. Besides, the intensifying of phosphorylation of catalytic subunit is not excluded also due to the activation of calcium-dependent protein kinase. In general, it is supposed that any influences, which cause the changes of concentration of cytosolic calcium, can potentially influence the activity of NADPH oxidase (Demidchik 2012).

One more mechanism of regulation of activity of NADPH oxidase consists in the binding with the catalytic subunit of phosphatidic acid (Marino et al. 2012), which is one of the products of transformation of membrane lipids with the participation of phospholipase D. So, it is revealed that the mutation on the arginine residue in N-terminal site of subunits of AtRBOHD and AtRBOHF leads to the loss of their ability to bind the phosphatidic acid and to be activated under its influence (Zhang et al. 2009). Naturally, the activity of NADPH oxidase, as well as the majority of other enzymes, can be regulated by the concentration of substrate (NADPH), the content of which depends on the oxidation-reduction processes in cytosol and on the processes happening in chloroplasts and mitochondria (Foyer and Noctor 2009). In particular, the reduction of electron transport chains in these organelles can serve to the increase of general pool of NADPH and activation of NADPH oxidase. Eventually, recently it is shown that NADPH oxidase can be activated by ROS (molecules of hydrogen peroxide) formed by the neighboring cells (Mittler et al. 2011).

Nitric oxide as one of the signaling molecules, apparently, participates in the regulation of activity of NADPH oxidase also. In the tissue culture of ginseng roots under the influence of NO donor, there were the activation of NADPH oxidase and increase of generation of superoxide anion radical (Tewari et al. 2008). The treatment of segments of wheat coleoptiles with the donor of nitric oxide sodium nitroprusside also caused the intensifying of generation of superoxide anion radical. This effect is bound, apparently, to the increase of activity of NADPH oxidase as it was suppressed with its inhibitor imidazole (Karpets et al. 2011).

On the other hand, there are data that NO can inhibit the NADPH oxidase due to S-nitrosylation of cysteine (Cys 890) (Yun et al. 2011). Thus, nitric oxide can participate both in activation and in inhibition of NADPH oxidase. If the NADPH oxidase inhibition by nitric oxide is bound to its direct interaction with specific sites of enzyme, then the activating effect of NO on the NADPH oxidase, most likely, is mediated by influence of NO on the calcium homeostasis (Glyan'ko and Ishcenko 2010).

NADPH oxidase is considered as the one of the main sources of signaling ROS (Kolupaev and Karpets 2013). During stress conditions, the NADPH oxidase also increases the level of activity of MAP kinase and takes part in the regulation of calcium homeostasis. The part of calcium channels (potential dependent, regulated by hyperpolarization) depends on ROS (Sagi and Fluhr 2001; Mori and Schroeder 2004; Kwak et al. 2006). It is supposed that mechanosensitive calcium channels can be also regulated by ROS (Mori and Schroeder 2004). Possibly, ROS, which appear owing to the activation of NADPH oxidase, can interplay with some functional groups of cytoskeleton proteins, which are associated with the plasma membrane. Thus, the state of mechanosensitive Ca^{2+} channels depends directly or indirectly on the state of these proteins (Mori and Schroeder 2004).

4.2.5 Cell Wall Compartmentalization

Cell wall is also a major site for ROS production. Peroxidase enzyme present in the cell wall is associated with the production of H_2O_2. Along with antioxidative activity, peroxidases can show also oxidase activity with the transfer of electrons from reductants (e.g., NADH) to the molecular oxygen (Bestwick et al. 1997).

Peroxidases, which are characteristics for this mechanism of action, are localized generally in cell walls in bounded state with their polymers by ionic or covalent bonds. It is considered that at "oxidative burst," the peroxidase of cellular walls can generate a large number of superoxide anion radical and, as the result,

H_2O_2 (Bestwick et al. 1997), though the process of ROS generation can be carried out by "tandem" peroxidase/NADPH oxidase. In the wheat roots, the presence of extracellular peroxidases with the molecular weight of 37, 40, and 136 kD is established. High sensitivity of generation of superoxide anion radical in wheat roots to the inhibitors of peroxidase is considered by authors as the evidence of participation of indicated peroxidases in ROS generation (Minibayeva et al. 2009). There are bases to believe that peroxidase of cellular walls is involved in the generation of signaling ROS, which activate the formation of adaptive responses of plants to the influence of abiotic stressors, in particular, hyperthermia and drought (Kolupaev and Karpets 2013).

4.2.6 Nuclear Compartmentalization

The most complex cell organelle in the eukaryotic cell is the nucleus. It is divided into various nuclear compartments which include nucleoli, PML bodies, histone local bodies, Cajal bodies, polycomb bodies, insulator bodies, and transcription and replication factories (Gavrilova and Razina 2014). For gene expression, the genome rearranges, thus creating dynamic compartments (Meldi and Brickner 2011). ROS are produced by reduction of molecular oxygen (O_2) by high-energy exposure or electron-transfer reactions in aerobic metabolism. The biotic and abiotic stresses also lead to the production of ROS in plants due to disturbances in the cellular homeostasis (Sharma et al. 2012). The redox potential of nuclei is more reduced than other organelles which are more oxidized, but ROS (hydrogen peroxide and NO) diffuse across membrane and thus cause potential damage to the DNA (Go and Jones 2008).

H_2O_2 can be produced in nuclei by stimulation of D-amino acid oxidase that increases glutathionylation in the nuclei (Halvey et al. 2007). ROS (H_2O_2) also cause oxidation of nuclear GSH but not of Trx-1 which is very important in regulating NF-κB activity. During the period of stress, cytoplasmic Trx-1 moves toward the nucleus as it is the main redox control system in the nucleus (Wei et al. 2000). Peroxiredoxins (Prx) proteins are found abundant in the nucleus, act as Trx-1-dependent peroxidases, and play an important role in redox signaling (Rhee et al. 2005). Prx prevents nuclear oxidation of both Trx-1 and NF-κB p50 (Jones and Go 2010). GSH is also needed for the maintenance of cysteine residues of intracellular proteins in reduced and functional state (Go and Jones 2008). Excessive ROS formation inhibits DNA-binding transcription factor, and therefore these cysteine residues need to be in reduced state (Kaludercic et al. 2014a). Thiols are the reduced form of cysteine residues (Schafer and Buettner 2001). Proteins which are redox sensitive within organelles can be identified by a variety of techniques (Kaludercic et al. 2014a). But it is relatively hard to find out redox potential inside the nucleus (Smith et al. 1996; Go and Jones 2002). Therefore, to protect genetic material from harmful effects of ROS, nuclear redox couples (GSH, Trx-1) are being maintained at high reduced values rather than cytoplasm. In reality, many studies reported oxidation in mitochondria and cytoplasm but were unable to detect any changes in nuclei exposed to oxidants (Jones and Go 2010).

4.2.7 Vacuolar Compartmentalization

Vacuole is the largest plant organelle that significantly contributes in redox homeostasis against ROS generated by various abiotic stresses (Andrev 2012). They are rich in antioxidants, which play a multifaceted role in plant metabolism (Gill and Tuteja 2010). Vacuoles also contribute to ROS accumulation during stress via class III peroxidases localized at the inner membrane of tonoplast. Peroxidases are generally ROS scavengers; they can attack H_2O_2-generating hydroxyl radicals via hydroxylic cycle. Proteomic studies also indicate localization of NADPH oxidases at tonoplast which can generate superoxide anions via cytosolic NADPH (Keunen et al. 2013). Vacuoles accumulate array of enzymatic and nonenzymatic antioxidant components for ROS scavenging during stress conditions (Andrev 2012). ROS are detoxified locally at their production sites, but when produced in excess during stress, they may diffuse to cytosol or other subcellular compartments like vacuoles where they are scavenged. H_2O_2 diffuses into vacuoles through tonoplast either directly or via aquaporins (Bienert et al. 2007). Vacuoles are abundant in flavonoids and ascorbate and also contain high content of peroxidase localized at inner surface of tonoplast (Gechev et al. 2006). Elevated levels of ascorbate in vacuoles play an important role in detoxification of H_2O_2 accumulated due to leakage from peroxisomes and chloroplast during high light stress (Heyneke et al. 2013) and drought stress (Koffler et al. 2014). Vacuoles though contain low levels of GSH, but in response to oxidative stress, the GSH accumulates as oxidized form GSSG in vacuoles (Queval et al. 2011). GSSG is directly transferred from cytosol to vacuoles through transporters like ABCCs (subclass C of ATP-binding cassette transporters). This transport is crucial for redox homeostasis and signaling through GSH-dependent system (Noctor et al. 2013). Soluble sugars (sucrose) and sugar alcohol (fructans) accumulate in vacuoles under oxidative stress which is a common consequence of abiotic stress. Vacuole of sugar-accumulating tissues such as taproots of sugar beet, phloem tissues, and sugarcane containing high levels of sucrose might protect tonoplast against ROS-mediated damage (Andreev 2012). Fructans that are also present in vacuoles protect membranes against freezing and drought stress by scavenging hydroxyl-free radicals generated at membrane surface by integrating with them leading in production of less toxic H_2O_2. This reaction also leads to transformed fructans that are less damaging which can be reduced by phenolics and/or GSH or AsA present in vacuoles. This suggests synergistic action of phenolics and fructans in scavenging of vacuolar H_2O_2 (Bolouri-Moghadden et al. 2010; Peshav et al. 2013). During stress, sucrose and fructans can be transported from vacuoles to apoplast as tonoplast exocytosis vesicles for stabilization of cell membrane (Valluru et al. 2008).

4.2.8 Apoplastic Compartmentalization

The region of plant cell outside of the plasma membrane is known as apoplast, also referred as extracellular space. It is one of the important sites of ROS generation under endogenous and exogenous influence. Apoplastic ROS are generated by the cell surface or enzymes like NADPH oxidases, amine oxidases, class III peroxidases, and oxalate oxidases which are localized apoplast (You and Chan 2015). Apoplastic ROS generation is essential for drought and salt stress tolerance responses such as cell wall strengthening, leaf cell elongation, and growth (Miller et al. 2010). NADPH oxidase genes (*Atrboh D* and *Atroboh F*) induced apoplastic ROS production that is crucial for ABA-mediated stomatal closure during drought and salinity stress (Kwak et al. 2003). Oxalate oxidase is implicated in acclimatization of root growth during drought stress (Noctor et al. 2014). Apoplastic ROS generation can also influence chloroplastic ROS production which in turn can modulate nuclear gene expression (Shapiguzou et al. 2012). Apoplast is also rich in antioxidants which regulate apoplastic redox state. Thioredoxins (Trxs) are members of multigenic proteins which help in regulation of redox state under various environmental conditions through thiol-disulfide interchange reactions also localized in apoplast. *OsTRXh1* H-type Trx in *Oryza sativa* regulates salt stress responses by regulating apoplastic ROS under salt stress (Zhang et al. 2011). Secretion of polyamines such as putrescine, triamine spermidine, and tetraamine spermine in apoplast during abiotic stress also regulates redox state in apoplast. However polyamine oxidases (PAOs) and amine oxidases (AOs) localized in apoplast catabolize PAs titers producing H_2O_2 which in turn may signal stress tolerance signals or lead to programmed cell death (PCD) syndrome depending upon the level of H_2O_2 produced (Toumi et al. 2010; Petrov et al. 2015).

4.2.9 Cytoplasmic Compartmentalization

Cytosol is not the major site for ROS production, but superoxide radicals are produced in cytosol by aldehyde oxidase and xanthine dehydrogenase (Yesbergenova et al. 2005; Zarepour et al. 2010). Superoxide anions that are produced by small ETC consisting of NADH and Cyt b present in peroxisomal membrane are also released into cytosol (Del Rio et al. 2002). Cytosol is an important site for redox signal integration, as signal from apoplast and other organelles passes through it for modification of gene expression in the nucleus. Cytoplasmic NADPH play a central role in redox signaling as it firstly generates reduced substrates for ROS-detoxifying antioxidative enzymes and secondly it also regulates thiol/disulfide redox couple status and supplies electrons to NADPH oxidases for ROS production (Fig. 4.4; Tripathy and Oelmüller 2012). Under stress such as high light and high-temperature leakage, diffusion or active transport of ROS from mitochondria or chloroplast to cytosol can cause disruption of normal metabolism and even cell death. Under such conditions, cytosolic ROS scavenging system maintains the redox homeostasis and modifies or buffers ROS leakage

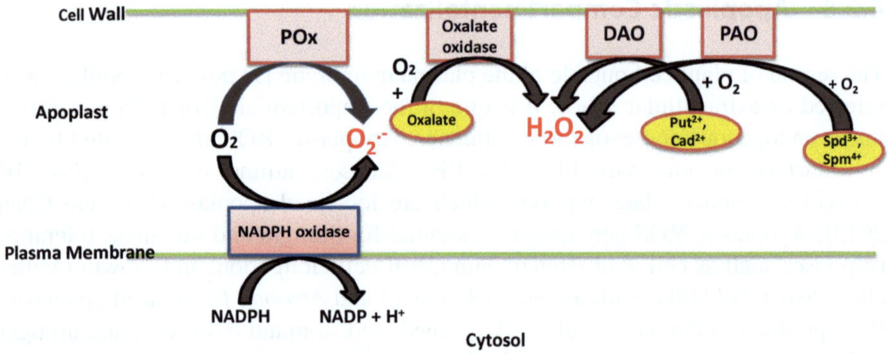

Fig. 4.4 ROS-generating enzymes in apoplast showing generation of superoxide anion and hydrogen peroxide. In addition to NADPH oxidase, oxalate oxidase and class III peroxidases (POx) are main sources of ROS production in the apoplast. Amine oxidases such as DAO (diamine oxidase) and PAO (polyamine oxidase) also generate H_2O_2 via catabolism of polyamines. DAO selectively oxidizes Put^{2+} (putrescine) and Cad^{2+} (cadaverine), while PAO catabolizes Spd^{3+} (spermidine) and Spm^{4+} (thermospermine). H_2O_2 can be further converted to most toxic ROS, i.e., $OH^{\bullet \cdot}$, by Haber-Weiss and Fenton reaction (Pottosin et al. 2014)

signals between other organelles like mitochondria or plastids and nucleus (Suzuki et al. 2013). Additionally various enzymatic and nonenzymatic antioxidants are localized in cytosol. Cytosolic APX and ascorbate-glutathione cycle plays an important role in scavenging H_2O_2 using ascorbate as an H donor (Asada 2000). Drought stress upregulated mRNA of cytosolic APX in alfalfa nodule (Naya et al. 2007). Moreover transgenics with overexpression of cytosolic APXs were more tolerant to salt stress than wild types (Lu et al. 2007). In rice under salt, cold, and drought stress, cytosolic APX 2 (OsAPX2) plays an important role in reproduction and growth of rice (Zhang et al. 2013). Cu/Zn SOD-type SOD metalloenzymes are also localized in cytosol in addition to chloroplast and peroxisomes; SOD enzymes dismutates superoxide radicals into hydrogen peroxide and oxygen (Del Rio et al. 1996; Chen et al. 2010). Cytosolic APXs and Cu/Zn SOD are involved in tolerance against drought stress (Faize et al. 2011). Cytosol is also abundant in GSH for reducing ROS; mutants with deficient cytosolic GSH metabolism in *Arabidopsis* exhibited ROS accumulation in cytosol suggesting its need for ROS removal during photooxidative stress (Ball et al. 2004).

4.3 Tools to Study ROS Compartmentalization

Production of different ROS species in cells needs to be analyzed to know the type, amount, location, as well as their role in cell disintegration. To measure redox potential, major redox couples such as GSH/GSSG and $NAD(P)H/NAD(P)^+$ were earlier used as tool to measure ROS production. Redox potential of a specific cell organelle or compartment can be identified or quantified using different techniques such as HPLC/UPLC (high-/ultra-performance liquid chromatography; Jones 2006;

Takimoto and Kaas 2007). The redox state of different proteins (Trx, Trx reductase, etc.) can be detected using techniques, such as mass spectroscopy (MS) and redox Western blotting along with tagging of free thiols. These techniques are highly specific and can detect both oxidized and reduced form of these redox molecules (Chen et al. 2006; Go et al. 2009; Go and Jones 2013). However, the major problem of using the above methods is that tissues analyzed need to undergo fractionation process, during which various other changes in tissues like redistribution and artificial oxidation/reduction can occur. In order to overcome this, the latest advanced techniques in molecular biology have come which use the epitope tagging of nuclear localization and nuclear export sequence inside of cell cytoplasm or nuclei having potential to overcome the fractionation (Go et al. 2010). Most of the redox molecules and ROS, nowadays, can be detected in vivo using fluorescent imaging. Genetically encoded fluorescent proteins and small molecule probes are the mostly recommended fluorescent imaging technique (Winterbourn 2014; Kaludercic et al. 2014a).

4.3.1 Small Molecule Redox Probes

Many small molecules (mitochondrial-targeted DHE (MitoSOX) Red, Peroxy Lucifer 1, 2′-7′-dichlorodihydrofluorescein (DCFH), dihydroethidium (DHE), the most popular ones, etc.) can be used as small molecule redox probes, but mostly few of them are commonly used due to limitation in sensitivity and selectivity. MitoSOX Red is most commonly used probe for measurement of superoxide formation in the mitochondria of live cells (Robinson et al. 2006; Zhou et al. 2011) and DHE is used as specific probe for detection of superoxide radical, DCF-DA can be used for monitoring H_2O_2 in plant tissues. The direct use of DHE and DCF is not much recommended as they lack in sensitivity and specificity as they will tag all peroxides and superoxides. For detection of singlet oxygen along with a singlet oxygen sensor green (SOSG) reagent has found to have high specificity to $^1O_2$ without the interference of hydroxyl radicals or superoxide (Yamamoto et al. 2002; Tarpey et al. 2004; Corpas et al. 2006).

The mitochondrial ROS are topped with either a 4-aminophenyl aryl ether (MitoAR) or a 4-hydroxy aryl ether group (MitoHR) on rhodamine-like fluorophores. The ether groups absorb the fluorophore emission by photoinduced electron transfer (PET). MitoAR tags mostly to OH• and HOCl, whereas MitoHR is most sensitive to •OH (Dickinson et al. 2010). In cultured cells, oxidation of hydroethidine (HE) is the usual reason for detection of superoxide by joining HE to hexyl triphenylphosphonium cation (Mito-HE); the HE can be localized in mitochondria (Robinson et al. 2006). To resolve the problem of oxidant-sensitive dyes, "non-redox" probes are used, which contain a coated fluorophore that is released without changing the oxidation state of fluorophore (Winterbourn 2014). Other example includes the boronate-masked groups, which are specially used for the detection of hydrogen peroxide (H_2O_2) (Miller et al. 2005). The formation of nitric oxide in mitochondria is detected by DAF-2DA in the endothelial cells.

Detection of other ROS such as hydrogen peroxide, $HOO^{\bullet}$, and $ONOO^{-}$ radicals has been accessed and reported using C-H2DFFDA (Rivera-Ingraham et al. 2016).

4.3.2 Fluorescent Protein-Based Redox Sensors

The limitation of oxidation, reversibility, specificity, and quantification has been overlooked by the targeting subcellular genetically encoded fluorescent redox-sensitive probes that have been developed. The advantage of these probes is that they have high specificity resulting in very accurate measurements of redox molecules and ROS even in the subcellular compartments (Miller et al. 2005; Kaludercic et al. 2014a). Other advantage of using genetically coded fluorescent probe is that the same probe can be used for different locations in various cell compartments as well as organelles (Rodriguez-Serrano et al. 2006). These genetically encoded redox probes are derived from green fluorescent proteins (GFP), which were then laterally advanced to redox-sensitive yellow FP (rxYFP), Hyper, redox-sensitive GFPs (roGFPs), etc. (Meyer and Dick 2010). Hyper is the most sensitive fluorescent probe for hydrogen peroxide labeling in peroxisomes (Belousov et al. 2006), and roGFP can be used for determination of H_2O_2 and thiol redox state depending on its state. The drawback of using these probes is the important factors such as pH and antioxidant activity, redox signaling, calcium oscillation, enzyme activities, and membrane potential (Hernandez-Barrera et al. 2013, Lukyanov and Belousov 2014; Jajic et al. 2015). To detect broad array of antioxidants, more advanced strategies such as reversible small molecule sensors have been developed (Ai 2015). A one- and two-photon excitable probe quinine-triazine-based excitation probe (TCA), which can be selectively and reversibly oxidized by superoxide to form TCAO (which has 1000× higher fluorescence), was reported. The reversibility of TCA fluorescence of HepG2 cells by chronological treatment of ROS inducer phorbol-12-myristate-13-acetate (PMA) and the reductant glutathione was studied by Zhang et al. (2015) and Kaur et al. (2016).

4.4 Conclusion and Future Prospects

ROS are produced by electron transport activities of cells and are compartmentalized in different organelles like mitochondria, chloroplast, and plasma membrane. ROS generation and localization in different compartments of cell is unavoidable. In normal conditions, ROS are produced in small amounts, whereas, under abiotic stress conditions, cellular homeostasis is disrupted and ROS production is enhanced. Each compartment of cell has diverse attribute toward ROS production, scavenging, and signaling. The systems within every compartment are continuously in unstable state. Those systems have evolved according to the function of each organelle, and each organelle has a different mode of ROS compartmentalization according to its line of function (Go and Jones 2008). For example, the ROS in mitochondria act as retrospective signaling molecules and make communication among mitochondria and other compartments (Lodi et al. 2006). The specific details of this communication still need explanation. Nuclear compartment is less prone to oxidative stress, whereas cytoplasm lacks the enzymes producing ROS, therefore, creating a buffer zone for other organelles from anomalous ROS generation (Schroeder et al. 2007). During abiotic stress, compartmentalization of ROS in different organelles makes the redox organization available which facilitates normal functioning. Potential advancement in metabolomics, genomics, and proteomics will make advances in understanding the biochemical complexes drawn in cellular reaction to oxidative stress.

References

Ai H (2015) Fluorescent-protein-based probes: general principles and practices. Anal Bioanal Chem 407:9–15

Alam MM, Hasanuzzaman M, Nahar K, Fujita M (2013) Exogenous salicylic acid ameliorates short-term drought stress in mustard (*Brassica juncea* L.) seedlings by upregulating the antioxidant defense and glyoxalase system. Aust J Crop Sci 7:1053–1063

Andreev M (2012) Role of the vacuole in the redox homeostasis of plant cells. Russ J Plant Physiol 59(5):611–617

Apel K, Hirt H (2004) Reactive oxygen species: metabolism, oxidative stress, and signal transduction. Ann Rev Plant Biol 55:373–399

Asada K (2000) The water-water cycle as alternative photon and electron sinks. Philoso Trans Royal Soc B: Biol Sci 355:1419–1431

Baker A, Graham AI (2002) Plant peroxisomes: biochemistry, cell biology and biotechnological applications. Kluwer Academic Publishers, Dordrecht

Ball L, Accotto GP, Bechtold U, Creissen G, Funck D, Jimenez A, Kular B, Leyland N, Mejia-Carranza J, Reynolds H, Karpinski S, Mullineaux PM (2004) Evidence for a direct link between glutathione biosynthesis and stress defense gene expression in *Arabidopsis*. Plant Cell 16(9):2448–2462

Bartoli CG, Gomez F, Martinez DE, Guiamet JJ (2004) Mitochondria are the main target for oxidative damage in leaves of wheat (*Triticum aestivum* L.) J Exp Bot 55:1663–1669

Belousov VV, Fradkov AF, Lukyanov KA, Staroverov DB, Shakhbazov KS, Terskikh AV, Lukyanov S (2006) Genetically encoded fluorescent indicator for intracellular hydrogen peroxide. Nat Methods 3:281–286

Bestwick CS, Brown IR, Bennett MHR, Mansfield JW (1997) Localization of hydrogen peroxide accumulation during the hypersensitive reaction of lettuce cells to *Pseudomonas syringae* pv. *Phaseolicola*. Plant Cell 9:209–221

Bienert GP, Møller ALB, Kristiansen KA, Schulz A, Møller IM, Schjoerring JK, Jahn TP (2007) Specific aquaporins facilitate the diffusion of hydrogen peroxide across membranes. J Biol Chem 282:1183–1192

Bolouri-Moghaddam MR, Le Roy K, Xiang L, Rolland F, Van den Ende W (2010) Sugar signalling and antioxidant network connections in plant cells. FEBS J 277:2022–2037

Bueso E, Alejandro S, Carbonell P, Perez-Amador MA, Fayos J, Belles JM (2007) The lithium tolerance of the *Arabidopsis* cat2 mutant reveals a cross talk between oxidative stress and ethylene. Plant J 52:1052–1065

Chen Y, Cai J, Jones DP (2006) Mitochondrial thioredoxin in regulation of oxidant induced cell death. FEBS Lett 580:6596–6602

Chen Q, Zhang M, Shen S (2010) Effect of salt on malondialdehyde and antioxidant enzymes in seedling roots of Jerusalem artichoke (Helianthus tuberosus L.) Acta Physiol Plant 33:273–278

Choudhury S, Panda P, Sahoo L, Panda SK (2013) Reactive oxygen species signaling in plants under abiotic stress. Plant Signal Behav 8:e23681
Cleland RE, Grace SC (1999) Voltammetric detection of superoxide production by photosystem II. FEBS Lett 457:348–352
Conde A, Chaves MM, Geros H (2011) Membrane transport, sensing and signaling in plant adaptation to environmental stress. Plant Cell Physiol 52(9):1583–1602
Corpas FJ, Fernandez-Ocana A, Carreras A, Valderrama R, Luque F, Esteban FJ, Rodriguez-Serrano M, Chaki M, Pedrajas JR, Sandalio LM (2006) The expression of different superoxide dismutase forms is cell-type dependent in olive (*Olea europaea* L.) leaves. Plant Cell Physiol 47:984–994
Cvetkovska M, Vanlerberghe GC (2013) Alternative oxidase impacts the plant response to biotic stress by influencing the mitochondrial generation of reactive oxygen species. Plant Cell Environ 36:721–732
Das P, Nutan KK, Singla-Pareek SL, Pareek A (2015) Oxidative environment and redox homeostasis in plants: dissecting out significant contribution of major cellular organelles. Front Environ Sci. doi:10.3389/fenvs.2014.00070
Del Rio LA, Palma JM, Sandalio LM, Corpas FJ, Pastori GM, Bueno P, Lopez-Huertas E (1996) Peroxisomes as a source of superoxide and hydrogen peroxide in stressed plants. Biochem Soc Trans 24:434–438
Del Rio LA, Corpas FJ, Sandalio LM, Palma JM, Gomez M, Barroso JB (2002) Reactive oxygen species, antioxidant systems and nitric oxide in peroxisomes. J Exp Bot 53:1255–1272
Del Rio LA, Sandalio LM, Corpas FJ, Palma JM, Barroso JB (2006) Reactive oxygen species and reactive nitrogen species in peroxisomes: production, scavenging and role in cell signalling. Plant Physiol 141:330–335
Demidchik, V. (2012) Reactive oxygen species and oxidative stress in plants. In: Shabala S (ed) Plant stress physiology. CAB International, p 24–58
Dickinson BC, Srikun D, Chang CJ (2010) Mitochondrial-targeted fluorescent probes for reactive oxygen species. Curr Opin Chem Biol 4(1):50–56
Elstner EF (1991) Mechanisms of oxygen activation in different compartments of plant cells. In: Pell EJ, Steffen KL (eds) Active oxygen/oxidative stress and plant metabolism. American Society of Plant Physiologists, Rockville, pp 13–25
Faize M, Burgos L, Faize L, Piqueras A, Nicolas E, Barba-Espin G, Clemente-Moreno MJ, Alcobendas R, Artlip T, Hernandez JA (2011) Involvement of cytosolic ascorbate peroxidase and Cu/Zn-superoxide dismutase for improved tolerance against drought stress. J Exp Bot 62:2599–2613
Foyer CH, Noctor G (2003) Redox sensing and signaling associated with reactive oxygen in chloroplast, peroxisomes and mitochondria. Physiol Plant 119:355–364
Foyer CH, Noctor G (2009) Redox regulation in photosynthetic organisms: signaling, acclimation, and practical implications. Antioxid Redox Signal 11:861–906
Gao C, Xing D, Li L, Zhang L (2008) Implication of reactive oxygen species and mitochondrial dysfunction in the early stages of plant programmed cell death induced by ultraviolet-C overexposure. Planta 227:755e767
Gavrilova AA, Razina SV (2014) Compartmentalization of the cell nucleus and spatial organization of the genome. Mol Biol 49(1):21–39
Gechev TS, Breusegem FV, Stone JM, Denev I, Laloi C (2006) Reactive oxygen species as signals that modulate plant stress responses and programmed cell death. BioEssays 28(11):1091–1101
Gill SS, Tuteja N (2010) Reactive oxygen species and antioxidant machinery in abiotic stress tolerance in crop plants. Plant Physiol Biochem 48:909–930
Gleason C, Huang S, Thatcher L, Foley RC, Anderson CR, Carroll AJ, Millar AH, Singh KB (2011) Mitochondrial complex II has a key role in mitochondrial-derived reactive oxygen species influence on plant stress gene regulation and defense. Proc Natl Acad Sci U S A 108:10768–10773

Glyan'ko AK, Ischenko AA (2010) Structural and functional characteristics of plant NADPH oxidase: a review. Appl Biochem Microbiol 46:463–471

Go YM, Jones DP (2002) Redox potential of GSH/GSSG couple: assay and biological significance. Methods Enzymol 348:93–112

Go YM, Jones DP (2008) Redox compartmentalization in eukaryotic cells. Biochim Biophys Acta 1780:1273–1290

Go YM, Jones DP (2013) The redox proteome. J Biol Chem 288:26512–26520

Go YM, Pohl J, Jones DP (2009) Quantification of redox conditions in the nucleus. Methods Mol Biol 464:303–317

Go YM, Park H, Koval M, Orr M, Reed M, Liang Y (2010) A key role for mitochondria in endothelial signalling by plasma cysteine/cystine redox potential. Free Radic Biol Med 48:275–283

Golldack D, Li C, Mohan H, Probst N (2014) Tolerance to drought and salt stress in plants: unraveling the signaling networks. Front Plant Sci 5:151–157

Grivennikova VG, Vinogradov AD (2013) Mitochondrial production of reactive oxygen species. Biochem Mosc 78(13):1490–1511

Gupta KJ, Igamberdiev AU (2015) Compartmentalization of reactive oxygen species and nitric oxide production in plant cells: an overview. In: Gupta KJ, Igamberdiev AU (eds) Reactive oxygen and nitrogen species signaling and communication in plants. Springer International Publishing, Switzerland. 2015. doi:10.1007/978-3-319-10079-1_1

Halvey PJ, Hansen JM, Johnson JM, Go YM, Samali A, Jones DP (2007) Selective oxidative stress in cell nuclei by nuclear-targeted D-amino acid oxidase. Antioxid Redox Signal 9(7):807–816

Hasanuzzaman M, Nahar K, Gil SS, Fujita M (2014) Drought stress responses in plants, oxidative stress and antioxidant defense. In: Gill SS, Tuteja N (eds) Climate change and plant abiotic stress tolerance. Wiley, Weinheim, pp 209–249

Hernandez-Barrera A, Quinto C, Johnson EA, Wu HM, Cheung AY, Cardenas L (2013) Using hyper as a molecular probe to visualize hydrogen peroxide in living plant cells: a method with virtually unlimited potential in plant biology. Methods Enzymol 527:275–290

Heyneke E, Luschin-Ebengreuth N, Krajcer I, Wolkinger V, Müller M, Zechmann B (2013) Dynamic compartment specific changes in glutathione and ascorbate levels in *Arabidopsis* plants exposed to different light intensities. BMC Plant Biol 13:104. doi:10.1186/1471-2229-13-104

Heyno E, Mary V, Schopfer P, Krieger-Liszkay A (2011) Oxygen activation at the plasma membrane: relation between superoxide and hydroxyl radical production by isolated membranes. Planta 234:35–45

Hideg E, Kalai T, Hideg K (2011) Direct detection of free radicals and reactive oxygen species in thylakoids. Methods Mol Biol 684:187–200

Hu YQ, Liu S, Yuan HM, Li J, Yan DW, Zhang JF (2010) Functional comparison of catalase genes in the elimination of photorespiratory H_2O_2 using promoter and 3 untranslated region exchange experiments in the *Arabidopsis cat2* photo respiratory mutant. Plant Cell Environ 33:1656–1670

Jacquot JP, Eklund H, Rouhier N, Schürmann P (2009) Structural and evolutionary aspects of thioredoxin reductases in photosynthetic organisms. Trends Plant Sci 14:336–343

Jajic I, Sarna T, Strzalka K (2015) Senescence, stress, and reactive oxygen species. Plant Sci 4:393–411

Jimenez A, Hernandez JA, Pastori G, del Rio LA, Sevilla F (1998) Role of the ascorbate-glutathione cycle of mitochondria and peroxisomes in the senescence of pea leaves. Plant Physiol 118:1327–1335

Jones DP (2006) Redefining oxidative stress. Antioxid Redox Signal 8:1865–1879

Jones DP, Go YM (2010) Redox compartmentalization and cellular stress diabetes. Diabetes Obes Metab 12(2):116–125

Kaludercic N, Carpi A, Nagayama T, Sivakumaran V, Zhu G, Lai EW (2014a) Monoamine oxidase B prompts mitochondrial and cardiac-dysfunction in pressure overloaded hearts. Antioxid Redox Signal 20:267–280

Kaludercic N, Deshwal S, Lisa FD (2014b) Reactive oxygen and redox compartmentalization. Front Physiol 5., Article:285. doi:10.3389/fphys.2014.00285

Karpets YV, Kolupaev YE, Yastreb TO (2011) Effect of sodium nitroprusside on heat resistance of wheat coleoptiles: dependence on the formation and scavenging of reactive oxygen species. Russ J Plant Physiol 58(6):1027–1033

Kaur A, Kolanowski JL, New EJ (2016) Reversible fluorescent probes for biological redox states. Angew Chem Int Ed 55:1602–1613

Kaye Y, Golani Y, Singer Y, Leshem Y, Cohen G, Ercetin M, Gillaspy G, Levine A (2011) Inositol polyphosphate 5-phosphatase7 regulates the production of reactive oxygen species and salt tolerance in Arabidopsis. Plant Physiol 157:229–241

Keunen E, Peshev D, Vangronsveld J, Van Den Ende W, Cuypers A (2013) Plant sugars are crucial players in the oxidative challenge during abiotic stress: extending the traditional concept. Plant Cell Environ 36(7):1242–1255

Khan MIR, Khan NA (2014) Ethylene reverses photosynthetic inhibition by nickel and zinc in mustard through changes in PS II activity, photosynthetic-nitrogen use efficiency and antioxidant metabolism. Protoplasma 251:1007–1019

Khan MIR, Asgher M, Khan NA (2014) Alleviation of salt-induced photosynthesis and growth inhibition by salicylic acid involves glycinebetaine and ethylene in mungbean (*Vigna radiata* L.) Plant Physiol Biochem 80:67–74

Khan MIR, Nazir F, Asgher M, Per TS, Khan NA (2015) Selenium and sulfur influence ethylene formation and alleviate cadmium-induced oxidative stress by improving proline and glutathione production in wheat. J Plant Physiol 178:9–18

Khan MIR, Iqbal N, Masood A, Mobin M, Anjum NA, Khan NA (2016a) Modulation and significance of nitrogen and sulfur metabolism in cadmium challenged plants. Plant Growth Regul 78:1–11

Khan MIR, Khan NA, Masood A, Per TS, Asgher M (2016b) Hydrogen peroxide alleviates nickel-inhibited photosynthetic responses through increase in use-efficiency of nitrogen and sulfur, and glutathione production in mustard. Front Plant Sci 7:44

Kimura S, Kaya H, Kawarazaki T, Hiraoka G, Senzaki E, Michikawa M, Kuchitsu K (2012) Protein phosphorylation is a prerequisite for the Ca^{2+}-dependent activation of Arabidopsis NADPH oxidases and may function as a trigger for the positive feedback regulation of Ca^{2+} and reactive oxygen species. Biochim Biophys Acta 1823:398–405

Koffler BE, Ebengreuth NL, Stabentheiner E, Muller M, Zechmann B (2014) Compartment specific response of antioxidants to drought stress in Arabidopsis. Plant Sci 227:133–144

Kolupaev YE, Karpets YV (2013) Participation of reactive oxygen species in formation of induced resistances of plants to abiotic stressors. In: Suzuki M, Yamamoto S (eds) Handbook on reactive oxygen species (ROS): formation mechanisms, physiological roles and common harmful effects. Nova Science Publishers, NY, pp 109–136

Kwak JM, Mori IC, Pei ZM, Leonhardt N, Torres MA, Dangl JL, Bloom RE, Bodde S, Jones JD, Schroeder JI (2003) NADPH oxidase Atrboh D and Atrboh F genes function in ROS dependent ABA signaling in Arabidopsis. EMBO J 22(11):2623–2633

Kwak JM, Nguyen V, Schroeder JI (2006) The role of reactive oxygen species in hormonal responses. Plant Physiol 141:323–329

Laloi C, Stachowiak M, Pers-Kamczyc E, Warzych E, Murgia I, Apel K (2007) Cross-talk between singlet oxygen- and hydrogen peroxide-dependent signaling of stress responses in *Arabidopsis thaliana*. Proc Natl Acad Sci USA 104:672–677

Lee KP, Kim C, Landgraf F, Apel K (2007) EXECUTER1- and EXECUTER2-dependent transfer of stress-related signals from the plastid to the nucleus of *Arabidopsis thaliana*. Proc Nat Acad Sci USA 104:10270–10275

Lee S, Seo PJ, Lee HJ, Park CM (2012) A nac transcription factor ntl4 promotes reactive oxygen species production during drought-induced leaf senescence in arabidopsis. Plant J Cell Mol Biol 70:831–844

Lodi R, Tonon C, Calabrese V, Schapira AH (2006) Friedreich's ataxia: from disease mechanisms to therapeutic interventions. Antioxid Redox Signal 8:438–443

Lu Z, Liu D, Liu S (2007) Two rice cytosolic ascorbate peroxidases differentially improve salt tolerance in transgenic Arabidopsis. Plant Cell Rep 26(10):1909–1917

Lukyanov KA, Belousov VV (2014) Genetically encoded fluorescent redox sensors. Biochim Biophys Acta 1840:745–756

Maheshwari R, Dubey RS (2009) Nickel-induced oxidative stress and the role of antioxidant defence in rice seedlings. Plant Growth Regul 59(1):37–49

Marino D, Dunand C, Puppo A, Pauly N (2012) A burst of plant NADPH oxidases. Trends Plant Sci 17:9–15

Maron BA, Michel T (2012) Subcellular localization of oxidants and redox modulation of endothelial nitric oxide synthase. Off J Japan Circ Soc. doi:10.1253/circj.CJ-12-1207

Meldi L, Brickner JH (2011) Compartmentalization of the nucleus. Trends Cell Biol 21 (12):701–708

Meyer AJ, Dick TP (2010) Fluorescent protein-based redox probes. Antioxid Redox Signal 13 (5):621–650

Meyer Y, Reichheld JP, Vignols F (2005) Thioredoxins in Arabidopsis and other plants. Photosynth Res 86:419–433

Miller EW, Albers AE, Pralle A, Isacoff EY, Chang CJ (2005) Boronate based fluorescent probes for imaging cellular hydrogen peroxide. J Am Chem Soc 127:16652–16659

Miller G, Suzuki N, Ciftci-Yilmaz S, Mittler R (2010) Reactive oxygen species homeostasis and signaling during drought and salinity stresses. Plant Cell Environ 33(4):453–467

Minibayeva F, Kolesnikov O, Chasov A, Beckett RP, Lüthje S, Vylegzhanina N, Buck F, Böttger M (2009) Wound-induced apoplastic peroxidase activities: their roles in the production and detoxification of reactive oxygen species. Plant Cell Environ 32:497–508

Mishra S, Jha AB, Dubey RS (2011) Arsenite treatment induces oxidative stress, upregulates antioxidant system, and causes phytochelatin synthesis in rice seedlings. Protoplasma 248 (3):565–577

Mittler R, Vanderauwera S, Gollery M, Van Breusegem F (2004) Reactive oxygen gene network of plants. Trends Plant Sci 9:490–498

Mittler R, Vanderauwera S, Suzuki N, Miller G, Tognetti VB, Vandepoele K, Gollery M, Shulaev V, Van Breusegem F (2011) ROS signaling: the new wave? Trends Plant Sci 16:300–309

Moller IM (2001a) Plant mitochondria and oxidative stress: electron transport, NADPH turnover, and metabolism of reactive oxygen species. Annu Rev Plant Physiol Plant Mol Biol 52:561–591

Moller IM (2001b) Plant mitochondria and oxidative stress: electron transport, NADPH turnover, and metabolism of reactive oxygen species. Ann Rev Plant Physiol Mol Biol 52:561e591

Mori IC, Schroeder JS (2004) Reactive oxygen species activation of plant Ca2+ channels. A signaling mechanism in polar growth, hormone transduction, stress signaling, and hypothetically mechanotransduction. Plant Physiol 135:702–708

Naya L, Ladrera R, Ramos J, González EM, Arrese-Igor C, Minchin FR, Becana M (2007) The response of carbon metabolism and antioxidant defenses of alfalfa nodules to drought stress and to the subsequent recovery of plants. Plant Physiol 144(2):1104–1114

Noctor G, Veljovic-Jovanovic S, Driscoll S, Novitskaya L, Foyer CH (2002) Drought and oxidative load in the leaves of C3 plants: a predominant role for photorespiration. Ann Bot 89:841–850

Noctor G, Mhamdi A, Queval G, Foyer CH (2013) Regulating the redox gatekeeper: vacuolar sequestration puts glutathione disulfide in its place. Plant Physiol 163(2):665–671

Noctor G, Mhamdi A, Foyer CH (2014) The roles of reactive oxygen metabolism in drought: not so cut and dried. Plant Physiol 164:1636–1648

Nühse TS, Bottrill AR, Jones AME, Peck SC (2007) Quantitative phosphoproteomic analysis of plasma membrane proteins reveals regulatory mechanisms of plant innate immune responses. Plant J 51:931–940

Oda T, Hashimoto H, Kuwabara N, Akashi S, Hayashi K, Kojima C, Wong HL, Kawasaki T, Shimamoto K, Sato M, Shimizu T (2010) Structure of the N-terminal regulatory domain of a plant NADPH oxidase and its functional implications. J Biol Chem 285:1435–1445

Ogasawara Y, Kaya H, Hiraoka G, Yumoto F, Kimura S, Kadota Y, Hishinuma H, Senzaki E, Yamagoe S, Nagata K, Nara M, Suzuki K, Tanokura M, Kuchitsu K (2008) Synergistic activation of the Arabidopsis NADPH oxidase AtrbohD by Ca^{2+} and phosphorylation. J Biol Chem 283:8885–8892

Pani G, Bedogni B, Colavitti R, Anzevino R, Borrello S, Gleotti T (2001) Cell compartmentalization in redox signaling. IUBMB Life 52:7–16

Peshev D, Vergauwen R, Moglia A, Hideg E, Van den Ende W (2013) Towards understanding vacuolar antioxidant mechanisms: a role for fructans? J Exp Bot 64(4):1025–1038

Petrov V, Hille J, Mueller-Roeber B, Gechev TS (2015) ROS-mediated abiotic stress-induced programmed cell death in plants. Front Plant Sci 6:69. doi:10.3389/fpls..00069

Piantadosi CA (2008) Carbon monoxide, reactive oxygen signaling, and oxidative stress. Free Radic Biol Med 45:562–569

Pottosin I, Velarde-Buendí AM, Bose J, Zepeda-Jazo I, Shabala S, Dobrovinskaya O (2014) Cross-talk between reactive oxygen species and polyamines in regulation of ion transport across the plasma membrane: implications for plant adaptive responses. J Exp Bot:1–13. doi:10.1093/jxb/ert423

Purvis AC (1997) Role of the alternative oxidase in limiting superoxide production by plant mitochondria. Physiol Plant 100:165–170

Quan LJ, Zhang B, Shi WW, Li HY (2008) Hydrogen peroxide in plants: a versatile molecule of the reactive oxygen species network. J Integr Plant Biol 50:2e18

Queval G, Jaillard D, Zechmann B, Noctor G (2011) Increased intracellular H2O2 availability preferentially drives glutathione accumulation in vacuoles and chloroplasts. Plant Cell Environ 34:21–32

Rhee SG, Chae HZ, Kim K (2005) Peroxiredoxins: a historical overview and speculative preview of novel mechanisms and emerging concepts in cell signalling. Free Radic Biol Med 38:1543–1552

Rhoads DM, Umbach AL, Subbaiah CC, Siedow JN (2006a) Mitochondrial reactive oxygen species. Contribution to oxidative stress and interorganellar signaling. Plant Physiol 141:357–366

Rhoads DM, Umbach AL, Subbaiah CC, Siedow JN (2006b) Mitochondrial reactive oxygen species: contribution to oxidative stress and interorganellar signaling. Plant Physiol 141:357–366

Rivera-Ingraham GA, Rocchetta I, Bickmeyer U, Meyer S, Abele D (2016) Spatial compartmentalization of free radical formation and mitochondrial heterogeneity in bivalve gills revealed by live-imaging techniques. Front Zool 13:4. doi:10.1186/s12983-016-0137-1

Robinson KM, Janes MS, Pehar M, Monette JS, Ross MF, Hagen TM (2006) Selective fluorescent imaging of super oxide in vivo using ethidium-based probes. Proc Natl Acad Sci USA 103:15038–15043

Rodriguez-Serrano M, Romero-Puertas MC, Zabalza A, Corpas FJ, Gomez M, Del R, Sandalio LA, L.M. (2006) Cadmium effect on oxidative metabolism of pea (Pisum sativum L.) roots. Imaging of reactive oxygen species and nitric oxide accumulation in vivo. Plant Cell Environ 29:1532–1544

Sagi M, Fluhr R (2001) Superoxide production by plant homologues of the gp91phox NADPH oxidase: modulation of activity by calcium and by tobacco mosaic virus infection. Plant Physiol 126:1281–1290

Sagi M, Fluhr R (2006) Production of reactive oxygen species by plant NADPH oxidases. Plant Physiol 141:336–340

Schafer FQ, Buettner GR (2001) Redox environment of the cell as viewed through the redox state of the glutathione disulfide/glutathione couple. Free Radic Biol Med 30:1191–1212

Schrader M, Fahimi HD (2006) Peroxisomes and oxidative stress. Biochim Biophys Acta 1763:1755–1766

Schroeder P, Popp R, Wiegand B, Altschmied J, Haendeler J (2007) Nuclear redox-signalling is essential for apoptosis inhibition in endothelial cells- important role for nuclear thioredoxin-1. Arterioscler Thromb Vasc Biol 27:2325–2331

Sewelam N, Kazan K, Schenk PM (2016) Global plant stress signaling: reactive oxygen species at the cross-road. Front Plant Sci 7., Article:187. doi:10.3389/fpls.2016.00187

Shapiguzov A, Vainonen JP, Wrzaczek M, Kangasjärvi J (2012) ROS-talk – how the apoplast, the chloroplast, and the nucleus get the message through. Front Plant Sci 3:292

Sharma P, Jha AB, Dubey RS, Pessarakli M (2012) Reactive oxygen species, oxidative damage, and antioxidative defense mechanism in plants under stressful conditions. Aust J Bot 2012:1–26

Smith CV, Jones DP, Guenthner TM, Lash LH, Lauterburg BH (1996) Compartmentation of glutathione: implications for the study of toxicity and disease. Toxicol Appl Pharmacol 140:1–12

Suzuki N, Koussevitzky S, Mittler R, Miller G (2012) ROS and redox signalling in the response of plants to abiotic stress. Plant Cell Environ 35:259–270

Suzuki N, Miller G, Salazar C, Mondal HA, Shulaev E, Cortes DF (2013) Temporal-spatial interaction between reactive oxygen species and abscisic acid regulates rapid systemic acclimation in plants. Plant Cell 25:3553–3569

Takeda S, Gapper C, Kaya H, Bell E, Kuchitsu K, Dolan L (2008) Local positive feedback regulation determines cell shape in root hair cells. Science 319:1241–1244

Takimoto E, Kass DA (2007) Role of oxidative stress in cardiac hypertrophy and remodeling. Hypertension 49:241–248

Tarpey MM, Wink DA, Grisham MB (2004) Methods for detection of reactive metabolites of oxygen and nitrogen: in vitro and in vivo considerations. Am J Phys Regul Integr Comp Phys 286:431–444

Tewari RK, Hahn EJ, Paek KY (2008) Function of nitric oxide and superoxide anion in the adventitious root development and antioxidant defence in Panax ginseng. Plant Cell Rep 27:563–573

Thakur P, Kumar S, Malik JA, Berger JD, Nayyar H (2010) Cold stress effects on reproductive development in grain crops: an overview. Environ Exp Bot 67(3):429–443

Torres MA, Dangl JL, Jones JDG (2002) Arabidopsis gp91phox homologues AtrbohD and AtrbohF are required for accumulation of reactive oxygen intermediates in the plant defense response. Proc Natl Acad Sci U S A 99:517–522

Toumi I, Moschou PN, Paschalidis KA, Bouamama B, Ben Salem-Fnayou A, Ghorbel AW, Mliki A, Roubelakis-Angelakis KA (2010) Abscisic acid signals reorientation of polyamine metabolism to orchestrate stress responses via the polyamine exodus pathway in grapevine. J Plant Physiol 167(7):519–525

Tripathy BC, Oelmüller R (2012) Reactive oxygen species generation and signaling in plants. Plant Signal Behav 7(12):1621–1633

Valluru R, Lammens W, Claupein W, Van den Ende W (2008) Freezing tolerance by vesicle-mediated fructan transport. Trends Plant Sci 13:409–414

Wanders RJ, Waterham HR (2006) Biochemistry of mammalian peroxisomes revisited. Annu Rev Biochem 75:295–332

Wang W, Vinocur B, Shoseyov O, Altman A (2004) Role of plant heat-shock proteins and molecular chaperones in the abiotic stress response. Trends Plant Sci 9(5):244–252

Waypa GB, Marks JD, Guzy R, Mungai PT, Schriewer J, Dockie D (2010) Hypoxia triggers subcellular compartment redox signaling in vascular smooth muscle cells. Circ Res 106:526–535

Wei SJ, Botero A, Hirota K, Bradbury CM, Markovina S, Laszlo A, Spitz DR, Goswami PC, Yodoi J, Gius D (2000) Thioredoxin nuclear translocation and interaction with redox factor-1 activates the activator protein-1 transcription factor in response to ionizing radiation. Cancer Res 60(23):6688–6695

Winterbourn CC (2014) The challenges of using fluorescent probes to detect and quantify specific reactive oxygen species in living cells. Biochim Biophys Acta 1840:730–738

Wong HL, Pinontoan R, Hayashi K, Tabata R, Yaeno T, Hasegawa K, Kojima C, Yoshioka H, Iba K, Kawasaki T, Shimamoto K (2007) Regulation of rice NADPH-oxidase by Rac GTPase to its N-terminal extension. Plant Cell 19:4022–4034

Yamamoto Y, Kobayashi Y, Devi SR, Rikiishi S, Matsumoto H (2002) Aluminum toxicity is associated with mitochondrial dysfunction and the production of reactive oxygen species in plant cells. Plant Physiol 128:63–72

Yesbergenova Z, Yang G, Oron E, Soffer D, Fluhr R, Sagi M (2005) The plant mo-hydroxylases aldehyde oxidase and xanthine dehydrogenase have distinct reactive oxygen species signatures and are induced by drought and abscisic acid. Plant J 42(6):862–876

You J, Chan Z (2015) ROS regulation during abiotic stress responses in crop plants. Front Plant Sci 6:1092. doi:10.3389/fpls.2015.01092

Yun BW, Feechan A, Yin M, Saidi NBB, Le Bihan T, Yu M, Moore JW, Kang JG, Kwon E, Kang JG, Spoel SH, Pallas JA, Loake GJ (2011) S-nitrosylation of NADPH oxidase regulates cell death in plant immunity. Nature 478:264–268

Zarepour M, Kaspari K, Stagge S, Rethmeier R, Mendel R, Bittner F (2010) Xanthine dehydrogenase *AtXDH1* from Arabidopsis thaliana is a potent producer of superoxide anions via its NADH oxidase activity. Plant Mol Biol 72:301–310

Zhang Y, Zhu H, Zhang Q, Maoyin L, Yan M, Wang R, Wang L, Welti R, Zhang W, Wang X (2009) Phospholipase Dα1 and phosphatidic acid regulate NADPH oxidase activity and production of reactive oxygen species in ABA-mediated stomatal closure in Arabidopsis. Plant Cell 21:2357–2377

Zhang CJ, Zhao BC, Ge WN, Zhang YF, Song Y, Sun DY, Guo Y (2011) An apoplastic h-type thioredoxin is involved in the stress response through regulation of the apoplastic reactive oxygen species in rice. Plant Physiol 157(4):1884–1899

Zhang Z, Zhang Q, Wu J, Zheng X, Zheng S, Sun X, Qiu Q, Lu T (2013) Gene knockout study reveals that cytosolic ascorbate peroxidase 2 (OsAPX2) plays a critical role in growth and reproduction in rice under drought, salt and cold stresses. PLoS One 8:e57472

Zhang W, Wang X, Li P, Huang F, Wang H, Zhang W, Tang B (2015) Elucidating the relationship between superoxide anion levels and lifespan using an enhanced two-photon fluorescence imaging probe. Chem Commun 51:9710–9713

Zhou L, Aon MA, Liu T, O'Rourke B (2011) Dynamic modulation of C^{a2+} sparks by mitochondrial oscillations in isolated guinea pig cardiomyocytes under oxidative stress. J Mol Cell Cardiol 51:632–639

Reactive Oxygen Species Production and Detoxification During Leaf Senescence

5

Alice Trivellini, Giacomo Cocetta, Alessandra Francini, and Antonio Ferrante

Abstract

The natural leaf senescence is the last stage of leaf development and is a highly genetically regulated process. Induced leaf senescence, instead, usually occurs in detached leaves or in plants placed in dark conditions under postharvest storage. In both cases, an oxidative burst is a typical feature of this phenomenon and includes a significant increment in reactive oxygen species (ROS), such as superoxide anion, hydrogen peroxide, hydroxyl radical, and singlet oxygen. Thus, ROS play a key role in leaf senescence process. Under normal (i.e., non-stressed) conditions, the levels of ROS are controlled and balanced by the antioxidant systems present in leaves. The leaf cells act in the defense response against the accumulation of ROS by enhancing the activity of some key enzymes such as catalase (CAT) and superoxide dismutase (SOD). In addition, the ascorbate–glutathione cycle, also known as Asada–Halliwell cycle, is the most important enzymatic antioxidant system acting in the leaf cells during senescence. The leaves are considered the main target of ROS production, and in particular, the photosynthetic apparatus and the related molecules act to support the light perception and photosynthesis activity (chlorophylls and carotenoids).

Keywords

Ascorbate • Ethylene • Fruit • Oxidative stress • Postharvest • Vegetables

A. Trivellini • A. Francini
Institute of Life Sciences, Scuola Superiore Sant'Anna, Piazza Martiri della Libertà 33, 56127 Pisa, Italy

G. Cocetta • A. Ferrante (✉)
Department Agricultural and Environmental Science – Production, Landscape, Agroenergy, Università degli Studi di Milano, Via Celoria 2, 20133 Milano, Italy
e-mail: antonio.ferrante@unimi.it

M.I.R. Khan, N.A. Khan (eds.), *Reactive Oxygen Species and Antioxidant Systems in Plants: Role and Regulation under Abiotic Stress*,
DOI 10.1007/978-981-10-5254-5_5

Abbreviations

1O2	Singlet oxygen
AAO	Aldehyde oxygenase
ABA	Abscisic acid
ACO	1-Aminocyclopropane-1-carboxylate oxidase
AOI	IAA oxidase
APX	Ascorbate peroxidase
AsA	Ascorbate
ATG	Affecting autophagy
CAT	Catalase
DHA	Dehydroascorbate
DHAR	Dehydroascorbate reductase
GR	Glutathione reductase
GSH	Glutathione
GSH	Reduced glutathione
GSSG	Oxidized form of glutathione
H2O2	Hydrogen peroxide
IAA	Indole acetic acid
JA	Jasmonic acid
MDHA	Monodehydroascorbate reductase
MeJA	Methyl jasmonate
NahG	Naphthalene oxygenase
NCED	9-cis-Epoxycarotenoid dioxygenase
NIT1-3	Nitrilases
$O2^-$	Superoxide anion
$OH^{\cdot}$	Hydroxyl radical
PAO	Pheophorbide a oxygenase
ROS	Reactive oxygen species
SA	Salicylic acid
SAGs	Senescence-associated genes
SARK	Senescence-associated receptor-like kinase
SOD	Superoxide dismutase
TSA1	Tryptophan synthase

5.1 Introduction

Leaf senescence is considered the last stage of leaf development, and it is genetically regulated but also influenced by environmental factors such as biotic and abiotic stresses. During leaf senescence, several degradation processes occur at the cellular and subcellular level. Many macromolecules such as chlorophylls, carotenoids, nucleic acids, proteins, and lipids are degraded, while nutrients are mobilized to storage sinks (Buchanan-Wollaston et al. 2003; Woo et al. 2016). When leaves are attached on plants, their senescence follow the plant physiology: (1) can naturally senesce on plants during the season changes (deciduous species) or (2) following the genetic programmed fate (aging), which occurs in the evergreen species. In deciduous plants, such as fruit and ornamental trees, leaf senescence is induced by the transition of the plant status from vegetative to winter dormancy. On the contrary, in annual plants, leaves can senescence with the whole plants and complete their cycle with whole plant life (Munné-Bosch and Alegre 2004; Lim et al. 2007).

Among vegetable crop production, some produces called baby leaf vegetables are commonly harvested at the early stage of plant development when plants are 10–15 cm height and have three to five fully expanded leaves. These vegetables are in the faster growth stage and are very perishable during postharvest. In this case, the leaf senescence is not genetically programmed, but it is induced by the harvest and postharvest conditions (storage or transportation), phytohormone unbalance, wounding (due to processing or mechanical damage), etc. In this case, the deterioration or senescence can be called postharvest-induced senescence and can involve different pathways, depending on developmental stage of leaves (Ferrante et al. 2004; Jibran et al. 2013).

Considering the ornamental sector, cut branches, which are used as filler in bouquets, can undergo senescence because they are detached from the plants. Leaves of cut branches do not receive any nutrients or hormones, and the transpiration can induce water stress enhancing the postharvest senescence (Ferrante et al. 2002). When leaf senescence occurs both in attached leaves and detached leaves placed in dark conditions under postharvest storage, an oxidative burst is observed, and it includes the accumulation of large quantities of reactive oxygen species (ROS) such as hydrogen peroxide (H_2O_2), hydroxyl radical ($OH^{\cdot}$), superoxide anion (O_2^{-}), and singlet oxygen ($^1O_2$), making these molecules a central component of the leaf senescence process. During leaf senescence, the pattern of ROS accumulation in cells is characterized by an initial increment followed by a decline (Fig. 5.1). Although ROS half-life approximately ranges from 2 μs to 1 ms (Gechev et al. 2006), these molecules cause severe cell damages that effectively compromise plants' functionality, leading to the degradation of proteins, lipids, carbohydrates, and nucleic acids (Mittler et al. 2004; Bhattacharjee 2005). Most of ROS are formed in the electron transport systems such as mitochondria, chloroplast, glyoxysome, and peroxisome; in particular, ROS in leaves are produced in the photosystem II (PSII), especially $^1O_2$, and this molecule can then move from the thylakoid membrane to the cytoplasm. The accumulation of ROS can be due also to the impairment of the antioxidant system of mitochondria, associated with membrane damage and alteration of electron transport with negative effect on photosynthesis, respiration, and growth of plants (Khan and Khan 2014; Khan et al. 2014, 2015, 2016a, b). Important sources of ROS are also represented by the cytochrome P450 and endoplasmic reticulum but also cell wall peroxidase (pH dependent) and NADH

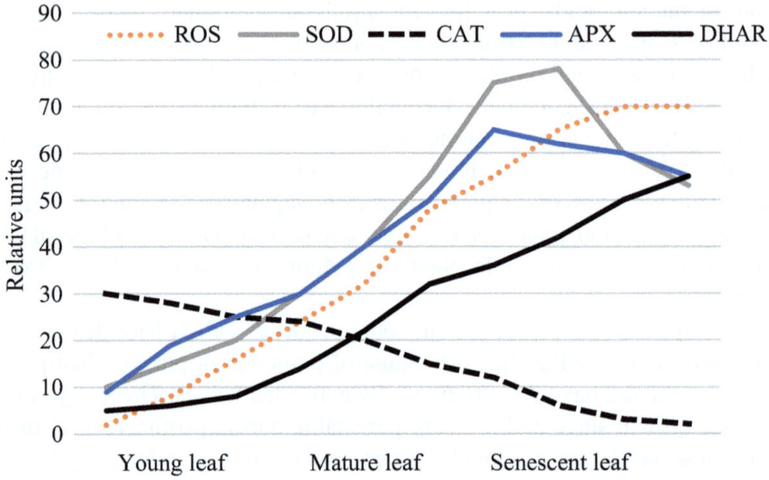

Fig. 5.1 Schematic representation of ROS accumulation and detoxification enzyme activities during leaf development in plant

oxidase. These molecules are also involved in the signal transduction and in the control of many physiological processes, including development and growth under biotic or abiotic stresses (Mittler et al. 2011).

The signal transduction function of ROS in cells is carried out at low concentrations, showing a transient increase. This ROS production is regulated by the respiratory burst or NADPH oxidase enzyme (Torres and Dangl 2005; Suzuki et al. 2011). The respiratory burst oxidase homologue (RBOH) genes in plants are directly involved in the signaling network in response to several environmental stresses (Sewelam et al. 2016).

5.2 Leaf Senescence and Oxidative Stress

Leaf senescence is closely correlated to the oxidative stress, membrane leaking, and lipid peroxidation (Navabpour et al. 2003). The senescence of the cell membrane through the lipid peroxidation generates high levels of ROS; as a direct consequence, the detoxification systems can be impaired or less efficient. In postharvest-induced senescence, the increase of peroxidation is due to tissue damage such as wounding (Ferrante et al. 2009; Cocetta et al. 2014) and high storage temperature (Spinardi and Ferrante 2012). The degradation of cell membrane lipids is also induced by lipoxygenase, phospholipase D, and lipase. In wounded baby leaf spinach, the increase of lipid peroxidation is followed by the increase of phospholipases C and D at 20 °C (Antonacci et al. 2011). Temperature is surely involved in the accumulation of ROS during senescence, especially in postharvest-induced senescence. In rocket leaves exposed at 20 °C in darkness, anion

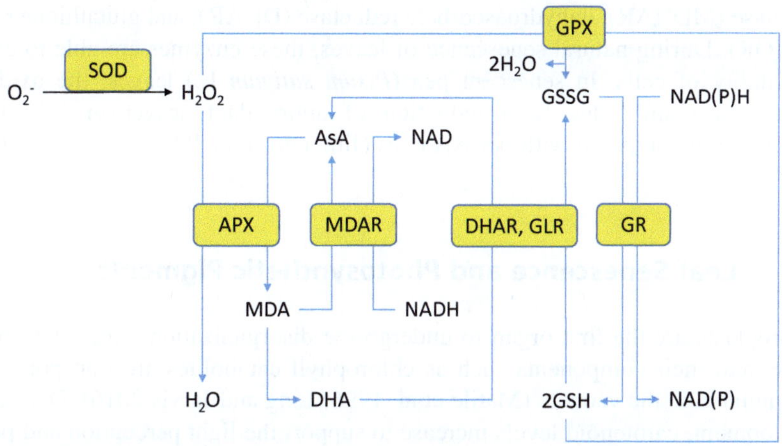

Fig. 5.2 Schematic overview of the ascorbate (AsA) glutathione (GSH) enzymes involved in reactive oxygen species scavenger system in plant cells. The superoxide radicals ($O_2^{\bullet -}$) are dismutated by superoxide dismutase (SOD) with the contemporary production of hydrogen peroxide (H_2O_2). H_2O_2 is converted to water by the reaction catalyzed by ascorbate peroxidase (APX) with the production of oxidized forms of AsA monodehyroascorbate and dehyroascorbate (MDHA and DHA). Monodehyroascorbate reductase (MDHAR) reduces the MDHA to AsA. However, part of the MDHA may be converted to AsA and DHA by nonenzymatic reactions. The DHA is converted to AsA by dehydroascorbate reductase (DHAR); contemporarily, the enzyme glutathione reductase (GR) catalyzes the last step of the cycle in which the oxidized form of glutathione (GSSG) produced by the action of DHAR is converted to the reduced form (GSH) with consumption of NADPH

superoxide levels increased. The hydrogen peroxide instead did not significantly increase but an increase of ascorbic acid was observed, which might counteract the ROS accumulation (Cavaiuolo et al. 2015).

Leaves are active and important organs for the plant; therefore, they have efficient systems to counteract the senescence and oxidative processes related to ROS production. During normal metabolism, many specific scavengers are able to control the accumulation of ROS (Liu and He 2016); unfortunately, during senescence the impairment of the turnover of the detoxification enzymes and antioxidant molecules compromises the cell functionality and vitality.

The antioxidant mechanism includes two important antioxidant protection systems (protecting enzyme system and ascorbate (AsA)–glutathione (GSH) cycle) to avoid the deleterious effect of ROS (Li et al. 2016). The ascorbate–glutathione cycle (Fig. 5.2) involves several enzymes which are highly coordinated and tightly connected, and their activation lead to efficient reduction in the ROS accumulation. The first enzyme able to convert the O_2^- to molecular oxygen and hydrogen peroxide is the superoxide dismutase (SOD). This enzyme is well known to play a key role in cell protection mechanism against ROS (Foyer et al. 1994). The hydrogen peroxide is then removed from the enzymes of the ascorbate glutathione (Foyer and Halliwell 1976; Bieker et al. 2012) such as ascorbate peroxidase (APX), monodehydroascorbate reductase (MDHAR), dehydroascorbate reductase (DHAR), and glutathione reductase (GR). During natural senescence of leaves, these enzymes are able to extend the vitality of cells. In senescent pea (*Pisum sativum* L.) leaves, the oxidative damage is mainly due to a reduction of antioxidant scavenger systems in mitochondria compared with peroxisomes (Jiménez et al. 1998; Jajic et al. 2015).

5.3 Leaf Senescence and Photosynthetic Pigments

Chloroplasts are the first organ to undergo the disorganization during leaf senescence, and their components such as chlorophyll catabolites are transported and accumulated in the vacuole (Matile et al. 1996; Ling and Jarvis 2016). During leaf development, carotenoid levels increase to support the light perception and photosynthesis activity. During leaf senescence, instead, they act as scavengers of ROS to contrast oxidative imbalance produced and consequently, their level declines (Biswal 1995). Thus, carotenoids content in leaves acts in the mechanism of chlorophyll photo-protection. Carotenoids are in fact involved in the dissipation of energy in excess by heat (xanthophyll cycle), protection against singlet oxygen generated from the photooxidation (Apel and Hirt 2004), indirect ABA biosynthesis through their degradation, and membrane stability (Cutler and Krochko 1999).

The understanding of the regulation of chlorophyll breakdown during senescence has been partly elucidated using stay-green mutants. These mutants do not show leaf yellowing and senescence is strongly delayed. The use of stay-green mutants in different crops revealed that chlorophyll is retained because the pheophorbide a oxygenase (PAO) enzyme lost its function; thus, the process of chlorophyll breakdown results to be impaired (Barry et al. 2009). The comparison between the wild-type and stay-green mutants evidenced that leaf senescence proceeds in both plants even though a marked difference in chlorophyll content was observed.

It has been demonstrated that ANAC046 (*Arabidopsis*, no apical meristem) transcription factor activates the expression of chlorophyll catabolic genes and SAGs. The ANAC046 is able to bind the promoter regions of non-yellow coloring1, stay-green1, stay-green2, and pheophorbide a oxygenase, regulating the expression of the related genes. Recent studies indicate that several transcription factors are able to bind directly the promoters of chlorophyll catabolic genes, regulating the leaf senescence process. However, the interplay among the different transcription factors is not completely understood (Oda-Yamamizo et al. 2016). The SAG21 (late embryogenesis abundant-like protein) is ROS-inducible and in *Arabidopsis* plants is able to modulate stress response and plays a role in mitigating the ROS effect at mitochondrial level (Salleh et al. 2012).

5.3.1 Senescence-Associated Genes

There are many genes related to ethylene biosynthesis that are directly or indirectly involved in leaf senescence and chlorophyll loss. The expression of genes that encode for 1-aminocyclopropane-1-carboxylate oxidase (ACO) enzyme has been found to increase along with leaf development and aging. In white clover (*Trifolium repens*), three ACO genes were isolated, and their expression was found to be spatially differentiated during leaf ontogeny (Hunter et al. 1999). Among the different ACO genes, the *TrACO1* and *TrACO2* were expressed in young leaves in the apical part, and the *TrACO3* gene was essentially expressed in senescing leaves. These results link the *TrACO3* gene activation to chlorophyll breakdown.

In *Arabidopsis thaliana*, many senescence-associated genes (SAGs) have been identified and related transcription factor has been characterized (Gepstein et al. 2003). SAGs are genes that are constitutively expressed in leaves and up-regulated or activated during senescence (Fig. 5.3).

One of the first SAG that has been isolated is the *see1* which was strongly induced by ethylene (Oh et al. 1996). The *see1* encode for cysteine proteases. The expression of this gene was inversely proportional to chlorophyll degradation. Other SAGs were also affected by ethylene (such as *SAG13*, *14*, and *20*), but their expression was less strong in younger leaves (Weaver et al. 1998).

At the molecular level, a membrane-bound NAC transcription factor NTL4 has been found in *Arabidopsis*, which is involved in the ROS accumulation under induced leaf senescence. The knockout mutants for NTL4 did not show a phenotype under natural senescence, while under stress the senescence progression was delayed and ROS was lowered (Fig. 5.4).

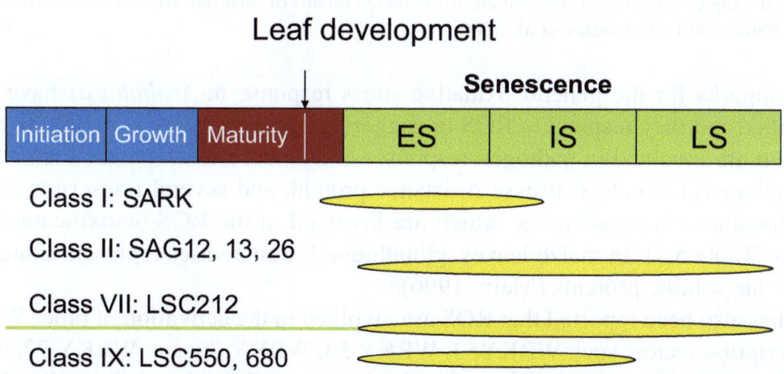

Fig. 5.3 Spatiotemporal expression of genes related to senescence, which was divided in classes as suggested by Smart (1994) and used by Buchanan-Wollaston (1997) for *Brassica napus*. For example, in this figure are reported senescence-associated receptor-like kinase (SARK), senescence-associated genes (SAGs), LSC212 (antifungal protein in *Bassica napus*), LSC550 (cytochrome P450 in *Bassica napus*), and LSC680 (ATP sulfurylase in *Brassica napus*) during leaf development (Smart 1994; Buchanan-Wollaston 1997; Gan and Amasino 1997; Hajouj et al. 2000)

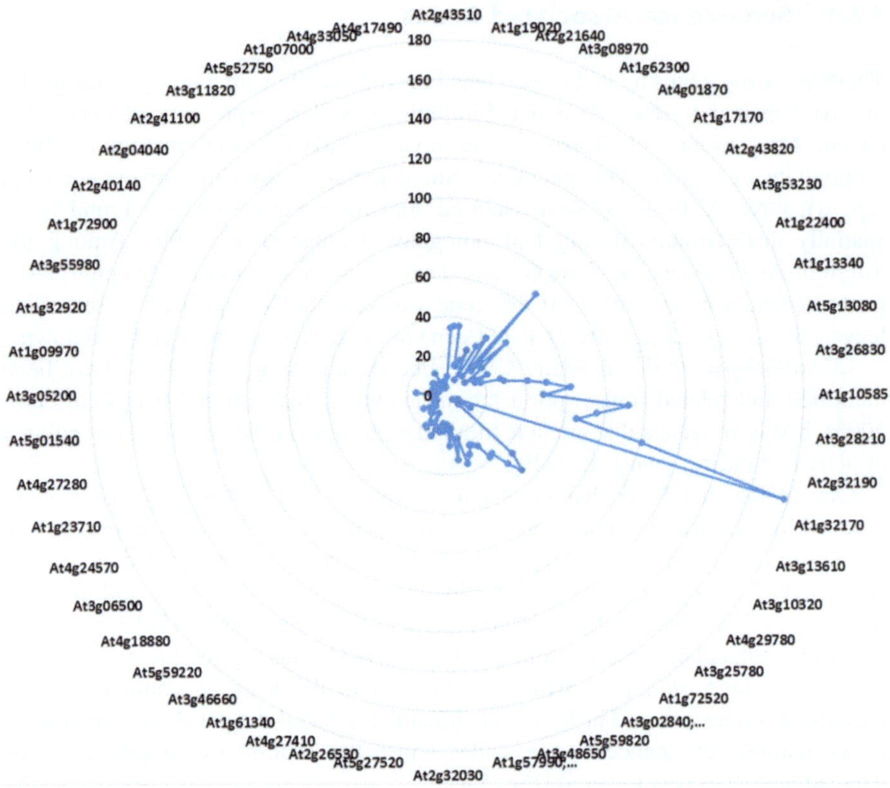

Fig. 5.4 Gene expression in *Arabidopsis thaliana* of the most affected genes exposed to ROS-inducing treatment. Values that are fold change means of different ROS-inducing treatments (Data elaborated form Gadjev et al. 2006)

Hallmarks for the general oxidative stress response in *Arabidopsis* have been identified and they respond to ROS or triggering events (Gadjev et al. 2006). Some of them are involved in pathogen responses and defense mechanisms such as toll-interleukin-1(TIR)-class disease resistance protein, and several transcripts encode for glutathione S-transferases, which are involved in the ROS detoxification processes (Table 5.1). In maize leaves, glutathione S-transferases represent more than 1% of the soluble proteins (Marrs 1996).

It has also been reported that ROS are involved in the activation of other WRKY transcription factors such WRKY53, WRKY 54, WRKY 70, and WRKY 72, which are all involved in the regulation of leaf senescence (Bakshi and Oelmüller 2014). In *Arabidopsis* the phytochrome-interacting factor4 (PIF4) was found to positively regulate ROS level during dark-induced leaf senescence (Song et al. 2014). In fact, mutants defective in PIF expression delayed inflorescence de-greening in dark

Table 5.1 Up-regulated genes in response to ROS in leaf of *Arabidopsis thaliana* and *Oryza sativa* during leaf senescence

Gene ID	Gene name	Description/function	Accession number	
	OsPLS1	*Oryza sativa* premature leaf senescence 1, vacuolar H+-ATPase A1-subunit		Song et al. (2014)
At5G45890	SAG12	Cysteine protease	NM_123957	Guo et al. (2004)
At4g02380	SAG21	Transcription factor, late embryogenesis abundant-like protein	NM_116471.3	Weaver et al. (1998)
At3G04060	ANAC046	Transcription factor	NW_003302553	Woo et al. (2016)
At3g10500	NTL4	Transcription factor	AY079345	He et al. (2002)
At2g43510	ATTI	ROS-responsive genes, trypsin inhibitor protein	AY096456	Gadjev et al. (2006)
At1g57630	TIR	ROS-responsive genes, transmembrane receptor activity	NM_104560	Ramel et al. (2013)
At1g19020		Unknown	AY117287	Gadjev et al. (2006)
At1g05340		Unknown	AY056369	Gadjev et al. (2006)
At2g21640		Unknown	BT005648	Gadjev et al. (2006)
At1g62300	AtWRKY6	Transcription factor	AY056172	Yoshida (2003)
At2g29470	Glutathione S-transferase	Detoxification	AK117612	Gadjev et al. (2006)
At5g13080	AtWRKY75	Transcription factor	BT006530	Rosenwasser et al. (2011)
At1g10585	bHLH	Transcription factor	BT015403	Gadjev et al. (2006)
At3G49620	Din11	Oxidoreductase	NM_114822	Balazadeh et al. (2011)
599593	AO	Ascorbate oxidase	D43624	

incubated inflorescences (Trivellini et al. 2012). The role of PIF in ROS regulation and leaf senescence requires further studies and may provide interesting insight in the leaf senescence regulatory network.

5.3.2 ROS, Leaf Senescence, and Plant Hormone Regulation

In the leaf during aging, the hormonal balance and signaling pathways of various phytohormones are critical components in this process. Cytokinins are the most effective senescence-retarding plant hormone, whereas ethylene, jasmonic acid (JA), abscisic acid (ABA), and salicylic acid are known to promote leaf senescence (Nooden 1988; Woo et al. 2013; Song et al. 2016). Gibberellins are also able to delay leaf senescence and in particularly leaf yellowing and avoid chlorophyll degradation. It has been demonstrated in *Pelargonium* cuttings that GA_3 reduced ROS accumulation and down-regulated the *PeSAG12-1* and PeWRKY6-1. In dark-induced senescence, GA_3 inhibited the ROS accumulation even if it was applied later when senescence was already activated (Rosenvasser et al. 2006). Cytokinins are well known to delay leaf senescence both down-regulating SAGs and up-regulating genes involved in the chlorophyll biosynthesis and turnover. In *Alstroemeria* cut flowers, the application of thidiazuron, a compound with cytokinin-like activity, effectively delayed leaf senescence for more than a month (Ferrante et al. 2002). Transgenic plants expressing P_{SAG12}-*IPT* showed a delay in senescence by autoregulatory senescence inhibition system (Guo and Gan 2014). The biosynthesis of cytokinins during senescence restores the leaf functionality and lowers the ROS accumulation as well as the SAG transcripts (Xu et al. 2016).

The ABA biosynthetic genes 9-cis-epoxycarotenoid dioxygenase (NCED) and aldehyde oxygenase (AAO1, AAO2) are up-regulated during developmental senescence (Gepstein and Thimann 1980). Exogenous applications of ABA determine an increase in H_2O_2 and in the antioxidant enzyme activities as well as a decrease in reduced form of glutathione and ascorbic acid (Hung and Kao 2004). ABA is involved in the cellular-protective and senescence activities, which are important for efficient remobilization of nutrients (Lim et al. 2003).

The role of auxin in senescence is complex because of its crucial role in many aspects of plant growth and development (Lim et al. 2007). However, some works reported a role for this hormone in developmental senescence. Endogenous level IAA (indole acetic acid) increased over twofold in senescent *Arabidopsis* leaves (Quirino et al. 1999), and the auxin biosynthetic genes encoding for a tryptophan synthase (TSA1), nitrilases (NIT1-3), and IAA oxidase (AOI) were strongly up-regulated during developmental senescence (Van der Graaff et al. 2006). Moreover, the exogenous supply of auxin determines a decrease in expression of many SAGs, particularly in SAG12 (Noh and Amasino 1999). These data indicated also a negative function for auxins in the regulation of senescence program and illustrate the complexity of auxin signaling pathway in the repression and/or activation of senescence.

Salicylic acid (SA) has long been recognized as a key signaling hormone in pathogen responses such as pathogen-mediated cell death. This hormone has a role in regulating gene expression that occurs in naturally senescing leaves (Morris et al. 2000). *Arabidopsis* plants with reduced SA content due to overexpression of bacterial naphthalene oxygenase (NahG) showed delayed yellowing (Morris et al. 2000), and the up- and down-regulation of many SAGs during developmental senescence is altered in the NahG mutants (Buchanan-Wollaston et al. 2005). It is interestingly to note that SA is a key player in regulating stress tolerance during the senescence process affecting autophagy (ATG) and production of ROS (Jibran et al. 2013). In fact the ATG mutants, atg5 and atg2, which completely lack ATG, increased SA and ROS contents and senesce earlier compared to control plants (Yoshimoto et al. 2009).

Jasmonates, including jasmonic acid (JA) and methyl jasmonate (MeJA), are derived from the fatty acid and linoleic acid and act to promote senescence. The exogenous supply of JA induces premature leaf senescence symptoms, and in senescing leaves of *Arabidopsis*, its levels were significantly enhanced (He et al. 2002). Treatment with JA induces rapid chlorophyll loss, reduces the photosynthetic efficiency, and induces the upregulation of many SAGs (Lim et al. 2007; Weidhase et al. 1987).

5.4 Conclusion and Future Prospects

The onset and progression of senescence is mediated by the interplay among various molecular entities such as hormones, enzymes, and metabolites. ROS are accumulated during senescence but these molecules play also an important role in the signaling network. The initial increase of ROS is responsible of plat responses to senescence. Unfortunately, the alteration of detoxification systems induces cell disorders and can lead to cell death. Senescence can be also regulated by source–sink relationship and lack of nutrients, particularly during the reproductive phase of growth. These internal factors are controlled by complex network of regulatory genes and can be modulated by intrinsic/environmental factors such as oxidative stress and postharvest stresses (dark, detachment, and drought). Further studies are required to understand the regulatory activities of ROS during leaf development.

References

Antonacci S, Natalini A, Cabassi G et al (2011) Cloning and gene expression analysis of the phospholipase C in wounded spinach leaves during postharvest storage. Postharvest Biol Technol 59:43–52

Apel K, Hirt H (2004) Reactive oxygen species: metabolism, oxidative stress, and signal transduction. Annu Rev Plant Biol 55:373–399

Bakshi M, Oelmüller R (2014) WRKY transcription factors: Jack of many trades in plants. Plant Signal Behav 9(2):e27700. doi:10.4161/psb.27700

Balazadeh S, Kwasniewski M, Caldana C et al (2011) ORS1, an H_2O_2-responsive NAC transcription factor, controls senescence in *Arabidopsis thaliana*. Mol Plant 4(2):346–360

Bhattacharjee S (2005) Reactive oxygen species and oxidative burst: roles in stress, senescence and signal. Curr Sci India 89:1113–1121

Bieker S, Riester L, Stahl M et al (2012) Senescence-specific alteration of hydrogen peroxide levels in *Arabidopsis thaliana* and oilseed rape spring variety *Brassica napus* L. cv. Mozart. J Integr Plant Biol 54:540–554
Biswal B (1995) Carotenoid catabolism during leaf senescence and its control by light. J Photochem Photobiol B Biol 30:3–13
Buchanan-Wollaston V (1997) The molecular biology of leaf senescence. J Exp Bot 48 (2):181–199
Buchanan-Wollaston V, Earl S, Harrison E et al (2003) The molecular analysis of leaf senescence– a genomics approach. Plant Biotechnol J 1(1):3–22
Buchanan-Wollaston V, Page T, Harrison E et al (2005) Comparative transcriptome analysis reveals significant differences in gene expression and signalling pathways between developmental and dark/starvation-induced senescence in Arabidopsis. Plant J 42:567–585
Cocetta G, Baldassarre V, Spinardi A et al (2014) Effect of cutting on ascorbic acid oxidation and recycling in fresh-cut baby spinach (*Spinacia oleracea* L.) leaves. Postharvest Biol Technol 88:8–16
Cutler AJ, Krochko JO (1999) Formation and breakdown of ABA. Trend Plant Sci 4:472–478
Ferrante A, Mensuali-Sodi A, Serra G et al (2002) Effects of ethylene and cytokinins on vase life of cut *Eucalyptus parvifolia* Cambage branches. Plant Growth Regul 38:119–125
Ferrante A, Incrocci L, Maggini R et al (2004) Colour changes of fresh-cut leafy vegetables during storage. J Food Agric Environ 2(3&4):40–44
Ferrante A, Martinetti L, Maggiore T (2009) Biochemical changes in cut vs. intact lamb's lettuce (*Valerianella olitoria*) leaves during storage. Int J Food Sci Tech 44:1050–1056
Foyer CH, Halliwell BT (1976) The presence of glutathione and glutathione reductase in chloroplasts: a proposed role in ascorbic acid metabolism. Planta 133:21–25
Foyer CH, Lelandais M, Kunert KJ (1994) Photooxidative stress in plants. Physiol Plant 92:696–717
Gadjev I, Vanderauwera S, Gechev TS et al (2006) Transcriptomic footprints disclose specificity of reactive oxygen species signaling in Arabidopsis. Plant Physiol 141:436–445
Gan S, Amasino RM (1997) Making sense of senescence (molecular genetic regulation and manipulation of leaf senescence). Plant Physiol 113(2):313–319
Gechev TS, Van Breusegem F, Stone JM et al (2006) Reactive oxygen species as signals that modulate plant stress responses and programmed cell death. BioEssays 28:1091–1101
Gepstein S, Thimann KV (1980) Changes in the abscisic acid content of oat leaves during senescence. Proc Natl Acad Sci 77:2050–2053
Gepstein S, Sabehi G, Carp MJ et al (2003) Large-scale identification of leaf senescence-associated genes. Plant J 36(5):629–642
Guo Y, Gan SS (2014) Translational researches on leaf senescence for enhancing plant productivity and quality. J Exp Bot 65(14):3901–3913
Guo Y, Cai Z, Gan S (2004) Transcriptome of Arabidopsis leaf senescence. Plant Cell Environ 27 (5):521–549
Hajouj T, Michelis R, Gepstein S (2000) Cloning and characterization of a receptor-like protein kinase gene associated with senescence. Plant Physiol 124(3):1305–1314
He YH, Fukushige H, Hildebrand DF, Gan SS (2002) Evidence supporting a role of jasmonic acid in Arabidopsis leaf senescence. Plant Physiol 128:876–884
Hung KT, Kao CH (2004) Hydrogen peroxide is necessary for abscisic acid-induced senescence of rice leaves. J Plant Physiol 161:1347–1357
Hunter DA, Yoo SD, Butcher SM (1999) Expression of 1-aminocyclopropane-1-carboxylate oxidase during leaf ontogeny in white clover. Plant Physiol 120(1):131–142
Jajic I, Sarna T, Strzalka K (2015) Senescence, stress, and reactive oxygen species. Plants 4 (3):393–411
Jibran R, Hunter DH, Dijkwel PP (2013) Hormonal regulation of leaf senescence through integration of developmental and stress signals. Plant Mol Biol 82:547–561

Jiménez A, Hernández JA, Pastori G et al (1998) Role of the ascorbate-glutathione cycle of mitochondria and peroxisomes in the senescence of pea leaves. Plant Physiol 118 (4):1327–1335

Khan MIR, Khan NA (2014) Ethylene reverses photosynthetic inhibition by nickel and zinc in mustard through changes in PS II activity, photosynthetic-nitrogen use efficiency and antioxidant metabolism. Protoplasma 251:1007–1019

Khan MIR, Asgher M, Khan NA (2014) Alleviation of salt-induced photosynthesis and growth inhibition by salicylic acid involves glycine betaine and ethylene in mungbean (*Vigna radiata* L.) Plant Physiol Biochem 80:67–74

Khan MIR, Nazir F, Asgher M, Per TS, Khan NA (2015) Selenium and sulfur influence ethylene formation and alleviate cadmium-induced oxidative stress by improving proline and glutathione production in wheat. J Plant Physiol 178:9–18

Khan MIR, Khan NA, Masood A, Per TS, Asgher M (2016a) Hydrogen peroxide alleviates nickel-inhibited photosynthetic responses through increase in use-efficiency of nitrogen and sulfur, and glutathione production in mustard. Front Plant Sci 7:44

Khan MIR, Iqbal N, Masood A, Mobin M, Anjum NA, Khan NA (2016b) Modulation and significance of nitrogen and sulfur metabolism in cadmium challenged plants. Plant Growth Regul 78:1–11

Li H, He J, Yang X et al (2016) Glutathione-dependent induction of local and systemic defense against oxidative stress by exogenous melatonin in cucumber (*Cucumis sativus* L.) J Pineal Res 60(2):206–216

Lim PO, Woo HR, Nam HG (2003) Molecular genetics of leaf senescence in Arabidopsis. Trends Plant Sci 8:272–278

Lim PO, Kim HJ, Nam HG (2007) Leaf senescence. Annu Rev Plant Biol 58:115–136

Ling Q, Jarvis P (2016) Plant signaling: ubiquitin pulls the trigger on chloroplast degradation. Curr Biol 26(1):R38–R40

Liu Y, He C (2016) Regulation of plant reactive oxygen species (ROS) in stress responses: learning from *AtRBOHD*. Plant Cell Rep 35(5):995–1007

Marrs KA (1996) The functions and regulation of glutathione S-transferases in plants. Annu Rev Plant Phys Plant Mol Biol 47:127–158

Matile P, Hörtensteiner S, Thomas H et al (1996) Chlorophyll breakdown in senescent leaves. Plant Physiol 112:1403–1409

Mittler R, Vanderauwera S, Gollery M et al (2004) Reactive oxygen gene network of plants. Trends Plant Sci 9(10):490–498

Mittler R, Vanderauwera S, Suzuki N et al (2011) ROS signaling: the new wave? Trends Plant Sci 16:300–309

Morris K, Mackerness SAH, Page T et al (2000) Salicylic acid has a role in regulating gene expression during leaf senescence. Plant J 23:677–685

Munné-Bosch S, Alegre L (2004) Die and let live: leaf senescence contributes to plant survival under drought stress. Funct Plant Biol 31(3):203–216

Navabpour S, Morris K. Allen R et al (2003) Expression of senescence- enhanced genes in response to oxidative stress J Exp Bot 54:2285–2292

Noh YS, Amasino RM (1999) Identification of a promoter region responsible for the senescence-specific expression of SAG12. Plant Mol Biol 41:181–194

Nooden LD (1988) The phenomena of senescence and aging. In: Nooden LD, Leopold AC (eds) Senescence and aging in plants. Academic Press, San Diego

Oda-Yamamizo C, Mitsuda N, Sakamoto S et al (2016) The NAC transcription factor ANAC046 is a positive regulator of chlorophyll degradation and senescence in Arabidopsis leaves. Sci Rep 9 (6):23609. doi:10.1038/srep23609

Oh SA, Lee SY, Chung IK et al (1996) A senescence-associated gene of Arabidopsis thaliana is distinctively regulated during natural and artificially induced leaf senescence. Plant Mol Biol 30(4):739–754

Quirino BF, Normanly J, Amasino RM (1999) Diverse range of gene activity during Arabidopsis thaliana leaf senescence includes pathogen-independent induction of defense-related genes. Plant Mol Biol 40:267–278

Ramel F, Ksas B, Akkari E et al (2013) Light-induced acclimation of the Arabidopsis chlorina1 mutant to singlet oxygen. Plant Cell 25(4):1445–1462

Rosenvasser S, Mayak S, Friedman H (2006) Increase in reactive oxygen species (ROS) and in senescence associated gene transcript (SAG) levels during dark induced senescence of Pelargonium cuttings, and the effect of gibberellic acid. Plant Sci 170:873–879

Rosenwasser S, Rot I, Sollner E (2011) Organelles contribute differentially to reactive oxygen species-related events during extended darkness. Plant Physiol 156(1):185–201

Salleh FM, Evans K, Goodall B (2012) A novel function for a redox-related LEA protein (SAG21/AtLEA5) in root development and biotic stress responses. Plant Cell Environ 35(2):418–429

Sewelam N, Kazan K, Schenk PM (2016) Global plant stress signaling: reactive oxygen species at the cross-road. Front Plant Sci 23(7):187

Smart CM (1994) Gene expression during leaf senescence. New Phytol 126:419–448

Song Y, Yang C, Gao S (2014) Age-triggered and dark-induced leaf senescence require the bHLH transcription factors PIF3, 4, and 5. Mol Plant 7(12):1776–1787

Song Y, Xiang F, Zhang G et al (2016) Abscisic acid as an internal integrator of multiple physiological processes modulates leaf senescence onset in *Arabidopsis thaliana*. Front Plant Sci 19(7):181

Spinardi A, Ferrante A (2012) Effect of storage temperature on quality changes of minimally processed of baby lettuce. J Food Agric Environ 10(1):38–42

Suzuki N, Miller G, Morales J et al (2011) Respiratory burst oxidases: the engines of ROS signaling. Curr Opin Plant Biol 14:691–699

Torres MA, Dangl JL (2005) Functions of the respiratory burst oxidase in biotic interactions, abiotic stress and development. Curr Opin Plant Biol 8:397–403

van der Graaff E, Schwacke R, Schneider A et al (2006) Transcription analysis of arabidopsis membrane transporters and hormone pathways during developmental and induced leaf senescence. Plant Physiol 141:776–792

Weaver LM, Gan S, Quirino B et al (1998) A comparison of the expression patterns of several senescence-associated genes in response to stress and hormone treatment. Plant Mol Biol 37 (3):455–469

Weidhase R, Kramell H, Lehmann J et al (1987) Methyl jasmonate-induced changes in the polypeptide pattern of senescing barley leaf segments. Plant Sci 51:177–186

Woo HR, Kim JH, Nam HG et al (2013) Plant leaf senescence and death – regulation by multiple layers of control and implications for aging in general. J Cell Sci 126:4823–4833

Woo HR, Koo HJ, Kim J et al (2016) Programming of plant leaf senescence with temporal and inter-organellar coordination of transcriptome in Arabidopsis. Plant Physiol 171(1):452–467

Xu Y, Burgess P, Zhang X et al (2016) Enhancing cytokinin synthesis by overexpressing ipt alleviated drought inhibition of root growth through activating ROS-scavenging systems in *Agrostis stolonifera*. J Exp Bot 67(6):1979–1992

Yoshida S (2003) Molecular regulation of leaf senescence. Curr Opin Plant Biol 6(1):79–84

Yoshimoto K, Jikumaru Y, Kamiya Y (2009) Autophagy negatively regulates cell death by controlling NPR1-dependent salicylic acid signaling during senescence and the innate immune response in Arabidopsis. Plant Cell 21:2914–2927

ROS-Induced Transcription Factors During Oxidative Stress in Plants: A Tabulated Review

6

Rashmi Kalia, Shelja Sareen, Avinash Nagpal, Jatinder Katnoria, and Renu Bhardwaj

Abstract

Different abiotic stresses lead to overproduction of reactive oxygen species (ROS) in plants which are highly reactive, toxic and cause damage to proteins, lipids, carbohydrates and DNA which ultimately results in oxidative stress. The ROS comprises of free radicals including superoxide radicals and hydroxyl radicals and molecular forms like hydrogen peroxide and singlet oxygen. Oxidative stress has been devastating effects on the overall growth of plant and it results in loss of vigour and germination. In recent years, it has become a well-known fact that reactive oxygen intermediates are produced in plants as signalling molecules to control metabolic processes such as programmed cell death, abiotic stress responses, pathogen defence and systemic signalling. In response to stress, plants respond through crosstalk between different signal transduction pathways (abscisic acid dependent and abscisic independent), involving transcription factors (TFs). Transcription factors are also called as regulons, having a pivotal role in regulation of gene expression in plants. In the present attempt, we present tabulated review of TFs, proteins, genes and enzymes which have a central role in combating oxidative stress through gene expression regulation.

Keywords

Transcription factors • Abiotic stress • Regulons • Reactive oxygen intermediates

R. Kalia • S. Sareen
BBK DAV College for Women, Amritsar, Punjab 143001, India

A. Nagpal • J. Katnoria • R. Bhardwaj (✉)
Department of Botanical and Environment Sciences, Guru Nanak Dev University, Amritsar, Punjab 143005, India
e-mail: renubhardwaj82@gmail.com

M.I.R. Khan, N.A. Khan (eds.), *Reactive Oxygen Species and Antioxidant Systems in Plants: Role and Regulation under Abiotic Stress*,
DOI 10.1007/978-981-10-5254-5_6

Abbreviations

ABA	Abscisic acid
ABF	ABRE-binding factor
ABRE	ABA-responsive element
APX	Ascorbate peroxidase
ARCK	ABA and osmotic stress-inducible receptor-like cytosolic kinase
AREB1	ABA-responsive element-binding protein
ATAF	*Arabidopsis* transcription activation factor
ATAF	*Arabidopsis* transcription activation factor
AtMYC2	*Arabidopsis thaliana* myelocytomatosis 2
BOS1	*Botrytis*-susceptible 1
bZIP	Basic leucine zipper
CaERFLP	*Capsicum annuum* ethylene-responsive factor-like protein
CAT	Catalases
CBF1C	Repeat-binding factor
CRK	Cysteine-rich receptor-like kinase
CRLK	Calcium-regulated RLK
CRT/DRE	C-repeat/dehydration-responsive element
CUC	Cup-shaped cotyledon
DEAR1	DREB and EAR motif protein 1
DHAR	Dehydroascorbate reductase
DREB	Dehydration-responsive element-binding protein
DREB	Dehydration-responsive element-binding protein
EAR	Ethylene response factor-associated amphiphilic repression
EREBP	Ethylene-responsive element-binding protein
ERF	Ethylene-responsive element-binding factors
ERF	Ethylene-responsive factor
ET	Ethylene
GHR	Guard cell hydrogen peroxide
GPX	Glutathione peroxidase
GR	Glutathione reductase
GSSG	Oxidized glutathione
H_2O_2	Hydrogen peroxide superoxide
HOS1	High expression osmotically responsive gene 1
HSFA2	Heat shock transcription factor A2
JA	Jasmonic acid
LOO	Lipid peroxyl radical
LOOH	Lipid hydroperoxide
MDHAR	Monodehydroascorbate reductase
MYB	Myeloblastosis
MYC	Myelocytomatosis
MYC	Myelocytomatosis
NAC	NAM(no apical meristem)

NADPH	Nicotinamide adenine dinucleotide phosphate
NTL6	NAC with transmembrane motif-like 6
O_2^{-}	Hydroxyl radicals
O_2	Singlet oxygen
OPBP1	Osmotin promoter-binding protein
OsBIERF	*Oryza sativa* benzothiadiazole-induced ethylene-responsive transcription factors
OSSIPK	*O. sativa* stress-induced protein kinase
PCD	Programmed cell death
PERK	Proline-rich extension-like kinase
PLP	Pyridoxal-5-phosphate
PSTOL	Phosphorus starvation tolerance
RD	Responsive to dehydration
ROS	Reactive oxygen species
RPK	Receptor-like protein kinase
SIRLK	Salt-induced receptor-like kinase
SLAIM	*Solanum lycopersicum* abscisic acid-induced Myb1
SOD	Superoxide dismutase
TaPIMP	*Triticum aestivum* pathogen-induced membrane protein
TFs	Transcriptional factors
TSi	Tobacco stress-induced gene
WRKY	W(tryptophan), R (arginine), K (lysine), Y(tyrosine)
ZAT	Zinc finger of *Arabidopsis thaliana*

6.1 Introduction

Oxidative stress, occurring due to devastating effects of reduced oxygen species, is an important phenomenon in many biological systems. Plants are constantly exposed to the adversity of extreme environmental conditions which is perceived as abiotic stress. It is one of the most deleterious effects on plants as it reduces photosynthetic and growth attributes, yield of crop, vigour and germination (Woodrow et al. 2012; Khan and Khan 2014; Khan et al. 2014, 2015). These stressful conditions result in the production of reactive oxygen species (ROS), resulting in oxidative stress (Ahmad et al. 2008, 2010; Yousuf et al. 2012; Khan et al. 2016a, b). Plants in response to adversity of environmental conditions overproduce ROS such as superoxide (O_2^{-}), hydrogen peroxide (H_2O_2) and hydroxyl radicals (OH^{-}). Although plants utilize ROS in a beneficial way (Inze and Van Montagu 1995), plants must exercise a control on the production of ROS as accumulation of ROS can trigger series of events which leads to cell death. Thus, equilibrium must be established by plants between two mechanisms where by one mechanism they regulate signalling pathways through ROS and by another mechanism overproduced ROS must be scavenged (Mittler 2002). ROS are produced during photosynthesis and aerobic respiration in mitochondria, chloroplast and peroxisomes (Gill and Tuteja 2010; Tuteja et al. 2009). Out of various plant responses, regulation of gene expression is an important plant response towards abiotic stresses such as salinity (Liao et al. 2008a, b; Ding et al. 2009; Fujita et al. 2004; Ciftci-Yilmaz et al. 2007), drought (Haake et al. 2002; Shinozaki et al. 2007; Vannini et al. 2007; Xiang et al. 2008), heat (Rizhsky et al. 2004a; Ciftci-Yilmaz et al. 2007), cold (Liu et al. 1998; Nakashima et al. 2007; Liao et al. 2008a; Ouyang et al. 2010) and heavy metals, and it involves unique as well as universal changes in transcript levels of genetic machinery. In lieu of the abiotic stress, plants are expected to respond with complex, interconnected signal transduction pathways which share proteins, cofactors and enzymes at various levels of tolerance. Transcription factors may regulate gene expression by binding to specific DNA sequences in the promoters of respective target genes. These factors are called as regulons (Saibo et al. 2009; Chinnusamy et al. 2004). All these factors involve abscisic acid (ABA)-dependent and ABA-independent systems for regulation of gene expression. Apart from this glutathione and glutathione reductase also have significant role in providing defence against abiotic stress in plants. An important and interesting question for plant biologist since a long time is that how do plants perceive these stresses? How are the signals transmitted so that plant can be prepared for adaptive responses? These questions and its answers are crucial in leading towards developing plant stress tolerance genotypes. Signal relays and cue perception are clear indications that a crosstalk does occur between different signal transduction pathways in plants (Chinnusamy et al. 2004). The present chapter focuses upon the tabulated review of TFs, various genes and enzymes along with their significant role involved in response to abiotic stress.

6.2 Reactive Oxygen Species and Its Types

The ROS are the partially reduced forms of oxygen. There are three main kinds of ROS which are produced in a stepwise manner depending upon the degree to which oxygen is reduced

Superoxide (O_2^-): It is the primary ROS which is formed when the molecular oxygen picks up one electron. NADPH is the electron donor and NADPH oxidase acts as the reaction catalyst. Formation of this radical has devastating effect upon photosystem I and photosystem II and the whole photosynthetic machinery (Gill and Tuteja 2010).

Hydrogen peroxide (H_2O_2): It is the other ROS which is dismutation of superoxide. Firstly, it forms peroxide (O_2^{2-}) which after neutralization by two protons forms hydrogen peroxide (Garg and Manchanda 2009). At low concentration, it relays signals that lead to tolerance against biotic and abiotic stress, and at higher concentrations, it triggers programmed cell death (Quan et al. 2008).

Hydroxyl radicals (OH^-): The third ROS which is the most reactive and most toxic are formed by Haber-Weiss reaction and Fenton reaction. This radical can react with all kinds of biological molecules, and because there is no mechanism in plants which can remove this radical, ultimately it leads to death of plant (Garg et al. 2009).

6.3 ROS Production in Cellular Organelles: An Overview

ROS are produced in different cell organelles mainly in mitochondria, chloroplast and peroxisomes. In dark conditions, the mitochondrion is the major producer and, in the presence of light chloroplast and peroxisomes, is considered to be the ROS producer (Foyer and Noctor 2003):

Chloroplast: In this organelle, PSI and PSII are the major sites for the production of ROS. Under abiotic stress conditions, there is overproduction of ROS (Gill and Tuteja 2010).

Peroxisomes: These small subcellular organelles produce both H_2O_2 and O_2- radicals as a part of normal metabolism. Peroxisome membrane and matrix are thought to be the site for the production of O_2- (Luis et al. 2002), while H_2O_2 is generated as an outcome of beta-oxidation of fatty acids and few other metabolic pathways (Corpas et al. 2001).

Mitochondria: The "power house" of the cell produces H_2O_2. It is produced in this organelle as the product of normal respiration, while under the abiotic stress conditions, it overproduces H_2O_2 (Gill and Tuteja 2010).

6.4 ROS Scavenging

Plants have evolved efficient mechanisms to detoxify ROS through various enzymatic and non-enzymatic components of antioxidative defence system. Table 6.1 presents an overview of enzymatic and non-enzymatic participants involved in scavenging of ROS.

6.5 Transcription Factors Are Key Players in Oxidative Stress

Various metabolic proteins like TFs and kinases are produced in response to abiotic stresses. There is a dense and co-ordinated network of TFs which regulate the genes responsible for stress tolerance. For the past two decades, constant advances have been made to understand which and how environmental cues bring transcriptional changes. A lot of work has been done to identify signalling proteins and TFs which regulate the stress-induced gene expression. Plant responses towards stress involve transcriptional activation of genes. Gene expression as a result is systemic as well as localized response involves hormones like salicylic acid (SA), ethylene (ET) and jasmonic acid (JA) which act as secondary signals for the gene expression (Rushton and Somssich 1998). Four different regulons that are active in response to abiotic stresses have been identified. Dehydration-responsive element-binding protein 1 (DREB1)/C-repeat-binding factor (CBF) and DREB2 regulons function in

Table 6.1 Antioxidant defence mechanism and their scavenging mechanisms

S. no.	Component	Site of production	ROS	Mechanism of action	References
Non-enzymatic components					
1.	Ascorbate (AsA)	Cytosol	H_2O_2	Scavenge ROS in hydrophilic environment. Two molecules of AsA are utilized by APX to reduce H_2O_2 to water	Hasanuzzaman et al. (2012)
2.	Glutathione (GSH)	Cytosol, mitochondria, peroxisomes, apoplast	H_2O_2, O_2^- and OH^-	It scavenges H_2O_2, O_2^-, and OH^- by producing AsA via AsA-GSH cycle	Mittler (2002), Jimenez (1998) and Foyer and Halliwell (1976)
3.	Tocopherol	Thylakoid membranes	O_2^- and OH^- lipid radicals	Physical quenching and lipid peroxidation by reducing LOO• (lipid peroxyl radical)	Hasanuzzaman et al. (2012) and Maeda et al. (2005)
4.	Proline	Mitochondria	O_2^- and OH^-	NADPH-mediated pentose phosphate pathway activity	Hare and Cress (1997)
5.	Carotenoids	Chloroplast	O_2^-	*Car*, a lipid-soluble antioxidant plays an important role in quenching O_2^- and also found to protect photosynthetic apparatus	Gill and Tuteja (2010)
6.	Flavonoids	Leaves, pollens and floral parts	H_2O_2, O_2^- and OH^-	Neutralization of radicals	Gill and Tuteja (2010)
Enzymatic components					
7.	SOD	Chloroplast, cytosol, mitochondria, peroxisome	O_2^-	Dismutation of O_2^- by reducing one O_2^- to H_2O_2 and another to O_2	Gill and Tuteja (2010)
	(a) Cu/Zn-SOD	Chloroplast and cytosol	–	Shows sensitivity towards KCN and H_2O_2	Luis et al. (2002)
	(b) Mn-SOD	Mitochondria and peroxisome	–	Resistant to both KCN and H_2O_2	Luis et al. (2003)
	(c) Fe-SOD	Chloroplast	–	Resistant to KCN and sensitivity towards H_2O_2	Alscher et al. (2002)

(continued)

Table 6.1 (continued)

S. no.	Component	Site of production	ROS	Mechanism of action	References
8.	CAT	Peroxisomes, glyoxysomes	H_2O_2	Scavenges H_2O_2 by oxidases through beta-oxidation of fatty acids and photorespiration	Gill and Tuteja (2010)
9.	APX	Chloroplast, cytosol, mitochondria, peroxisome	H_2O_2	AsA is utilized as substrate and electrons are transferred to H_2O_2, producing water and DHA	Raven (2000) and Pang and Wang (2010)
10.	MDHAR	Chloroplast and cytosol, mitochondria, peroxisome	H_2O_2	A FAD enzyme is reduced to MDHA by oxidation of AsA. In mitochondria and peroxisomes, it need APX to scavenge H_2O_2	del Río et al. (2002)
11.	DHAR	Chloroplast and cytosol, mitochondria, peroxisome	H_2O_2	A critical component in AsA recycling system and regenerates AsA from DHA	Martínez and Araya (2010)
12.	GPX	Chloroplast and cytosol, mitochondria, endoplasmic reticulum	H_2O_2 and LOOH	GPX reduces GSH to H_2O_2. Also it plays an important role in oxidative signal transduction	Miao et al. (2006) and Noctor et al. (2002a)
13.	GR	Chloroplast and cytosol, mitochondria,	H_2O_2	It scavenges H_2O_2 by maintaining GSH/GSSHG ratio. It's a key enzyme in AsA-GSH cycle	Pang and Wang (2010) and Hasanuzzaman et al. (2012)

abscisic acid (ABA)-independent gene expression, whereas the ABA-responsive element (ABRE)-binding protein (AREB)/ABRE-binding factor (ABF) regulon functions in ABA-dependent gene expression (Saibo et al. 2009). Regulons, including the NAC (or NAM, no apical meristem) and myeloblastosis/myelocytomatosis (MYB/MYC), also play significant role in stress-responsive gene expression. TFs proved to be a game-changing element for genetic engineering in abiotic stress resistance in crop plants, and a lot has been done in this pretext since the last two decades (Woodrow et al. 2012).

6.5.1 Regulons

DREB/CBF regulon: This regulon has a significant role in cold stress response and has cosmopolitan distribution in the plant kingdom, including non-acclimatized plants like rice and tomato. The response towards cold stress is quick, and once the gene is expressed, it sets a cascade of activation for other proteins acting as osmoprotectants and antioxidants. Expression of DREB1/CBF has shown improved tolerance in plants against drought and low temperatures (Kasuga et al. 1999). DREB2 (DREB2A and DREB2B genes) are induced by salinity and dehydration stress (Liu et al. 1998; Akhtar et al. 2012; Lata and Prasad 2011).

AREB/ABF regulon: AREB/ABF is bZIPs which binds to ABRE motif, a cis-element which triggers ABA-dependent gene expression. ABFs are multi-functional with different genes showing tolerance towards salinity (ABF2, ABF3, ABF4), drought (ABF2, ABF4), cold (ABF1, ABF4) and heat (ABF2) stress conditions (Lata et al. 2011).

MYB/MYC regulon: A universal group of proteins is found in both plants and animals and participates in ABA-dependent gene expression. Usually, they play important role in seed development, meristem formation and secondary metabolism. Significant tolerance against stress conditions like drought, heat and salinity has been observed in various transgenic plants (Lata et al. 2011).

NAC regulon: These regulons are cis-regulatory sequences and play important role in both temporal and spatial gene expression. They provide tolerance against both abiotic and biotic stress through ABA-dependent and ABA-independent pathways. They can prove to be powerful targets for production of economically and stress-tolerant plants (Puranik et al. 2012). The summary of literature for transcription factors, genes and enzymes responsible for stress resistance is given in the Table 6.2.

6.6 Conclusion and Future Prospects

With the advent of novel genetic tools in molecular biology, noticeable efforts are made to develop transgenic plants which have shown tolerance against abiotic stress and pathogen defence. An evident crosstalk between various signalling pathways has opened the gateway of understanding and unravelling the role of TFs in abiotic stress-induced gene expression. The development of transgenic plants with altered enzymatic mechanisms has shown considerable tolerance against abiotic stress, and improved yield is being noticed. Earlier ROS accumulation was perceived as toxic, but now it is a well-established fact that they mediate crosstalk between various signalling pathways to influence various developmental processes and making them adaptable towards extreme abiotic stresses. Recently, many transgenic plants with overexpressed genes have shown promising role in

Table 6.2 Tabulated review of transcription factors/genes and enzymes involved in oxidative stress resistance

S. no.	Family	Plant species	Gene/TF/enzymes	Mechanism of action	Stress (abiotic/biotic)	References
1	Brassicaceae	*Arabidopsis thaliana*	*CBF1/DREB1*	Enhances freezing and osmotic tolerance of plants	Abiotic (cold and osmotic stress)	Guo et al. (2004)
			CBF3	Freezing tolerance increased upon exposure to low temperature. COR (cold responsive) genes are activated in response to low temperature in transgenic plants which enhance cold tolerance in plants. Plants expressing CBF3 have high level of proline, sucrose, glucose and fructose. Increased proline levels are expected to be due to expression of enzyme delta (1)-pyrroline-5-carboxylate synthase	Abiotic (cold stress)	Gilmour et al. (2000)
			CBF4	Elevated drought resistance is observed. Overexpression in transgenic plants activates DREB element which provides tolerance against cold and drought	Abiotic (cold and drought stress)	Haake et al. (2002)
			DREB1A and DREB2A	DREB1A and DREB2A (dehydration-responsive element) play a pivotal role in dehydration and cold stress-induced gene expression in *Arabidopsis* plants	Abiotic (cold stress)	Liu et al. (1998)

(continued)

Table 6.2 (continued)

S. no.	Family	Plant species	Gene/TF/enzymes	Mechanism of action	Stress (abiotic/biotic)	References
			BOS1	Infection of plant with fungus *Botrytis cinerea* induces the expression of BOS1 (*Botrytis* susceptible1) gene. There is strong evidence of interaction of BOS1 gene with jasmonate signalling pathway. Moreover, it has been found to encode a TF protein named R2R3MYB which is speculated to be a mediator in signalling pathways	Biotic (resistance against fungal pathogen *B. cinerea* and necrotrophic pathogen *Alternaria brassicicola*); abiotic (BOS1 gene encoding R2R3MYB TF)	Mengiste et al. (2003)
			DEAR1 DREB and EAR (ethylene response factor associated-amphiphilic repression motif protein1)	Regulator of freezing tolerance pathways. Also DEAR1 accumulates in response to *Pseudomonas syringae* infection. Moreover, this TF also regulates crosstalk between various signalling pathways	Abiotic (freezing tolerance) and biotic (resistance against gram-negative bacterium *P. syringae*)	Tsutsui et al. (2009)
			ABF2 (*ABRE* binding bZIP factor 2)	ABF2 member of bZIP proteins is involved in drought stress response. Overexpression of the gene in mutants has shown tolerance against various stresses. This gene is a key factor in normal functioning of glucose	Abiotic (ABA/stress response)	Kim et al. (2004)

			ABF3 (abscisic acid-responsive element-binding factor 3)	Key component in ABA signalling. Overexpression has shown enhanced tolerance against drought	Abiotic (ABA signalling pathway)	Kang et al. (2002)
			ABF4	Plays an important role in ABA signalling and promotes resistance against drought. ABF3 and ABF4 together mediate stress-responsive ABA signalling	Abiotic (ABA signalling pathway)	Kang et al. (2002)
			bZIP44 (basic leucine zipper), *bZIP62*, *bZIP78*	These genes are identified in soybean plant. Transgenic *Arabidopsis* plants overexpressing these genes have lesser ABA sensitivity. Genes confer tolerance against salinity and cold	Abiotic (cold and salinity tolerance)	Liao et al. (2008a)
			bZIP132	Reduced ABA sensitivity has been found in transgenic *Arabidopsis* plants overexpressing bZIP132 and increased water deficit. Overexpression of this gene has been found to increase salt tolerance	Abiotic (salt tolerance)	Liao et al. (2008b)
			bZIP60	FOX (full length cDNA over-expressor) gene hunting system was developed to identify stress-inducible transcription factors. Various salt stress-resistant lines were	Abiotic (stress signal transduction)	Fujita et al. (2007)

(continued)

Table 6.2 (continued)

S. no.	Family	Plant species	Gene/TF/enzymes	Mechanism of action	Stress (abiotic/biotic)	References
				screened, and it was found to possess *F39* transgene which encodes bZIP TF identical to *bZIP60*, and it is believed to be involved in stress signal transduction and is associated with endoplasmic reticulum stress response		
			MYB2 (myeloblastosis 2)	Regulation of ABA- and JA-inducible gene expression in drought-resistant plants. There is an improved osmotic tolerance in plants	Abiotic (drought and osmotic tolerance)	Shinozaki et al. (2007)
			MYC2 (myelocytomatosis)	MYC2 along with MYB2 activate the transcription of beta-glucuronidase reporter gene in transgenic *Arabidopsis* plants. Both the genes are found to play a key role in providing drought tolerance through ABA signal transduction	Abiotic (drought tolerance)	Abe et al. (2003)
			MYB4 (myeloblastosis 4)	Antagonistic regulation of two MYB TFs have been found, MYB12, a positive regulator of flavanol pathway, and MYB4, a negative regulator.	Abiotic (crosstalk involving two MYB TFs, MYB12 and MYB4, and biotic (synthesis of pathogen-related defence compounds like lignin,	Schenke et al. (2011)

				Overexpression of MYB4 is known to synthesize pathogen defence compounds	phytoalexin and chemalexin)	
			MYB15 (myeloblastosis 15)	Overexpression of this gene provides tolerance against salt and drought stress through elevated levels of stress-tolerant proteins and genes involved in ABA biosynthesis	Abiotic (salt and drought tolerance)	Ding et al. (2009)
			MYB96 (myeloblastosis 96)	One of the most versatile TF MYB96 plays an important role in ABA-dependent drought resistance; enhanced plant's pathogen resistance by SA biosynthesis and a molecular link that interconnects ABA-SA pathways	Abiotic (drought tolerance); biotic (pathogen resistance)	Seo et al. (2010), (2011)
				MBY96 TF also encodes enzymes involved in synthesis of cuticular wax, and there is direct correlation of deposition of cuticular wax in drought conditions		

(continued)

Table 6.2 (continued)

S. no.	Family	Plant species	Gene/TF/enzymes	Mechanism of action	Stress (abiotic/biotic)	References
			ATAF2 (*Arabidopsis* transcription activation factor 2)	Drought resistance and wounding defence and enhanced resistance to *Fusarium oxysporum*; responsible for necrosis	Abiotic (drought tolerance) and biotic (increased susceptibility towards soil-borne fungal pathogen)	Delessert et al. (2005)
			ATAF1 (*Arabidopsis* transcription activation factor 1)	Repressor of ABA pathway during biotic stress and discourage the penetration of fungus Bgh (*Blumeria graminis* f. sp. *hordei)*	Abiotic (ATAF1 is a negative regulator of ABA) and Biotic (susceptibility against fungus *Blumeria graminis*)	Jensen et al. (2008)
			NTL6 (NAC with transmembrane motif-like 6)	Cold stress resistance and SA (salicylic acid)-independent induction of PR (pathogenesis related) genes, like *PR1*, *PR2* and *PR5* by NTL6 gene which enhances pathogen resistance	Abiotic (tolerance to cold) and biotic (quick response of plant to pathogens in cold conditions regulated by *PR* genes	Seo et al. (2010)
			AtMYC2 (*Arabidopsis thaliana* myelocytomatosis2)	Drought resistance and cross-connection between ABA and JA pathways. Increased resistance to *Fusarium oxysporum* has been noticed	Abiotic (drought tolerance) and biotic (resistance against fungus *Fusarium oxysporum)*	Anderson et al. (2004)

			RD26 (responsive to dehydration)	RD26 function as transcriptional activator in ABA-dependent pathways and is drought- and salinity-induced gene. In vivo results suggest that the RD26 protein functions as a transcription activator involved in the ABA-responsive expression of the GLY (glyoxalase I family protein) gene. Also plays a key role in defence against pathogen infection and senescence	Abiotic (drought and salinity tolerance) and biotic (resistance against pathogen and senescence)	Fujita et al. (2004)
			HSFA2	A positive regulator of defence genes (ascorbate peroxidase 2 and galactinol synthase 1 and 2) and enhanced stress tolerance to abiotic factors	Abiotic (tolerance against environmental stress) and biotic (important regulator in stimulation of defence system)	Nishizawa et al. (2006)
			ZAT7 (zinc finger of *Arabidopsis thaliana*)	Resistance against salinity, temperature and crosstalk between JA and SA signalling pathways	Abiotic (salinity, temperature and tolerance)	Ciftci-Yilmaz et al. (2007)
			ZAT12 (zinc finger of *Arabidopsis thaliana*)	ZAT7, ZAT12, and WRKY25 {W (tryptophan), R (arginine), K (lysine), Y (tyrosine)} work in tandem	Abiotic (heat shock, salt, cold, wounding) and biotic (pathogen defence)	Rizhsky et al. (2004a)

(continued)

Table 6.2 (continued)

S. no.	Family	Plant species	Gene/TF/enzymes	Mechanism of action	Stress (abiotic/biotic)	References
				to reduce H_2O_2 stress conditions, and ZAT7, ZAT12, WRKY25 along with Apx1(ascorbate peroxidase) work during oxidative stress. Overexpression of ZAT		
			GHR 1(guard cell hydrogen peroxide-resistant 1)	Localized on the plasma membrane, this gene in *Arabidopsis thaliana* plants mediate ABA and H_2O_2 regulation. It is also involved in stomatal movement	Abiotic (critical component of ABA signalling)	Hua et al. (2012)
			PERK4(proline-rich extension-like receptor kinase)	PERK4 belonging to family of PRELRK (proline-rich extension-like receptor kinase) family, localized in the plasma membrane, is a $Ca2^+$-activated protein. It is found to be an ABA signalling pathway regulator. Mutant varieties have shown lesser sensitivity towards ABA for growth and germination. Moreover, it functions at early stage of ABA signalling to inhibit root cell elongation	Abiotic (ABA signalling)	Bai et al. (2009)

			RPK1 (receptor-like protein kinase)	RPK1 is LRR (leucine-rich repeat) receptor kinase which is localized in the plasma membrane. It acts as a regulator of ABA signalling	Abiotic (ABA-mediated signalling)	Osakabe et al. (2005), (2010)
				There is increased drought tolerance, less loss of water during transpiration and reduction in ROS accumulation		
			RLK7 (receptor-like kinase)	RLK7 belong to LRR (leucine-rich receptor); RLKs have a pivotal role in oxidative stress tolerance as well as regulator of germination rate. Using mutants, one without cytoplasmic kinase domain (RLK7Δkin) and other one without 14 LRR repeats (RLK7ΔLRR) results, indicating that they have a central role in controlling the speed of germination	Abiotic (oxidative stress tolerance and germination speed regulator)	Pitorre et al. (2010)
			CRK36 (cysteine-rich receptor-like kinase)	CRK36 is RLK; a drought and salinity stress-inducible gene was screened by studying its interaction with six RLKs. It was found to interact with ARCK1, and the bimolecular fluorescence	Abiotic (drought and salinity stress)	Tanaka et al. (2012)

(continued)

Table 6.2 (continued)

S. no.	Family	Plant species	Gene/TF/enzymes	Mechanism of action	Stress (abiotic/biotic)	References
				complementation assay (BiFC) was done to confirm the findings. Its role as a negative regulator ABA and osmotic stress signalling was established		
			ARCK1 (ABA and osmotic stress-inducible receptor-like cytosolic kinase)	ARCK1 is an RLCK expressed under abiotic stress and it encodes cytosolic protein kinase. Comparison of mutant and wild-type varieties has shown that mutants have lower sensitivity towards osmotic stress and ABA, making it a negative regulator of osmotic and ABA signal transduction	Abiotic (osmotic stress)	Tanaka et al. (2012)
			OsRLCK253 (receptor-like cytoplasmic kinase)	Drought and salt stress resistance. Protection against loss of yield is also reported in transgenic plants	Abiotic (drought and salt)	Giri et al. (2011)
			CRLK1 calcium/ calmodulin-regulated receptor-like kinase	*CRLK1* is found to play a central role in increasing tolerance for cold in plants. No phenotypic difference has been found among wild-type and *crlk1* mutants but mutants have shown increased tolerance towards freezing temperature	Abiotic (freezing tolerance)	Yang et al. (2010a)

			NAC2 NAM domain (no apical meristem), ATAF (*Arabidopsis* transcription activation factor), CUC(cup-shaped cotyledon), *NAC019*, *NAC055*	Three cDNA clones were created using yeast one-hybrid system, encoding NAC2, NAC019 and NAC055; members of NAC TF family. Protein expression is found to be activating the beta-glucuronidase gene resulting in providing the enhanced tolerance against drought	Abiotic (drought tolerance)	Tran et al. (2004)
			WRKY45, *WRKY21*, *WRKY54*	WRKY21 provides tolerance against cold, while WRKY54 provides tolerance against salt and drought stress. WRKY45 confer pathogenic and drought resistance	Abiotic (salt and drought tolerance) and biotic (pathogen defence)	Qiu and Yu (2009) and Zhou et al. (2008)
2.	Gramineae	*Oryza sativa*	*MYB4*	Overexpression of MYB4 has shown significant tolerance against freezing climatic conditions. Moreover, MYB4 overexpression also activates cold-inducible promoters like PAL2, ScD9 SAD and COR15a	Abiotic (freezing tolerance) and biotic (increased level of amino acids involved in stress response)	Vannini et al. (2004), (2006), (2007)
				Osmyb4 encodes MYB TF which is involved in cold		

(continued)

Table 6.2 (continued)

S. no.	Family	Plant species	Gene/TF/enzymes	Mechanism of action	Stress (abiotic/biotic)	References
				acclimatization. This gene's expression has been found to increase levels of certain amino acids which play an important role in stress response		
			OsBIERF1-4 (*Oryza sativa* benzothiadiazole-induced ethylene-responsive transcription factors)	Four genes, OsBIERF1, OsBIERF2, OsBIERF3, and OsBIERF4, were isolated and studied and were found to be a positive regulator of abiotic stress tolerance and also seem to be involved in various signalling pathways	Abiotic (mediator of signalling pathways)	Cao et al. (2006)
			OsNAC6	Biotic (wounding and blast disease) and abiotic (cold, drought, high salinity) stress response regulator	Abiotic (cold, drought) and biotic (blast disease)	Nakashima et al. (2007)
			OsWRKY45	Inducer of ABA-responsive abiotic stress-tolerant genes. Enhanced expression of PR genes, improved resistance to the bacterial pathogen *Pseudomonas syringae*, salt and drought stresses	Abiotic (ABA-dependent pathway) and biotic (resistance against bacteria)	Qiu et al. (2009)
			OsWRKY82	Versatile stress-inducible gene crosstalk between JA and ET signalling pathway and resistance to both abiotic and biotic stress	Abiotic (heat shock, wounding) and biotic (bacterial blight diseases and fungal blast disease)	Peng et al. (2011)

			OsWRKY89	OsWRKY89 confers tolerance against UV-B radiation and significant resistance against rice blast fungus and white-backed plant hopper	Abiotic (UV-B tolerance) and biotic (resistance against fungal pathogen)	Wang et al. (2007)
			OsSIK1	Salt and drought resistance and decreased ROS buildup	Abiotic (salinity and drought tolerance)	Ouyang et al. (2010)
			PSTOL1 (phosphorus starvation tolerance)	PSTOL1, a recombinant protein kinase, is an enhancer which makes plants get more phosphorus and other nutrients through early growth of roots	Abiotic (increased uptake of phosphorus)	Gamuyao et al. (2012)
			OsAB15	Gene isolated from *O. sativa* is responsible for regulation of plant stress response and fertility. An ABA-inducible gene, if overexpressed, provides high sensitivity to salinity stress and if repressed confers stress tolerance and low fertility	Abiotic (salinity stress tolerance)	Zou et al. (2008)
			OsbZIP23	A TF in rice regulate the expression of various stress-related genes though ABA-dependent pathway. It provides tolerance against drought and high salinity	Abiotic (drought tolerance)	Xiang et al. (2008)

(continued)

Table 6.2 (continued)

S. no.	Family	Plant species	Gene/TF/enzymes	Mechanism of action	Stress (abiotic/biotic)	References
		Triticum aestivum	*TaNAC4* (*Triticum aestivum* NAC4)	Transcriptional activator of signalling pathways and tolerance against wounding, low temperature and high salinity	Abiotic (freezing tolerance and salinity)	Xia et al. (2010)
3.	Solanaceae	*Solanum lycopersicum*	*SIAIM1* (*Solanum lycopersicum* abscisic acid-induced Myb1)	This gene encodes R2R3MYB TF which regulate pathogen resistance and abiotic stress tolerance through regulation of ABA signalling pathway	Abiotic (ABA-dependent regulation)	AbuQamar et al. (2009)
			AREB1 (ABA-responsive element)	Two TFs SLAREB1 (*Solanum lycopersicum AREB*) and SLAREB2 regulate abiotic stress and pathogen defence through ABA-dependent pathways. Both of them have been found to enhance drought and salt resistance in tomato plants. Moreover, involvement of bZIP in ABA pathway has been speculated which mediate abiotic stress tolerance	Abiotic (salinity)	Orellana et al. (2010)
			MYB4	Significant tolerance against viral diseases and drought stress	Abiotic (drought tolerance) and biotic (resistance to viral diseases)	Vannini et al. (2007)

		Nicotiana tabacum	*TSi1* (tobacco stress-induced gene)	Activation of several pathogen-related genes providing enhanced resistance against salt and pathogen resistance against salt tolerance	Abiotic (salinity stress tolerance)	Park et al. (2001)
			OPBP1 (osmotin promoter-binding protein)	OPBP1 is a TF of tobacco plant. Overexpression of this gene in transgenic tobacco plants has shown high buildup of PR-1a and PR-5d genes which have important role in combating infection by *Pseudomonas* and *Phytophthora* pathogens. High resistance against salt stress has been noticed	Abiotic (salinity tolerance) and biotic (resistance to fungal pathogens)	Zhang et al. (2009)
			Wlip19 (Wheat low temperature-induced protein)	This gene plays an important role in making the plant cold acclimatized. It provides tolerance against freezing conditions and drought	Abiotic (cold stress tolerance)	Kobayashi et al. (2008)
			OSISAP1 (*O. sativa* stress-induced protein kinase)	A wide spectrum of stress and ABA-induced gene isolated from rice plant, which when overexpressed in tobacco plant provides tolerance against cold, desiccation, salt and heavy metal stress	Abiotic (salinity, drought tolerance)	Mukhopadhyay et al. (2004)

(continued)

Table 6.2 (continued)

S. no.	Family	Plant species	Gene/TF/enzymes	Mechanism of action	Stress (abiotic/biotic)	References
			SCOF-1 (soya bean cold-inducible factor-1)	ABA and low temperature-inducible gene. High tolerance is due to *COR* gene expression, making it a positive regulator of *COR* gene expression	Abiotic (freezing stress tolerance)	Kim et al. (2001)
4.	Leguminosae	*Glycine max*	*GmERF3* (*Glycine max* ethylene-responsive factor)	GmERF3, a TF from soya bean, has been found to provide increased resistance against abiotic (high salinity and dehydration) as well as biotic (against pathogens like *Alternaria alternata* and TMV) stress tolerance	Abiotic (dehydration) and biotic (pathogen-related stress resistance)	Zhang et al. (2009)
5	Piperaceae	*Capsicum annuum*	*CaERFLP1* (*Capsicum annuum* ethylene-responsive factor-like protein)	Positive regulator of genes in salt tolerance and defence against pathogens. In transgenic tobacco plants, ectopic expression of gene resulted in enhanced tolerance against *Pseudomonas syringae*	Abiotic (salt stress tolerance)	Lee et al. (2004)

combating ROS-induced stress, but the future belongs to the development of a more novel transgenic plants with improved genetic mechanisms to achieve plethora of abiotic stresses (Hasanuzzaman et al. 2012).

References

Abe H, Urao T, Ito T, Seki M, Shinozaki K, Yamaguchi-Shinozaki K (2003) Arabidopsis AtMYC2 (bHLH) and AtMYB2 (MYB) function as transcriptional activators in abscisic acid signaling. Plant Cell 15(1):63–78

Abu Qamar S, Luo HL, Laluk K, Mickelbart MV, Mengiste T (2009) Crosstalk between biotic and abiotic stress responses in tomato is mediated by the AIM1 transcription factor. Plant J 58:347–360

Ahmad P, Sarwat M, Sharma S (2008) Reactive oxygen species, antioxidants and signaling in plants. J Plant Biol 51(3):167–173

Ahmad P, Jaleel CA, Salem MA, Nabi G, Sharma S (2010) Roles of enzymatic and nonenzymatic antioxidants in plants during abiotic stress. Crit Rev Biotechnol 30(3):161–175

Akhtar M, Jaiswal A, Taj G, Jaiswal JP, Qureshi MI, Singh NK (2012) DREB1/CBF transcription factors: their structure, function and role in abiotic stress tolerance in plants. J Genet 91 (3):385–395

Alscher RG, Erturk N, Heatrhj LS (2002) Role of superoxide dismutases (SODs) in controlling oxidative stress in plants. J Exp Bot 53(372):1331–1341

Anderson JP, Badruzsaufari E, Schenk PM, Manners JM, Desmond OJ, Ehlert C, Maclean DJ, Ebert PR, Kazan K (2004) Antagonistic interaction between abscisic acid and jasmonate–ethylene signaling pathways modulates defense gene expression and disease resistance in Arabidopsis. Plant Cell 16:3460–3479

Bai L, Zhang G, Zhou Y, Zhang Z, Wang W, Du Y, Wu Z, Song CP (2009) Plasma membrane-associated proline-rich extensin-likereceptor kinase 4, a novel regulator of Ca signalling, is required for abscisic acid responses in *Arabidopsis thaliana*. Plant J 60:314–327

Cao YF, Song FM, Goodman RM, Zheng Z (2006) Molecular characterization of four rice genes encoding ethylene-responsive transcriptional factors and their expressions in response to biotic and abiotic stress. J Plant Physiol 163:1167–1178

Chinnusamy V, Schumaker K, Zhu JK (2004) Molecular genetic perspectives on cross-talk and specificity in abiotic stress signalling in plants. J Exp Bot 55(395):225–236

Ciftci-Yilmaz S, Morsy MR, Song LH, Coutu A, Krizek BA, Lewis MW, Warren D, Cushman J, Connolly EL, Mittler R (2007) The EAR-motif of the Cys2/His2-type zinc finger protein Zat7 plays a key role in the defense response of Arabidopsis to salinity stress. J Biol Chem 282:9260–9268

Corpas FJ, Barroso JB, del Río LA (2001) Peroxisomes as a source of reactive oxygen species and nitric oxide signal molecules in plant cells. Trends Plant Sci 6(4):145–150

Delessert C, Kazan K, Wilson IW, Van Der Straeten D, Manners J, Dennis ES, Dolferus R (2005) The transcription factor ATAF2 represses the expression of pathogenesis-related genes in *Arabidopsis*. Plant J 43:745–757

del Río LA, Corpas FJ, Sandalio LM, Palma JM, Gómez M, Barroso JB (2002) Reactive oxygen species, antioxidant systems and nitric oxide in peroxisomes. J Exp Bot 53:1255–1272

Ding Z, Li S, An X, Liu X, Qin H, Wang D (2009) Transgenic expression of MYB15 confers enhanced sensitivity to abscisic acid and improved drought tolerance in *Arabidopsis thaliana*. J Genet Genomics 36(1):17–29

Foyer CH, Halliwell B (1976) The presence of glutathione and glutathione reductase in chloroplasts: a proposed role in ascorbic acid metabolism. Planta 133(1):21–25

Foyer CH, Noctor G (2003) Redox sensing and signaling associated with reactive oxygen in chloroplasts, peroxisomes and mitochondria. Physiol Plant 119(3):355–364

Fujita M, Fujita Y, Maruyama K, Seki M, Hiratsu K, Ohme-Takagi M, Tran LSP, Yamaguchi-Shinozaki K, Shinozaki K (2004) A dehydration-induced NAC protein, RD26, is involved in a novel ABA-dependent stress-signaling pathway. Plant J 39:863–876

Fujita M, Mizukado S, Fujita Y, Ichikawa T, Nakazawa M, Seki M, Matsui M, Yamaguch-Shinozaki K, Shinozaki K (2007) Identification of stress-tolerance-related transcription-factor genes via mini-scale full length cDNA over-eXpressor (FOX) gene hunting system. Biochem Biophys Res Commun 364(2):250–257

Gamuyao R, Chin JH, Pariasca-Tanaka J, Pesaresi P, Catausan S, Dalid C, Slamet-Loedin I, Tecson-Mendoza EM, Wissuwa M, Heuer S (2012) The protein kinase Pstol1 from traditional rice confers tolerance of phosphorus deficiency. Nature 488:535–539

Garg N, Manchanda G (2009) ROS generation in plants: boon or bane? Plant Biosyst 143(1):81–96

Gill SS, Tuteja N (2010) Reactive oxygen species and antioxidant machinery in abiotic stress tolerance in crop plants. Plant Physiol Biochem 48(12):909–930

Gilmour SJ, Sebolt AM, Salazar MP, Everard JD, Thomashow MF (2000) Overexpression of Arabidopsis CBF3 transcriptional activator mimics multiple biochemical changes associated with cold acclimation. Plant Physiol 124(4):1854–1865

Giri J, Vij S, Dansanam PK, Tyagi AK (2011) Rice A20/AN1 zincfinger containing stress-associated proteins (SAP1/11) and a receptor like cytoplasmic kinase (OsRLCK253) interact via A20 zinc-finger and confer abiotic stress tolerance in transgenic Arabidopsis plants. New Phytol 191:721–732

Guo ZJ, Chen XJ, Wu XL, Ling JQ, Xu P (2004) Overexpression of the AP2/EREBP transcription factor OPBP1 enhances disease resistance and salt tolerance in tobacco. Plant Mol Biol 55 (4):607–618

Haake V, Cook D, Riechmann JL, Pineda O, Thomashow MF, Zhang JZ (2002) Transcription factor CBF4 is a regulator of drought adaptation in Arabidopsis. Plant Physiol 130(2):639–648

Hare PD, Cress WA (1997) Metabolic implications of stress induced proline accumulation in plants. Plant Growth Regul 21(2):79–102

Hasanuzzaman M, Hossain MA, da Silva JAT, Fujita M (2012) Plant response and tolerance to abiotic oxidative stress: antioxidant defense is a key factor. In: Crop stress and its management: perspectives and strategies. Springer, Dordrecht, pp 261–315

Hua D, Wang C, He J, Liao H, Duan Y, Zhu Z, Guo Y, Chen Z, Gong Z (2012) A plasma membrane receptor kinase, GHR1, mediates abscisic acid- and hydrogen peroxide-regulated stomatal movement in Arabidopsis. Plant Cell 24:2546–2561

Inze D, Van Montagu M (1995) Oxidative stress in plants. Curr Opin Biotechnol 6(2):153–158

Jensen MK, Hagedorn PH, de Torres-Zabala M, Grant MR, Rung JH, Collinge DB, Lyngkjaer MF (2008) Transcriptional regulation by an NAC (NAM-ATAF1,2-CUC2) transcription factor attenuates ABA signalling for efficient basal defence towards Blumeria graminis f. sp hordei in Arabidopsis. Plant J 56:867–880

Jimenez A, Hernandez JA, Pastori G, Luis A, Sevilla F (1998) Role of the ascorbate- glutathione cycle of mitochondria and peroxisomes in the senescence of pea leaves. Plant Physiol 118 (4):1327–1335

Kang JY, Choi HI, Im MY, Kim SY (2002) Arabidopsis basic leucine zipper proteins that mediate stress-responsive abscisic acid signaling. Plant Cell 14:343–357

Kasuga M, Liu Q, Miura S, Yamaguchi-Shinozaki K, Shinozaki K (1999) Improving plant drought, salt, and freezing tolerance by gene transfer of a single stress-inducible transcription factor. Nat Biotechnol 17(3):287–291

Khan MIR, Khan NA (2014) Ethylene reverses photosynthetic inhibition by nickel and zinc in mustard through changes in PS II activity, photosynthetic-nitrogen use efficiency and antioxidant metabolism. Protoplasma 251:1007–1019

Khan MIR, Asgher M, Khan NA (2014) Alleviation of salt-induced photosynthesis and growth inhibition by salicylic acid involves glycinebetaine and ethylene in mungbean (*Vigna radiata* L.) Plant Physiol Biochem 80:67–74

Khan MIR, Nazir F, Asgher M, Per TS, Khan NA (2015) Selenium and sulfur influence ethylene formation and alleviate cadmium-induced oxidative stress by improving proline and glutathione production in wheat. J Plant Physiol 178:9–18

Khan MIR, Iqbal N, Masood A, Mobin M, Anjum NA, Khan NA (2016a) Modulation and significance of nitrogen and sulfur metabolism in cadmium challenged plants. Plant Growth Regul 78:1–11

Khan MIR, Khan NA, Masood A, Per TS, Asgher M (2016b) Hydrogen peroxide alleviates nickel-inhibited photosynthetic responses through increase in use-efficiency of nitrogen and sulfur, and glutathione production in mustard. Front Plant Sci 7:44

Kim JC, Lee SH, Cheong YH, Yoo CM, Lee SI, Chun HJ, Yun D-J, Hong JC, Lee SY, Lim CO, Cho MJ (2001) A novel cold-inducible zinc finger protein from soybean, SCOF-1, enhances cold tolerance in transgenic plants. Plant J 25(3):247–259

Kim S, Kang JY, Cho DI, Park JH, Kim SY (2004) ABF2, an ABRE-binding bZIP factor, is an essential component of glucose signaling and its overexpression affects multiple stress tolerance. Plant J 40(1):75–87

Kobayashi F, Maeta E, Terashima A, Kawaura K, Ogihara Y, Takumi S (2008) Development of abiotic stress tolerance via bZIP-type transcription factor LIP19 in common wheat. J Exp bot 59(4):891–905

Lata C, Prasad M (2011) Role of DREBs in regulation of abiotic stress responses in plants. J Exp Bot 62(14):4731–4748

Lata C, Yadav A, Prasad M (2011) Role of plant transcription factors in abiotic stress tolerance. In: Shanker A (ed) Abiotic stress response in plants – physiological, biochemical and genetic perspectives. ISBN: 978-953-307-672-0, InTech. Available from: http://www.intechopen.com/books/abiotic-stress-response-in-plants-physiological-biochemical-and-geneticperspectives/role-of-plant-transcriptionfactors-in-abiotic-stress-tolerance

Lee JH, Hong JP, Oh SK, Lee S, Choi D, Kim WT (2004) The ethylene-responsive factor like protein 1 (CaERFLP1) of hot pepper (Capsicum annuum L.) interacts in vitro with both GCC and DRE/CRT sequences with different binding affinities: possible biological roles of CaERFLP1 in response to pathogen infection and high salinity conditions in transgenic tobacco plants. Plant Mol Biol 55:61–81

Liao Y, Zou H, Wei W, Hao YJ, Tian AG, Huang J, Liu Y-F, Zhang J-S, Chen S-Y (2008a) Soybean *GmbZIP44*, *GmbZIP62* and *GmbZIP78* genes function as negative regulator of ABA signaling and confer salt and freezing tolerance in transgenic *Arabidopsis*. Planta 228 (2):225–240

Liao Y, Zhang J-S, Chen S-Y, Zhang W-K (2008b) Role of soybean GmbZip132 under abscisic acid and salt stresses. J Integr Plant Biol 50(2):221–230

Liu Q, Kasuga M, Sakuma Y, Abe H, Miura S, Yamaguchi-Shinozaki K, Shinozaki K (1998) Two transcription factors, DREB1 and DREB2, with an EREBP/AP2 DNA binding domain separate two cellular signal transduction pathways in drought-and low-temperature-responsive gene expression, respectively, in Arabidopsis. Plant Cell 10(8):1391–1406

Luis A, Corpas FJ, Sandalio LM, Palma JM, Gómez M, Barroso JB (2002) Reactive oxygen species, antioxidant systems and nitric oxide in peroxisomes. J Exp bot 53(372):1255–1272

Luis A, Sandalio LM, Altomare DA, Zilinskas BA (2003) Mitochondrial and peroxisomal magnese superoxide dismutase: differential expression during leaf sense scene. J Exp Bot 54:923–933

Maeda H, Sakuragi Y, Bryant DA, DellaPenna D (2005) Tocopherols protect Synechocystis sp . strain PCC 6803 from lipid peroxidation. Plant Physiol 138:1422–1435

Martínez JP, Araya H (2010) Ascorbate–glutathione cycle: enzymatic and non-enzymatic integrated mechanisms and its biomolecular regulation. In: Ascorbate-glutathione pathway and stress tolerance in plants. Springer Netherlands, pp 303–322

Mengiste T, Chen X, Salmeron J, Dietrich R (2003) The BOTRYTIS SUSCEPTIBLE1 gene encodes an R2R3MYB transcription factor protein that is required for biotic and abiotic stress responses in Arabidopsis. Plant Cell 15(11):2551–2565

Miao Y, Lv D, Wang P, Wang XC, Chen J, Miao C, Song CP (2006) An Arabidopsis glutathione peroxidase functions as both a redox transducer and a scavenger in abscisic acid and drought stress responses. Plant Cell 18(10):2749–2766

Mittler R (2002) Oxidative stress, antioxidants and stress tolerance. Trends Plant Sci 7(9):405–410

Mukhopadhyay A, Vij S, Tyagi AK (2004) Overexpression of a zinc-finger protein gene from rice confers tolerance to cold, dehydration, and salt stress in transgenic tobacco. Proc Natl Acad Sci U S A 101(16):6309–6314

Nakashima K, Tran LSP, Van Nguyen D, Fujita M, Maruyama K, Todaka D, Ito Y, Hayashi N, Shinozaki K, Yamaguchi-Shinozaki K (2007) Functional analysis of a NAC-type transcription factor OsNAC6 involved in abiotic and biotic stress-responsive gene expression in rice. Plant J 51:617–630

Nishizawa A, Yabuta Y, Yoshida E, Maruta T, Yoshimura K, Shigeoka S (2006) Arabidopsis heat shock transcription factor A2 as a key regulator in response to several types of environmental stress. Plant J 48:535–547

Noctor G, Gomez L, Vanacker H, Foyer CH (2002) Interactions between biosynthesis, compartmentation and transport in the control of glutathione homeostasis and signalling. J Exp Bot 53 (372):1283–1304

Orellana S, Yanez M, Espinoza A, Verdugo I, Gonzalez E, Ruiz-Lara S, Casaretto JA (2010) The transcription factor SlAREB1 confers drought, salt stress tolerance and regulates biotic and abiotic stress-related genes in tomato. Plant Cell Environ 33:2191–2208

Osakabe Y, Maruyama K, Seki M, Satou M, Shinozaki K, Yamaguchi-Shinozaki K (2005) Leucine-rich repeat receptor-like kinase1 is a key membrane-bound regulator of abscisic acid early signaling in Arabidopsis. Plant Cell 17:1105–1119

Osakabe Y, Mizuno S, Tanaka H, Maruyama K, Osakabe K, Todaka D, Fujita Y, Kobayashi M, Shinozaki K, Yamaguchi-Shinozaki K (2010) Overproduction of the membrane-bound receptor-like protein kinase 1, RPK1, enhances abiotic stress tolerance in Arabidopsis. J Biol Chem 285:9190–9201

Ouyang SQ, Liu YF, Liu P, Lei G, He SJ, Ma B, Zhang WK, Zhang JS, Chen SY (2010) Receptor-like kinase OsSIK1 improves drought and salt stress tolerance in rice (Oryza sativa) plants. Plant J 62:316–329

Pang CH, Wang BS (2010) Role of ascorbate peroxidase and glutathione reductase in ascorbate–glutathione cycle and stress tolerance in plants. In: Ascorbate-glutathione pathway and stress tolerance in plants. Springer, Dordrecht, pp 91–113

Park JM, Park CJ, Lee SB, Ham BK, Shin R, Paek KH (2001) Overexpression of the tobacco TSI1 gene encoding an EREBP/AP2- type transcription factor enhances resistance against pathogen attack and osmotic stress in tobacco. Plant Cell 13:1035–1046

Peng XX, Tang XK, Zhou PL, Hu YJ, Deng XB, He Y, Wang HH (2011) Isolation and expression patterns of rice WRKY82 transcription factor gene responsive to both biotic and abiotic stresses. Agric Sci China 10:893–901

Pitorre D, Llauro C, Jobet E, Guilleminot J, Brizard JP, Delseny M, Lasserre E (2010) RLK7, a leucine-rich repeat receptor-like kinase, is required for proper germination speed and tolerance to oxidative stress in Arabidopsis thaliana. Planta 232:1339–1353

Puranik S, Sahu PP, Srivastava PS, Prasad M (2012) NAC proteins: regulation and role in stress tolerance. Trends Plant Sci 17(6):369–381

Qiu YP, Yu DQ (2009) Over-expression of the stress-induced OsWRKY45 enhances disease resistance and drought tolerance in Arabidopsis. Environ Exp Bot 65:35–47

Quan LJ, Zhang B, Shi WW, Li HY (2008) Hydrogen peroxide in plants: a versatile molecule of the reactive oxygen species network. J Integr Plant Biol 50(1):2–18

Raven EL (2000) Peroxidase-catalyzed oxidation of ascorbate. Structural, spectroscopic and mechanistic correlations in ascorbate peroxidase. Subcell Biochem 35:317–349

Rizhsky L, Davletova S, Liang HJ, Mittler R (2004a) The zinc finger protein Zat12 is required for cytosolic ascorbate peroxidase 1 expression during oxidative stress in Arabidopsis. J Biol Chem 279:11736–11743

Rushton PJ, Somssich IE (1998) Transcriptional control of plant genes responsive to pathogens. Curr Opin Plant Biol 1(4):311–315

Saibo NJ, Lourenço T, Oliveira MM (2009) Transcription factors and regulation of photosynthetic and related metabolism under environmental stresses. Ann Bot 103(4):609–623

Schenke D, Boettcher C, Scheel D (2011) Crosstalk between abiotic ultraviolet-B stress and biotic (flg22) stress signalling in Arabidopsis prevents flavonol accumulation in favor of pathogen defence compound production. Plant Cell Environ 34(11):1849–1864

Seo PJ, Park CM (2010) MYB96-mediated abscisic acid signals induce pathogen resistance response by promoting salicylic acid biosynthesis in Arabidopsis. New Phytol 186(2):471–483

Seo PJ, Lee SB, Suh MC, Park MJ, Go YS, Park CM (2011) The MYB96 transcription factor regulates cuticular wax biosynthesis under drought conditions in Arabidopsis. Plant Cell 23:1138–1152

Shinozaki K, Yamaguchi-Shinozaki K (2007) Gene networks involved in drought stress response and tolerance. J Exp bot 58(2):221–227

Tanaka H, Osakabe Y, Katsura S, Mizuno S, Maruyama K, Kusakabe K, Mizoi J, Shinozaki K, Yamaguchi-Shinozaki K (2012) Abiotic stress-inducible receptor-like kinases negatively control ABA signaling in Arabidopsis. Plant J 70:599–613

Tran L-SP, Nakashima K, Sakuma Y, Simpson SD, Fujita Y, Maruyama K, Fujita M, Seki M, Shinozaki K, Yamaguchi-Shinozaki K (2004) Isolation and functional analysis of *Arabidopsis* stress-inducible NAC transcription factors that bind to a drought responsive *cis*-element in the early responsive to dehydration stress promoter. Plant Cell 16(9):2481–2498

Tsutsui T, Kato W, Asada Y, Sako K, Sato T, Sonoda Y, Ichikawa T (2009) DEAR1, a transcriptional repressor of DREB protein that mediates plant defense and freezing stress responses in Arabidopsis. J Plant Res 122(6):633–643

Tuteja N, Ahmad P, Panda BB, Tuteja R (2009) Genotoxic stress in plants: shedding light on DNA damage, repair and DNA repair helicases. Mutat Res 681(2):134–149

Vannini C, Locatelli F, Bracale M, Magnani E, Marsoni M, Osnato M, Mattana M, Baldoni E, Coraggio I (2004) Overexpression of the rice Osmyb4 gene increases chilling and freezing tolerance of *Arabidopsis thaliana* plants. Plant J 37:115–127

Vannini C, Iriti M, Bracale M, Locatelli F, Faoro F, Croce P, Pirona R, Di Maro A, Coraggio I, Genga A (2006) The ectopic expression of the rice Osmyb4 gene in Arabidopsis increases tolerance to abiotic, environmental and biotic stresses. Physiol Mol Plant Pathol 69:26–42

Vannini C, Campa M, Iriti M, Genga A, Faoro F, Carravieri S, Rotino GL, Rossoni M, Spinardi A, Bracale M (2007) Evaluation of transgenic tomato plants ectopically expressing the rice Osmyb4 gene. Plant Sci 173:231–239

Wang H, Hao J, Chen X (2007) Overexpression of rice WRKY89 enhances ultraviolet B tolerance and disease resistance in rice plants. Plant Mol Biol 65(6):799–815

Woodrow P, Pontecorvo G, Ciarmiello LF, Annunziata MG, Fuggi A, Carillo P (2012) Transcription factors and genes in abiotic stress. In: Crop stress and its management: perspectives and strategies. Springer, Dordrecht, pp 317–357

Xia N, Zhang G, Liu XY, Deng L, Cai GL, Zhang Y, Wang XJ, Zhao J, Huang LL, Kang ZS (2010) Characterization of a novel wheat NAC transcription factor gene involved in defense response against stripe rust pathogen infection and abiotic stresses. Mol Biol Rep 37:3703–3712

Xiang Y, Tang N, Du H, Ye H, Xiong L (2008) Characterization of OsbZIP23 as a key player of the basic leucine zipper transcription factor family for conferring abscisic acid sensitivity and salinity and drought tolerance in rice. Plant Physiol 148(4):1938–1952

Yang T, Chaudhuri S, Yang L, Du L, Poovaiah BW (2010a) A calcium/calmodulin-regulated member of the receptor-like kinase family confers cold tolerance in plants. J Biol Chem 285:7119–7126

Yousuf, P. Y., Hakeem, K. U. R., Chandna, R., Ahmad, P., (2012). Role of glutathione reductase in plant abiotic stress. In: Abiotic stress responses in plants. Springer, New York, p 149–158

Zhang GY, Chen M, Li LC, Xu ZS, Chen XP, Guo JM, Ma YZ (2009) Overexpression of the soybean GmERF3 gene, an AP2/ERF type transcription factor for increased tolerances to salt, drought, and diseases in transgenic tobacco. J Exp Bot 60:3781–3796

Zhou QY, Tian AG, Zou HF, Xie ZM, Lei G, Huang J, Wang CM, Wang HW, Zhang JS, Chen SY (2008) Soybean WRKY-type transcription factor genes, GmWRKY13, GmWRKY21, and GmWRKY54, confer differential tolerance to abiotic stresses in transgenic Arabidopsis plants. Plant Biotechnol J 6:486–503

Zou M, Guan Y, Ren H, Zhang F, Chen F (2008) A bZIP transcription factor, OsABI5, is involved in rice fertility and stress tolerance. Plant Mol Biol 66(6):675–683

ROS-Induced Signaling and Gene Expression in Crops Under Salinity Stress

7

Vinay Kumar, Tushar Khare, Mansi Sharma, and Shabir H. Wani

Abstract

Salinity is one of the most serious environmental factors limiting the global productivity and superior quality of the agricultural crops, thus jeopardizing the capability of agronomic production to withstand the ever-increasing human population. Hypersaline conditions show multiple effects on crops such as water stress, ionic imbalance and toxicity, nutrition deficiency, oxidative stress, metabolic alterations, osmotic stress, and genotoxicity, which collectively reduce growth, development, and survival rate of the stressed crops. As a consequence, declined photosynthetic electron transport takes place, leading to the generation of reactive oxygen species (ROS) such as singlet oxygen, superoxide radical, hydrogen peroxide, and hydroxyl radical, albeit various forms and in different cellular compartments. These ROS are considered as key components for oxidative injury to lipid membrane and other vital cellular macromolecules, including nucleic acids, pigments, and proteins, eventually provoking cell death. To counteract the deleterious effects, these stress-induced ROS need to be scavenged by antioxidant machinery activated by the cells. Reports have advocated the ROS influenced expression of number of genes and signal transduction pathways, which implies the cellular strategies to use ROS as stimuli and signals that trigger and regulate numerous stress-responsive genetic

V. Kumar • T. Khare • M. Sharma
Department of Biotechnology, Modern College (Savitribai Phule Pune University), Ganeshkhind, Pune 411016, India

S.H. Wani (✉)
Division of Genetics and Plant Breeding, Faculty of Agriculture WADURA, Sher-e-Kashmir University of Agricultural Sciences and Technology, Srinagar, Jammu and Kashmir 191121, India

Department of Plant, Soil and Microbial Sciences, Michigan State University, East Lansing, MI 48824-1325, USA
e-mail: shabirhussainwani@gmail.com

M.I.R. Khan, N.A. Khan (eds.), *Reactive Oxygen Species and Antioxidant Systems in Plants: Role and Regulation under Abiotic Stress*,
DOI 10.1007/978-981-10-5254-5_7

networks. Therefore, in contrast to their acknowledged role as destructive species, ROS also act as signaling molecules in the regulation of salinity-induced alterations. The present chapter focuses on the role of ROS as signaling molecules and inducers of gene expression under the influence of saline conditions. We discuss herein the process of salinity-induced ROS generation, types of ROS, cellular damage by ROS, role of ROS as signaling molecules, and recent evidences of differential expression of antioxidants under saline conditions.

Keywords
Abiotic stress • Antioxidants • Genes • ROS • Salinity • Tolerance

Abbreviations

$^1O_2$	Singlet oxygen
ABA	Abscisic acid
APX	Ascorbate peroxidase
AsA	Ascorbate
AsA-GSH	Ascorbate-glutathione
AtMPK3	*Arabidopsis thaliana* mitogen-activated kinase 3
AtMPK6	*Arabidopsis thaliana* mitogen-activated kinase 6
AtrbohD	Arabidopsis respiratory burst oxidase homolog D
AtrbohF	Arabidopsis respiratory burst oxidase homolog F
Ca2+	Calcium ions
CaMV 35S	Cauliflower mosaic virus 35S promoter
Car	Carotene
Cars	Carotenoids
CAT	Catalase
CBL	Calcineurin B-like protein
CIPK	CBL-interacting protein kinase
Cu/ZnSOD	Copper/zinc superoxide dismutase
Cys	Cysteine
DHAR	Dehydroascorbate reductase ()
DNA	Deoxyribonucleic acid
ERF6	Ethylene regulatory factor
FeSOD	Iron superoxide dismutases
GB	Glycine betaine
GDH	Glutamate dehydrogenase
GPOX	Guaiacol peroxidase
GPX	Glutathione peroxidase

GR	Glutathione reductase
GSH	Glutathione
GST	Glutathione-S-transferase
H2O2	Hydrogen peroxide
HNE	4-Hydroxy-2-nonenal
K+	Potassium ion
LePHGPx	*Lycopersicon esculentum* phospholipid hydroperoxide glutathione peroxidase
MAPK	Mitogen-activated protein kinase
MAPKK	MAP-kinase kinase
MAPKKK	MAP-kinase kinase-kinase
MDA	Malondialdehyde
Mg2+	Magnesium ion
mM	Millimolar
MnSOD	Manganese superoxide dismutase
mRNA	Messenger ribonucleic acid
NaCl	Sodium chloride
NADPH	Nicotinamide adenine dinucleotide phosphate
NaMnSOD	Natrinema altunense manganese superoxide dismutase
NO	Nitric oxide
NO•	Reactive nitric oxide
O2$^{\cdot-}$	Superoxide radical
OH$^{\cdot}$	Hydroxyl radicals
P5CDH	Delta-1-pyrroline-5-carboxylate dehydrogenase (*Hordeum vulgare*)
P5CR	Pyrroline-5-carboxylate reductase
P5CS	Pyrroline-5-carboxylate synthase
PaSOD	Superoxide dismutase gene from *Potentilla atrosanguinea*
Polyols	Sugar alcohols
POX	Peroxidases
ProDH	Proline dehydrogenase
PSI	Photosystem I
PSII	Photosystem II
RaAPX	APX gene from rheum austral
Rboh	Respiratory burst oxidase homolog
ROS	Reactive oxygen species
SODs	Superoxide dismutases
TBARs	Thiobarbituric acid-reactive substances
TF	Transcription factor

7.1 Introduction

Salinity is one of the most severe threats for crop production. Hypersaline conditions not only limit the global crop productivity but also affect the quality of produce. Salinity exerts threefold effect as reduced plant growth and yield take place through osmotic stress, ion toxicities and imbalances, and oxidative stress (Khare et al. 2015). As a consequence, decreased stomatal conductivity under low water potential, over-reduction of electron transport in cellular organelles, and overall declined photosynthetic electron transport take place, leading to excessive generation of reactive oxygen species (ROS) such as singlet oxygen, superoxide radical, hydrogen peroxide, and hydroxyl radical. These ROS cause serious damages to lipid membranes and other essential macromolecules including proteins and nucleic acids and ultimately induce cell death (Qureshi et al. 2013; Khare et al. 2015). However, notably, they have the abilities to work as signaling molecule and are critical in regulating the responses of development as well as various aspects of stress (Ismail et al. 2014). There are recent reports of ROS-induced aberrant expression of multiple genes and altered signal transduction pathways, which implies the plant cellular strategies to use ROS as stimuli and signals that trigger and regulate numerous stress-responsive genetic networks in stressed plants. However, plants have evolved crucial biochemical strategies to keep a check on excess ROS generation to counter the deleterious effects of abiotic stress including salinity, which principally includes the induction of enzymatic and nonenzymatic antioxidant machinery for ROS detoxification or scavenging (Khan and Khan 2014; Khan et al. 2014, 2015, 2016a, b; Qureshi et al. 2013; Yildiztugay et al. 2014). Owing to their significance, the antioxidant entities have been altered to engineer the salt tolerance in various crops through overexpression of their pathway genes (Kumar et al. 2010; Kumar and Khare 2016; Wani et al. 2016). Through this chapter we are presenting herein a comprehensive and critical assessment and discussion of salinity stress-induced generation of different types of ROS in plant cells and tissues, cellular damages by the ROS, role of ROS as signaling molecules, and recent reports on differential expression of antioxidants under saline conditions, besides ROS biotechnology for improved crop salt tolerance.

7.2 Salinity-Induced ROS Accumulation

ROS includes radical and non-radical oxygen species which are formed due to partial reduction of oxygen in most of the aerobic organisms. The ROS typically explored are superoxide radical, hydroxyl radical, alkoxyl, and peroxyl as radical molecules and hydrogen peroxide, singlet oxygen, ozone, and hypochlorous acid as non-radical. These molecules are produced under normal physiological condition in various metabolic pathways as normal by-product (Halliwell 2006; del Rio 2015). Plants have evolved superior adaptive or survival strategies to counter fluctuating environmental conditions such as salinity. These include systemic and retrograde signaling, programmed cell death, in which ROS play valuable roles as signal

(Baxter et al. 2014; Petrov et al. 2015). All remote and unstressed plant tissues are alert by these systemic signals in the presence of abiotic or biotic stress elements and help in activation of acclimation pathways or resistance in plants against the stress condition (Szechynska-Hebda et al. 2010; Suzuki et al. 2013). Intricate responses of the plants against salinity are highly complicated and include metabolic and genetic changes (Fig. 7.1). The degree of such response depends on the salt tolerance nature of the plant and the treatment and exposure of the stress.

During reduction of the oxygen, the first step involves formation of superoxide radical ($O_2^{\cdot-}$) which is immediately converted to hydrogen peroxide (H_2O_2) by superoxide dismutases (SODs) (Rodrigo-Moreno et al. 2013). This species can be further responsible for formation of highly toxic hydroxyl radicals ($OH^{\cdot}$) or singlet oxygen ($^1O_2$). Cell does not hold the proper mechanism for detoxification of $OH^{\cdot}$ with the help of enzymes; hence either prevention of the $OH^{\cdot}$ formation or detoxification by means of nonenzymatic antioxidants is the strategy employed by the plant cells. However, water has the potential to scavenge $^1O_2$ (Asada 2006). Among various ROS, only H_2O_2 can diffuse across the neighboring cells and adjoining subcellular sections because of higher stability of H_2O_2. Transport of the H_2O_2 is facilitated by special types of aquaporins which are referred as peroxiporins (Bienert et al. 2006, 2007; Sies 2014). Half-life of H_2O_2 is in minutes,

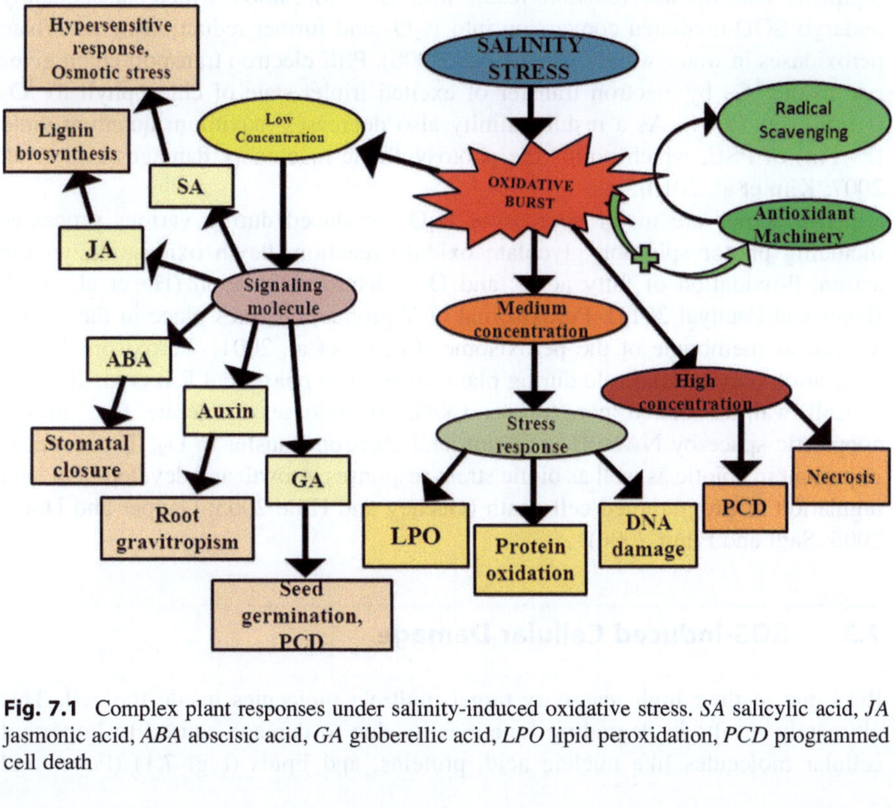

Fig. 7.1 Complex plant responses under salinity-induced oxidative stress. *SA* salicylic acid, *JA* jasmonic acid, *ABA* abscisic acid, *GA* gibberellic acid, *LPO* lipid peroxidation, *PCD* programmed cell death

while the same of the other species ($^1O_2$, $OH^\cdot$ and $O_2^{\cdot-}$) is in microseconds (Bienert et al. 2006; Pitzschke et al. 2006). During the course of salinity, the cellular ROS production is due to some prime ROS producer compartments including chloroplast, mitochondria, and peroxisomes. Additionally cell wall peroxidases, nicotinamide adenine dinucleotide phosphate (NADPH) oxidases, and oxalate oxidases are causative agents for ROS production in the apoplastic space (Mittler et al. 2004; Asada 2006).

Dysfunction in respiratory electron transport chain due to ionic/osmotic imbalance leads to the increase in mitochondrial ROS production with subsequent leakage of electrons to oxygen (Hernandez et al. 2001). Two key locations for ROS production in mitochondria are complexes I and III in mitochondrial electron transport chain, where $O_2^{\cdot-}$ is engendered and promptly catalyzed into H_2O_2 (Sweetlove and Foyer 2004). Salinity-induced ROS generation in mitochondria is reported in pea plant (Marti et al. 2011) and tobacco (Andronis and Roubelakis-Angelakis 2010). The degree of formation of ROS in mitochondria is lesser than that in peroxisomes and chloroplast whenever green tissues in plants are considered. But, in non-green cells, mitochondria act as main ROS source, for example, in root tissue (Rhoads et al. 2006).

In chloroplast, photosynthetic photosystem I and II (PSI and PSII) reaction centers are core sites of ROS production (Asada 2006). In case of PSI, antenna pigments and Mehler reaction result into $O_2^{\cdot-}$ formation which subsequently undergo SOD-mediated conversion into H_2O_2 and further reduction by ascorbate peroxidases in water-water cycle (Asada 2006). PSII electron transport chain gives rise to the $^1O_2$ by electron transfer of excited triplet state of chlorophyll to $^3O_2$ (Laloi et al. 2004). As a result, salinity also decreases maximum quantum yield (Fv/Fm) of PSII, which points the photosynthetic machinery damage (Lim et al. 2007; Kim et al. 2010).

Peroxisomes are major sources of H_2O_2 produced during various processes including photorespiration glycolate oxidase reaction, flavin oxidase enzymatic action, β-oxidation of fatty acids, and $O_2^{\cdot-}$ disproportionation (Hu et al. 2012; Baker and Paudyal 2014). Peroxisomal $O_2^{\cdot-}$ production takes place in the matrix as well as membrane of the peroxisome (Corpas et al. 2001). Peroxisomal ROS generation plays a vital role during plant senescence phase (del Rio et al. 2006).

Cell wall-associated peroxidases (NADPH oxidases) generate $O_2^{\cdot-}$ at the apoplastic space by NADPH oxidation and electron transfer to O_2. This event is important in abiotic as well as biotic stress responses, growth and development, and regulation of programmed cell death (Gechev and Hille 2005; Gapper and Dolan 2006; Sagi and Fluhr 2006).

7.3 ROS-induced Cellular Damage

ROS due to their high reactivity target multiple molecules inside the cell. This ultimately results in degrading functionality due to damage caused to important cellular molecules like nucleic acid, proteins, and lipids (Fig. 7.1) (Foyer and

Noctor 2005a, b). Therefore, ROS serves as a major cause of loss of productivity in crop plant throughout the globe (Mittler 2002; Apel and Hirt 2004; Mahajan and Tuteja 2005; Tuteja 2010; Gill et al. 2011). During photosynthesis ROS are continuously generated in chloroplasts; therefore, the prime site for superoxide radical ($O_2^{\bullet -}$) is primary electron acceptor of photosystem I in thylakoid membrane (Gill and Tuteja 2010). It is found that superoxide radical is the first ROS to be produced from O_2 in cell (Gill and Tuteja 2010), which triggers production of more reactive $OH^{\bullet}$, followed by $^1O_2$ (Halliwal 2006) and finally H_2O_2 by Fenton reaction (Gill and Tuteja 2010; Kumar et al. 2013).

7.3.1 Lipid Peroxidation

Cell membrane serves as one of the targets of ROS damage, which is composed of lipids. The degrading effect of ROS attack on lipids is considered worst among all (Gill and Tuteja 2010). During damage the small hydrocarbon molecules are formed like ketones and malondialdehyde (MDA) from polyunsaturated molecules (Garg and Manchanda 2009). The cellular and organelle membranes both are affected by lipid peroxidation, therefore hampering the cell function and accelerating the oxidative stress by generating lipid-developed radicals (Montillet et al. 2005). Initiation, progression, and termination are the three different steps involved in lipid peroxidation (Gill and Tuteja 2010).

Linoleic acid and linolenic acid are more prone to the damage caused by $^1O_2$ and $HO^{\bullet}$, thus forming a combination of hydroxyperoxides (Gill et al. 2011). Fluidity of the membrane decreases with increase in peroxidation of lipids, thereby making them leaky and secondarily damaging proteins (Moller et al. 2007). Various aldehydes are formed at the end of peroxidation process, which on further breakdown interacts with DNA and proteins. Cytoplasmic male sterility is caused in maize by aldehydes generated in mitochondria because restorer gene is responsible for mitochondrial aldehyde dehydrogenase in maize (Moller et al. 2007). Under salinity stress, the elevated levels of lipid peroxidation were recorded by Khan and Panda (2008) in *Oryza sativa* cultivars with higher magnitude in Begunbitchi cultivar than in Lunishree, which was attributed to activated antioxidant machinery in later. High lipid peroxidation was also observed in roots of *Cicer arietinum* under salt stress (Kukreja et al. 2005). MDA is an indirect measure of lipid peroxidation in cells which is measured using thiobarbituric acid-reactive substances (TBARS) method, where TBA reacts with MDA to give colored products. Pan et al. (2006) accounted the increase in MDA concentration in *Glycyrrhiza uralensis* under salinity and drought stress. Khare et al. (2015) suggested that the lipid peroxidation increases with increase in NaCl concentration in *Oryza sativa* cultivars.

7.3.2 Protein Oxidation

Under ROS attack covalent modifications occur in proteins, termed as protein oxidation. Except some proteins with sulfur-containing amino acids, the damage is irreversible (Ghezzi and Bonetto 2003). Carbonylation of proteins is broadly used to mark their oxidation (Job et al. 2005; Moller et al. 2007). The amino acids like arginine, histidine, lysine, proline, threonine, and tryptophan after oxidation provide free carbonyl groups which restrict their activities as well as make them more sensitive toward proteolytic attack (Moller et al. 2007). Amino acids like cysteine and methionine are more prone to damage by $OH^{\bullet}$ and $^1O_2$ (Gill and Tuteja 2010). H atom from cysteine can be removed by reactive oxygen which results in thiyl radical which interlinks with other thiyl radical-forming disulfide bridges (Hancock et al. 2006); on the other hand, methionine sulfoxide derivatives are formed by addition of oxygen to methionine (Sadanandom et al. 2000). Mitochondria are more prone to damage by ROS than that of chloroplast, and peroxisomes was reported by Bartoli et al. (2004) as protein carbonylation was seen higher in former than in latter. The lipid peroxidation end products like 4-hydroxy-2-nonenal (HNE) can also attack proteins (Miller and Leaver 2000; Taylor et al. 2005).

7.3.3 DNA Damage

Endogenously produced ROS under stress conditions can readily attack plant DNA. Though plant DNA is stable in nature, it cannot protect itself from ROS ($OH^{\bullet}$, $O_2^{\bullet-}$, and $NO^{\bullet}$) attack (Tuteja et al. 2009; Valko et al. 2006). $^1O_2$ was found to attack guanine (Wiseman and Halliwell 1996), whereas all apparatus of DNA like purines, pyrimidines, and deoxyribose backbone gets attacked by $OH^{\bullet}$, making it highly reactive against DNA (Halliwell and Gutteridge 1999). The damage caused by ROS to DNA is of multiple sorts like fragmentation of DNA, base modification by alkylation or oxidation, cross-linking, base removal, and dimer formation (Tuteja et al. 2001; Tuteja and Tuteja 2001; Kumar et al. 2013). As DNA is the primary coding center in cell, its damage triggers the cascade secondary responses. These responses include decreased protein synthesis, cell membrane damage, and destruction of photosynthetic proteins which, in turn, greatly affects growth and development (Britt 1999).

7.4 ROS as Signaling Molecules Under Stress and Optimal Conditions

The cellular ROS production occurs in response to many stimuli; however, the equilibrium between the synthesized and scavenged ROS needs to be maintained, which often gets disturbed by environmental stresses like drought, salinity, pathogen attack, etc. Interestingly, it was estimated that around 2% of O_2 taken up by plants is utilized for ROS production (Bhattacharjee 2005). In a cell, multiple pathways work at a time where some signaling molecules switch on different pathways, whereas others switch them off, as and when required. Therefore, the production of signaling molecule has to be tightly regulated and timed with the desired quantity to perform the function but soon needs to be removed once they serve the purpose in the cell (Neill et al. 2003). Soon, the plants experience the stress. The stress response mechanisms come in the picture with production of ROS, which are in turn highly reactive, and their overaccumulation can target lipids, proteins, and nucleic acids. The H_2O_2 is considered as relatively stable ROS and can pass the membranes through aquaporins which altogether leads to systemic responses (Mubarakshina et al. 2010; Borisova et al. 2012, Sewelam et al. 2016). The efficient antioxidant machinery gets activated when it senses ROS in the cell; these altogether make the ROS an effective signaling molecule. Stress conditions induce cellular imbalance which are converted to redox signaling, therefore in turn activating the specific responses at multiple levels (Scheibe and Dietc 2012). ROS responses generated under stress environments are vital as they regulate metabolism, cell functions, as well signaling network (Noctor 2006; Suzuki et al. 2012). The ROS signaling contributes to the responses like influx of Ca^{2+}, mitogen-activated protein kinase (MAPK) activation, and protein phosphorylation (Benschop et al. 2007) and further triggers long-term responses on growth, development, and survival (Pesaresi et al. 2009; Dubiella et al. 2013). All these responses happen due to ROS burst. The ROS production leading to activation of antioxidant machinery under salt stress was observed in *Physcomitrella patens* as revealed via genomic and proteomic approaches (Wang et al. 2008). Glutamate dehydrogenase (GDH) is an omnipresent enzyme and plays a role in accumulation of ammonia in plant cell under stress condition. It was found that ROS directs the transcription of gene that codes for alpha-subunit of GDH. It was also found that this leads to accumulation of ammonia, therefore synthesizing glutamate for proline synthesis in tobacco (Skopelitis et al. 2006; Rejeb et al. 2014). Baxter et al. (2014) found out that systemic-acquired acclimation of plants to abiotic stress was controlled by temporal-spatial communication of ROS wave with stress-specific hormone or amino acid signals.

Due to its uncharged nature, H_2O_2 can easily be diffused through membranes; also, as it is stable yet reactive makes it an appropriate candidate for signaling molecule (Sewelam et al. 2016). Chemoselective oxidation of Cys residue of a signaling molecule (glutathione, thio-redoxins) by H_2O_2 arbitrates intracellular signal transduction (Paulsen and Carroll 2010). It was found that the transcription factors in some way get activated by ROS, which is intervened by MAPK cascades (Asai et al. 2000). The responses which H_2O_2 deliver are site specific, for example, when generated in chloroplast, they are involved in activation of transcription factor and biosynthetic genes which lead to production of secondary messengers, but when produced in peroxisomes, they initiate transcripts ultimately involved in protein repair responses (Sewelam et al. 2014).

When the cellular ROS homeostasis gets disturbed, the specific roles are delivered by the ROS as signaling molecules, thus resulting in appropriate responses by the cell (Foyer and Noctor 2005a). Among many macromolecules, proteins are major ROS targets, which after oxidation produce peptides, which in turn work as secondary messengers ultimately regulating the ROS signaling during different stresses including biotic and abiotic (Moller and Sweetlove 2010). Chung et al. (2008) reported that controlling of SOS pathway is critically controlled by ROS signaling from NADPH oxidase.

Specific feedback and feed-forward responses are produced by ROS through working up and down the line of secondary messengers which is a precise mesh of interlinked signals which ultimately results in different developmental and environmental signals (Sewelarn et al. 2016). The deep insights about interlinking large-scale signal transduction and expression of many genes are yet not very clear.

In stress conditions, the susceptible targets in plant cell are thylakoid membranes and redox complexes of photosynthesis assembly (especially PSII). During the stress environment, the excess photosynthetic light leads to ROS generation (Glatz et al. 1999). This synthesized ROS can lead to membrane damage by attacking lipids, which in turn triggers feedback signal for activation of stress genes by sensing the condition of membrane (Glatz et al. 1999).

Further, it was found that many membrane-bound receptor protein genes and signaling elements get activated by H_2O_2 from chloroplast (Sewelam et al. 2014). The oxidative stress produced in response to environmental pressure results in synthesis of lipid hydroxyperoxides from polyunsaturated fatty acids due to ROS attack, resulting in declined membrane fluidity (Moller et al. 2007). This further activates the downstream signaling intermediates (Sewelam et al. 2016). Respiratory burst oxidase homologs (Rboh) are plant NADPH oxidases which catalyze the generation of superoxide radicals (Torres and Dangal 2005; Sagi and Fluhr 2006). The concomitant increase in levels of abscisic acid (ABA) and ROS was observed during multiple stresses, which further imparts cross tolerance to various stresses (Verslues and Zhu 2004). The reports suggest that the ROS generated work downstream of ABA resulting in stomatal closing under stress conditions (Verslues and Zhu 2004). ABA is an abiotic stress interceding plant hormone found to synthesize AtrbohD and AtrbohF mediated reactive oxygen species under salinity stress, these ROS work as signaling molecules to regulate NA^+/K^+ homeostasis (Ma et al. 2012). *ERF6*, an ethylene regulatory factor, is a ROS regulator transcription factor in *Arabidopsis thaliana*, whose expression was regulated by *RbohD* (Sewelam et al. 2013). The calcium ions (Ca^{2+}) work as secondary messengers in a very effective way due to their peculiar features like complimentary ionic radius and hydration status in all life forms (Stael et al. 2012). The biotic and abiotic stresses induce ROS generation inside cells, resulting in increases of Ca^{2+} levels inside cells, which in turn works as secondary messenger (McAinsh and Pittman 2009; Dodd et al. 2010). Interlinking between calcineurin B-like protein (CBL)-CBL-interacting protein kinase (CIPK)-arbitrated Ca^{2+} and ROS signaling is found for synergistic activation of NADPH oxidase RbohF by direct Ca^{2+} binding to EF hands (Drerup et al. 2013).

ROS and NO (nitric oxide) are generated concurrently in plants under various stress conditions, as well as they cooperate with each other to trigger defense responses inside the cell (Neill et al. 2002; Molassiotis and Fotopoulos 2011). A proteomic study conducted by Tanou et al. (2009, 2010) in citrus plants under salinity stress revealed that when pretreated with H_2O_2 or NO, overlap between their respective signaling pathways was observed. MAPKs are enzymes which have been conserved in plants. The three interlinked proteins are involved in MAPK-mediated signaling: MAP-kinase kinase-kinase (MAPKKK), MAP-kinase kinase (MAPKK), and MAP kinase (MAPK) (Rodriguez et al. 2010; Sinha et al. 2011). Buildup of H_2O_2 inside cell triggers activation MAPKs which further induces H_2O_2-induced oxidative burst (Pitzschke et al. 2009; Petrov and Van Breusegem 2012). Kovtun et al. (2000) found that stress-induced H_2O_2 activates MAPKKK, ANP1 which activates downstream kinases, AtMPK3, and AtMPK6 using *A. thaliana* protoplast.

Signaling pathways to work efficiently need transient coordination between different phytohormones. It was observed that salicylic acid, jasmonic acid, and ethylene work along with ROS and play a role in defense responses (Mur et al. 2006; Mhamdi et al. 2010). Salicylic acid production is increased due to ROS synthesized out of the cell, and this results in stomatal closure in *A. thaliana* (Khokon et al. 2011). Jasmonic acid and methyl jasmonate play an important role in plant development, and MYC2 is a transcription factor (TF) engaged in jasmonic acid-mediated response. The ROS generated by *RbohD* and *RbohF* enzymes are crucial for genes controlled by MYC2 (Maruta et al. 2011). Lignin biosynthesis during cell damage is also controlled by interaction of jasmonic acid and reactive oxygen species, where lignin gets accumulated due to feedback inhibition by jasmonic acid isoleucine and ROS (Denness et al. 2011). The ROS-related microarray study conducted by Gadjev et al. (2006) revealed that out of 1500 TFs in *A. thaliana*, 500 were either activated or suppressed due to different ROS including H_2O_2 and $OH^{\bullet}$. ROS are found to be crucial in controlling the TFs, the most potent being H_2O_2. They do so at multiple levels like TF uprolling and elevation in stability of mRNA and translation; reducing binding with ubiquitin E3 ligase complex or blocking it, which degrades the proteins; camouflaging the nuclear localization signal hence controlling nucleocytoplasmic traffic; DNA binding and nuclear transactivation which involves affinity of TFs toward DNA, and aiming definite areas of chromatin to induce expression of certain genes (Marinho et al. 2014).

7.5 ROS Scavenging by Plants: Differential Gene Expression

Plant response to saline environment is amalgamation of combating mechanisms rather than a separate scavenger for single damaging entity. The developed strategies include selective accumulation, exclusion, and/or compartmentalization of ions at cell as well as whole plant level and stimulation of enzymatic and nonenzymatic antioxidative system for ROS scavenging (Qureshi et al. 2013; Yildiztugay et al. 2014). Plants hold a range of enzymes as their antioxidants to cope the injurious effects of ROS. All the enzymes work in coordination to scavenge different ROS. These enzymes can be broadly classified in two classes:

(1) enzymes that directly react with ROS and lower their level (SOD, CAT, and POX) and (2) enzymes that generate oxidized forms of antioxidants – ascorbate peroxidase and glutathione reductase (APX, GR) (Kumar et al. 2009; Khare et al. 2015).

Among all the antioxidative enzymes, SOD, a class of metalloenzymes, has an inimitable role as $O_2^{\cdot -}$ scavenger to convert it to H_2O_2. As at the cellular level, $O_2^{\cdot -}$ is the main ROS produced as a product of respiration and photosynthesis process, and the SOD activity is the first step in scavenging of $O_2^{\cdot -}$. Hence SODs are referred as first line of defense against stress in plants (Alscher et al. 2002). The correlation between the SOD activity and salinity stress tolerance shows both positive and negative connections in case of glycophytes (Maksimovic et al. 2013), whereas halophytes show the tremendous capacity to utilize SOD for protection against salinity-induced cellular damage (Ozgur et al. 2013). Depending on required metal cofactors, SODs can be FeSOD (present in chloroplast), MnSOD (present in mitochondria and peroxisome), or Cu/ZnSOD (present in cytosol and chloroplast). Apart from these different types, SODs also occur in different isoforms (Alscher et al. 2002; Jithesh et al. 2006). Sometimes the synthesis of a particular isoform can be dependent on the type, degree, and duration of the stress as well as the developmental stage of the plant (Ozgur et al. 2013).

Rapid generation of H_2O_2 is observed due to higher SOD detoxification activity. The generated H_2O_2 further results into more oxidative damage, as it shows more permeability across the cell membrane (Mittler et al. 2004). Hence, another set of scavenger enzymes is generated to detoxify the H_2O_2, which includes catalases, ascorbate peroxidases, glutathione peroxidase, and peroxidases (Gill and Tuteja 2010). Catalase (CAT) is a heme-containing enzyme, which is also an important scavenger of H_2O_2 (Mittler 2002; Foyer and Noctor 2005b). In plants, catalases are primarily present in peroxisomes with a principal role in cellular redox homeostasis (Corpas et al. 2001). Entry of two H_2O_2 molecules is necessary in the catalase active site; hence, catalases show lesser affinity toward the H_2O_2 than the peroxidases (Willekens et al. 1997). Catalases act as direct degraders of H_2O_2 without contribution of an electron donor which lead to the formation of water and oxygen (Ahmad et al. 2010).

Another class of enzyme responsible for H_2O_2 degradation is peroxidases (POX) which reduced H_2O_2 to water by simultaneous oxidation of series of substrates. POX shows many isoforms in a variety of cellular functions including growth and development, differentiation, lignification, senescence, and auxin catabolism with utilization of a range of phenolic substrates (Passardi et al. 2004). Ascorbate peroxidase (APX) is considered as a class I peroxidase enzyme which uses ascorbate (AsA) as an electron donor as H_2O_2 scavenger. APX is supported by the efficient ascorbate-glutathione cycle for AsA regeneration and helps to decrease toxic levels of H_2O_2. Guaiacol peroxidase (GPOX) is located in the cytosol, the vacuole, and the cell wall, which consumes H_2O_2 to produce phenoxy compounds. These compounds further polymerize and form some cell wall components such as lignins (Mishra et al. 2006). Glutathione peroxidase (GPX) is another type of peroxidase enzyme which scavenges the H_2O_2 by utilization of GSH as electron donor (Ursini et al. 1997). Collectively APX and GPX play key roles as enzymatic regulators in the ascorbate-glutathione cycle. Another enzyme from the same cycle is GR which helps in generation of GSH from oxidized glutathione to maintain the homeostasis cycle (Noctor and Foyer 1998). GR mainly catalyze the rate-limiting step of the ascorbate-glutathione cycle and is present in chloroplast, mitochondria, peroxisomes, and cytosol (Yannarelli et al. 2007).

All these enzymatic scavengers work collectively to combat the saline toxicities for better physiological performance and growth. Though the expression levels of all these enzymatic antioxidants indicate better cure for the damaging ROS, their expression levels differ in accordance with the extent of the salinity, time of the exposure, and the developmental stage of the plant. Many of the reports indicate differential behavior of these enzymes. For example, Yildiztugay and coworkers (2014) studied the enzymatic profile of *Apera intermedia* by exposing the plants at different salinity levels (0, 150, 300, 600 mM). The SOD and APX activities were increased at 150 mM NaCl treatment, but catalase (CAT) and peroxidase (POX) showed reduced activity in concentration in a time-dependent manner. The higher salinity concentrations (300 and 600 mM) were able to induce only APOX and GR which were unable to scavenge the overproduced H_2O_2. Similarly, the expression level of these enzymes was examined on two different *Plantago* species, namely, *Plantago maritima* (salt tolerant) and *P. media* (salt sensitive) by Turkan et al. (2013) under combined stress of salinity (200 mM) and waterlogging. They observed that the POX, APX, CAT, and GR activities in salt-tolerant species were high, whereas only POX and CAT in salt-sensitive species showed increased activities just by salinity treatment. However, in combined treatment, POX, APX, and GR levels showed increment without any effect in CAT level, in both the species. On the similar origin, same species were examined by Sekman et al. (2007) under the different levels of salinity (0, 100, and 200 mM of NaCl for 7 days). *P. media* showed reduced levels of SOD, CAT, and GR with increase in level of the salinity. But APX showed increment in the activity at 100 mM of salinity but decrease in the same at 200 mM salinity. On the other hand, activities of CAT, POX, and GR in *P. maritima* amplified under 200 mM NaCl, whereas the activities of the same were unchanged at 100 mM NaCl. SOD activity showed gradual increase with increase in salinity level. Changes in the antioxidant enzymes were also studied in *Glycine max* by Weisany et al. (2012), where the plantlets were treated with 0, 33, 66, and 99 mM of NaCl. CAT, POX, and APX showed a marked increase in the activity under all the mentioned NaCl concentrations. Qureshi et al. (2013) have demonstrated the long-term salinity effects on the cellular antioxidants in Sweet Annie (*Artemisia annua*) by treating the plants with different levels of salinity (0–160 mM). Salinity-induced increase was observed in SOD, APX, GR, and CAT levels. The differential expression of these enzymes was also studied (Uzilday et al. 2014) in *Eutrema parvulum*, an extreme halophyte and a very close relative of *A. thaliana*. The plants were grown for 30 days and then treated with 50, 200, and 300 mM NaCl. Compared to the control plants (without salt application), SOD, APX, and GR activities showed increase in the activities, in whole leaf level as well as in the isolated chloroplasts. Only the POX activity showed contrasting results where chlorophyll showed reduced or unchanged activity of POX, whereas whole-cell preparation showed increment in the same activity. All the above discussed reports indicated that these enzymes show dissimilar activities in different plants and growth conditions.

Among the different ROS discussed previously, $OH^{\cdot}$ and $^1O_2$ cannot scavenge by enzymatic antioxidants, and detoxification of these radicals is solely reliant on nonenzymatic low-molecular-weight antioxidants. This group involves reduced forms of AsA and glutathione (GSH), proline, betaine, carotenoids (Cars), sugar alcohols (polyols), and α-tocopherols (Krasuska and Gniazdowska 2012; Demidchik 2015). AsA is one of the most vital entities as it directly scavenges the H_2O_2 and acts as electron donor of APX (Foyer and Noctor 2005). Similarly, GSH is also responsible for excess H_2O_2 and works as an important partner molecule of AsA in the AsA-GSH cycle. Lipid peroxidation of the membrane is prevented by α-tocopherols by scavenging lipid peroxyl radicals. They also scavenge the $^1O_2$ in chloroplast. Regeneration of tocopherols is also enhanced by presence of AsA which boost the removal of lipid peroxyls to lessen the lipid peroxidation (Munne-Bosch 2005). Carotenoids are also low-molecular-weight compounds playing a central role in scavenging of $^1O_2$ in chloroplast and in prevention of $^1O_2$ and peroxyl-induced cellular damage. Proline is considered as a crucial component in cellular osmotic adjustments required in withstanding the abiotic stress environment. Elevated accumulation of free proline is frequently observed under abiotic stress conditions such as salinity (Deinlein et al. 2014). Proline and glycine betaine (GB) are two major osmo-protectants with affirmative effect on enzyme and cell membrane integrity required for osmotic adjustments. All these compounds ordinarily get accumulated in different plant tissues to combat the water-deficit environment formed due to disturbed osmobalance during the course of salinity (Ashraf and Foolad 2007).

Chenopodium album plants were treated with different salt concentrations (50 and 300 mM) to check the response of different antioxidants by Yao et al. (2010). When plants were treated with low salinity level (50 mM), there were no noticeable changes in AsA concentration, whereas carotene (Car) showed slightly high concentration. But contrastingly, at 300 mM of salt concentration, amount of AsA marked an increase, and Car accumulation was slightly low when compared to the control plants. Proline accumulation and expression of key enzymes involved in the proline biosynthesis (P5CS, P5CR, ProDH, and P5CDH) are always an important exploration in context of salinity tolerance and sustenance of the plant (Kumar et al. 2015). Celik and Unsal (2013) have studied Pro metabolism in *Glycine max* salt-tolerant mutants under the 90 mM of salinity treatment. Under the course of salinity, all the mutant lines showed high proline accumulation. Also, the quantitative reverse transcriptase PCR showed the augmented expression pattern of all the Pro biosynthesis genes. Increased proline accumulation in roots of *Oryza sativa* was also observed by Hien et al. (2003). In Sweet Annie, subjected to the salt treatment (Qureshi et al. 2013), total AsA content reduced under all the considered salinity levels (40, 80, and 160 mM), whereas total GSH content increased for the same treatment. Similarly, proline content showed the drastic increment at 40 mM salinity level, but for the remaining two treatments, the gradual decrease in proline was observed. Though there was gradual decrease, all the three salinity levels showed higher proline accumulation than the control plants. In the study involving the effect of long-term salinity on two wheat cultivars with contrasting salinity tolerance levels, Sairam et al. (2005) demonstrated that the extent of the accumulation of compatible solutes depends on the salt tolerance nature of the cultivar. Under the selected salinity level (1.7, 5.4, and 10.6 ds/m), both cultivars (Kharchia 65, salt tolerant; and KRL 19, moderately salt tolerant) showed decreased Car levels and increased proline and glycine-betaine levels. The only difference was the extent of increase in proline, and GB content was higher in the tolerant cultivar, whereas the extent of decrease was higher in the moderately tolerant cultivar. To conclude, the nonenzymatic antioxidants work very positively to decline the salinity-induced cellular damage. Hence their exogenous application is also one of the ways to improve the performance and yield of the plants under the saline environment.

7.6 ROS Biotechnology for Crop Improvement Against Salt Stress

The last couple of decades has provided the crop physiologists with insightful understanding of different roles played by ROS in context of their damaging effects as well as their pivotal role in cellular signaling in plants (Mylona and Polidoros 2010). Excess production of ROS leads to the toxic effects on cellular metabolism which in turn compromises the plant sustenance and eventually the yield, whereas low concentration of ROS acts as redox signals to maintain the many signal transduction pathways (Foyer and Noctor 2005). Biotechnological advancements in the field of gene discovery and functional genomics have discovered many tools and lead to the identification of many possible gene targets, which could confer adaptation and improved productivity in hostile environments (Kumar et al. 2012). Genes encoding either ROS-scavenging enzymatic antioxidants or enzymes which control the cellular antioxidant ability from several sources have been proven effective (Zhang et al. 2013). The primary goal of these approaches is betterment of crop species leading to minimization of massive loss in productivity. Such transgenic experimentation also helps in understanding the role of each and every scavenger imparting the tolerance to various abiotic factors including salinity. It also helps in understanding the coordinated mechanism between enzymatic and nonenzymatic antioxidant interactions in the complex signaling network. Hence as the total information about such coordinated network is available, it has been reported that rather than producing a transgenic line with a single transgene, it is beneficial to generate a transgenic line with co-expression of several antioxidant defense genes (Diaz-Vivancos et al. 2013; Xu et al. 2014). Hence the following are some reports discussed about transgenic lines generated with improved salt tolerance.

Overexpression of Cu/ZnSOD gene from *Kandelia candel* in tobacco has showed reduction in ROS formation specifically in plastids and improved salinity tolerance (Jing et al. 2015). Shafi et al. (2015) have demonstrated the co-expression of SOD gene from *Potentilla atrosanguinea* (*PaSOD*) and APX gene from *Rheum austral* (*RaAPX*) in transgenic *Arabidopsis* line. This resulted into enhanced lignin deposition and biomass production (yield) under salinity stress. Overexpression of yeast and pea mitochondrial *MnSOD* in plastids of rice conferred salt and oxidative stress tolerance (Wang et al. 2005). Also, manganese superoxide dismutase gene from halophilic archaeon, *Natrinema altunense* (*NaMnSOD*), is introduced into Nipponbare rice via *Agrobacterium*-mediated transformation. The resultant transgenic line showed increment in total SOD and CAT activity and enhanced elimination of the ROS under salt stress (Chen et al. 2013). Transgenic lines of plum (*Prunus domestica* cv. Claudia Verde) have risen via *Agrobacterium*-mediated transformation under the control of *CaMV35S* promoter with cytosolic Cu/ZnSOD from *Spinacia oleracea* and cytosolic APX cDNA from *Pisum sativum* as transgenes. Transgenic plants showed higher accumulation of the AsA and GSH and lower H_2O_2 accumulation with enhanced enzymatic and nonenzymatic antioxidative activities to confer the salinity tolerance (Diaz-Vivancos et al. 2013). Overexpression of *OsAPXa* or *OsAPXb* in *Arabidopsis* enhanced salinity tolerance to dissimilar levels. Overexpression of *OsAPXb* improved and sustained APX expression to greater extent than *OsAPXa* (Lu et al. 2007). APX gene from *Puccinellia tenuiflora* (*PutAPX*) was overexpressed in transgenic *Arabidopsis* by Guan et al. (2015). This showed the reduction in lipid peroxidation levels and higher chlorophyll content in transgenic seedlings with significantly enhanced salinity tolerance. Peroxisomal APX from *Salicornia brachiate* was overexpressed in which it conferred not only the salinity tolerance *Arachis hypogaea* but also the improved vegetative growth and germination rate in transgenic tobacco lines (Singh et al. 2014a, b). Wheat GPXs when expressed in *Arabidopsis* plastids showed greater growth rate and survival rate during salt stress. Transgenic lines also showed augmented peroxide scavenging capacity and enhanced tolerance to H_2O_2 (Zhai et al. 2013). Transient overexpression of the *Lycopersicon esculentum* phospholipid like GPX (*LePHGPx*) in tobacco leaves showed suppressed apoptotic feature during severe salt and high-temperature stresses to impart the better tolerance (Chen et al. 2004). Homoplastic chloroplast transformants of tobacco were generated by Le Martret et al. (2011), with either glutathione-S-transferase (GST) or dehydroascorbate reductase (DHAR) or combination of GST-GR/DHAR-GR. Transgenic progeny showed the better maintained redox state with respect to AsA and GSH content and improved tolerance to various abiotic stresses including salt, cold, and heavy metals. Overexpression of *Zea mays* CAT and SOD in plastids of *Brassica campestris* not only improved tolerance to salinity and sulfur dioxide with enhanced endogenous K^+, Mg^{2+}, and Ca^{2+}, which help to maintain ionic balance (Tseng et al. 2007). Tobacco plants that concurrently express *Cu/ZnSOD*, APX, and DHAR in their chloroplasts showed improved tolerance to salinity and oxidative damage as compared to the transgenic lines with single- or double-gene transfer (Lee et al. 2007). All the abovementioned cases clearly specify that overexpression of diverse isoforms of different genes amended the detoxification of ROS and imparted superior salinity tolerance in transgenic host system. Table 7.1

summarizes the successful attempts made to enhance the salt tolerance of several transgenic plants via overexpressing the ROS-detoxifying antioxidant genes.

Table 7.1 Summary of transgenic plants produced via over-expressing the ROS-detoxifying antioxidant genes and the improved phenotypic characters of resultant transgenics

Gene	Source organism	Transgenic plant	Improved characters	References
Cu/ ZnSOD	*Kandelia candel*	*Nicotiana tabacum*	Reduced ROS formation in plastids, improved salt tolerance	Jing et al. (2015)
PaSOD, *RaAPX*	*Potentilla atrosanguinea*, *Rheum austral*	*Arabidopsis thaliana*	Enhanced lignin deposition and yield under salinity stress	Shafi et al. (2015)
MnSOD	*Saccharomyces cerevisiae*, *Pisum sativum*	*Oryza sativa*	Enhanced salt and oxidative stress	Wang et al. (2005)
NaMnSOD	*Natrinema altunense*	*Oryza sativa*	Enhanced total SOD and CAT activity and ROS elimination under salt stress	Chen et al. (2013)
Cu/ ZnSOD, *APX* cDNA	*Spinacia oleracea*, *Pisum sativum*	*Prunus domestica*	Higher accumulation of the AsA and GSH and lower H_2O_2 accumulation, enhanced salinity tolerance	Diaz-Vivancos et al. (2013)
OsAPXa or *OsAPXb*	*Oryza sativa*	*Arabidopsis thaliana*	Enhanced salinity tolerance	Lu et al. (2007)
PutAPX	*Puccinellia tenuiflora*	*Arabidopsis thaliana*	Reduced lipid peroxidation, increased chlorophyll content, and enhanced salinity tolerance	Guan et al. (2015)
APX	*Salicornia brachiate*	*Arachis hypogaea*, *Nicotiana tabacum*	Enhanced vegetative growth and germination rate, improved salinity tolerance	Singh et al. (2014a, b)
GPX	*Triticum aestivum*	*Arabidopsis thaliana*	Enhanced growth rate and survival rate, increased peroxide scavenging capacity, enhanced tolerance to H_2O_2	Zhai et al. (2013)
LePHGPx	*Lycopersicon esculentum*	*Nicotiana tabacum*	Suppressed apoptotic during severe salt and high-temperature stresses imparting tolerance	Chen et al. (2004)
GST or *DHAR* or *GST-GR*/ *DHAR-GR*	*Oryza sativa*, *E. coli*	*Nicotiana tabacum*	Better maintained redox status, improved tolerance to various abiotic stresses including salt, cold, heavy metals	Le Martret et al. (2011)
CAT, *SOD*	*Zea mays*	*Brassica campestris*	Enhanced salinity tolerance, improved endogenous K^+, Mg^{2+}, and Ca^{2+} levels to maintain ionic balance	Tseng et al. (2007)

7.7 Conclusion and Future Perspective

A substantial amount of research has revealed that ROS generation in plant cells and tissues is an unfortunate consequence of all the biotic and abiotic stress conditions and these ROS have multiple negative implications on plant growth and development. Though there is a recent advocacy for their significance as signaling molecules, critical for cellular functions and stress responses, however, greater insights are yet to be revealed in these areas and need further efforts from the scientific community. Salinity stress-induced ROS triggers signaling cascades and alters the expression of several genes in plant cells particularly the ones related with antioxidants. Several attempts have been made in the recent past to develop salt-tolerant plants via altering the expression levels of major antioxidant pathway genes. Therefore, ROS biotechnology presents a solid platform for developing salt-tolerant crops.

Acknowledgment VK gratefully acknowledges the financial support from the Science and Engineering Research Board, Government of India, in the form of a Young Scientist Project [SERB-OYS; grant number SR/FT/LS-93/2011]. Authors acknowledge the support from the Department of Science and Technology, Government of India (DST-FIST Program).

References

Ahmad P, Jaleel CA, Salem MA, Nabi G, Sharma S (2010) Roles of enzymatic and nonenzymatic antioxidants in plants during abiotic stress. Crit Rev Biotechnol 30:161–175

Alscher RG, Erturk N, Heath LS (2002) Role of superoxide dismutases (SODs) in controlling oxidative stress in plants. J Exp Bot 53:1331–1341

Andronis EA, Roubelakis-Angelakis KA (2010) Short-term salinity stress in tobacco leads to the onset of animal-like PCD hallmarks in planta in contrast to long-term stress. Planta 23:437–448

Apel K, Hirt H (2004) Reactive oxygen species: metabolism, oxidative stress, and signal transduction. Annu Rev Plant Biol 55:373–399

Asada K (2006) Production and scavenging of reactive oxygen species in chloroplasts and their functions. Plant Physiol 141:391–396

Asai T, Stone JM, Heard JE, Kovtun Y, Yorgey P, Sheen J, Ausubel FM (2000) Fumonisin B1-induced cell death in Arabidopsis protoplasts requires jasmonate-, ethylene-, and salicylate-dependent signalling pathways. Plant Cell 12:1823–1836. doi:10.2307/3871195

Ashraf M, Foolad MR (2007) Roles of glycine betaine and proline in improving plant abiotic stress resistance. Environ Exp Bot 59:206–216

Baker A, Paudyal R (2014) The life of the peroxisome: from birth to death. Curr Opin Plant Biol 22:39–47

Bartoli CG, Gómez F, Martinez DE, Guiamet JJ (2004) Mitochondria are the main target for oxidative damage in leaves of wheat (*Triticum aestivum* L.) J Exp Bot 55:1663–1669

Baxter A, Mittler R, Suzuki N (2014) ROS as key players in plant stress signalling. J Exp Bot 65 (5):1229–1240

Benschop JJ, Mohammed S, O'Flaherty M, Heck AJ, Slijper M, Menke FL (2007) Quantitative phosphoproteomics of early elicitor signalling in Arabidopsis. Mol Cell Proteomics 6:1198–1214

Bhattachrjee S (2005) Reactive oxygen species and oxidative burst: roles in stress, senescence and signal transduction in plant. Curr Sci 89:1113–1121

Bienert GP, Schjoerring JK, Jahn TP (2006) Membrane transport of hydrogen peroxide. Biochim Biophys Acta 1758:994–1003

Bienert GP, Moller AL, Kristiansen KA, Schulz A, Moller IM, Schjoerring JK, Jahn TP (2007) Specific aquaporins facilitate the diffusion of hydrogen peroxide across membranes. J Biol Chem 282:1183–1192. doi:10.1074/jbc.M603761200

Borisova MM, Kozuleva MA, Rudenko NN, Naydov IA, Klenina IB, Ivanov BN (2012) Photosynthetic electron flow to oxygen and diffusion of hydrogen peroxide through the chloroplast envelope via aquaporins. Biochem Biophys Acta 1817:1314–1321. doi:10.1016/j.bbabio.2012.02.036

Britt AB (1999) Molecular genetics of DNA repair in higher plants. Trends Plant Sci 4:20–25

Çelik Ö, Ünsal SG (2013) Expression analysis of proline metabolism-related genes in salt-tolerant soybean mutant plants. Plant Omics 6:364

Chen S, Vaghchhipawala Z, Li W, Asard H, Dickman MB (2004) Tomato phospholipid hydroperoxide glutathione peroxidase inhibits cell death induced by Bax and oxidative stresses in yeast and plants. Plant Physiol 135:1630–1641

Chen Z, Pan YH, An LY, Yang WJ, Xu LG, Zhu C (2013) Heterologous expression of a halophilic archaeon manganese superoxide dismutase enhances salt tolerance in transgenic rice. Russ J Plant Physiol 60:359–366

Chung JS, Zhu JK, Bressan RA, Hasegawa PM, Shi H (2008) Reactive oxygen species mediate Naþ-induced SOS1 mRNA stability in Arabidopsis. Plant J 53:554–565

Corpas FJ, Barroso JB, del Río LA (2001) Peroxisomes as a source of reactive oxygen species and nitric oxide signal molecules in plant cells. Trends Plant Sci 6:145–150

Deinlein U, Stephan AB, Horie T, Luo W, Xu G, Schroeder JL (2014) Plant salt-tolerance mechanisms. Trends Plant Sci 19:371–379

del Rio LA (2015) ROS and RNS in plant physiology: an overview. J Exp Bot 66:2827–2837

del Rio LA, Sandalio LM, Corpas FJ, Palma JM, Barroso JB (2006) Reactive oxygen species and reactive nitrogen species in peroxisomes. Production, scavenging, and role in cell signaling. Plant Physiol 141:330–335

Demidchik V (2015) Mechanisms of oxidative stress in plants: from classical chemistry to cell biology. Environ Exp Bot 109:212–228

Denness L, McKenna JF, Segonzac C, Wormit A, Madhou P, Bennett M, Mansfield J, Zipfel C, Hamann T (2011) Cell wall damage-induced lignin biosynthesis is regulated by a reactive oxygen species-and jasmonic acid-dependent process in Arabidopsis. Plant Physiol 156:1364–1374. doi:10.1104/pp.111.175737

Diaz-Vivancos P, Faize M, Barba-Espin G, Faize L, Petri C, Hernández JA, Burgos L (2013) Ectopic expression of cytosolic superoxide dismutase and ascorbate peroxidase leads to salt stress tolerance in transgenic plums. Plant Biotechnol J 11:976–985

Dodd AN, Kudla J, Sanders D (2010) The language of calcium signalling. Annu Rev Plant Biol 61:593–620. doi:10.1146/annurev-arplant-070109- 104628

Drerup MM, Schlücking K, Hashimoto K, Manishankar P, Steinhorst L, Kuchitsu K, Kudla J (2013) The calcineurin B-like calcium sensors CBL1 and CBL9 together with their interacting protein kinase CIPK26 regulate the Arabidopsis NADPH oxidase RBOHF. Mol Plant 6:559–569. doi:10.1093/mp/sst009

Dubiella U, Seybold H, Durian G, Komander E, Lassig R, Witte CP, Schulze WX, Romeis T (2013) Calcium-dependent protein kinase/NADPH oxidase activation circuit is required for rapid defense signal propagation. Proc Natl Acad Sci U S A 110:8744–8749

Foyer CH, Noctor G (2005a) Redox homeostasis and antioxidant signaling: a metabolic interface between stress perception and physiological responses. Plant Cell 17:1866–1875

Foyer CH, Noctor G (2005b) Oxidant and antioxidant signalling in plants: a re-evaluation of the concept of oxidative stress in a physiological context. Plant Cell Environ 28:1056–1107

Gadjev I, Vanderauwera S, Gechev TS, Laloi C, Minkov IN, Shulaev V, Apel K, Inzé D, Mittler R, Van Breusegem F (2006) Transcriptomic footprints disclose specificity of reactive oxygen species signalling in Arabidopsis. Plant Physiol 141:436–445

Gapper C, Dolan L (2006) Control of plant development by reactive oxygen species. Plant Physiol 141:341–345

Garg N, Manchanda G (2009) ROS generation in plants: boon or bane? Plant Biosyst 143:8–96

Gechev TS, Hille J (2005) Hydrogen peroxide as a signal controlling plant programmed cell death. J Cell Biol 68:17–20

Ghezzi P, Bonetto V (2003) Redox proteomics: identification of oxidatively modified proteins. Proteomics 3:1145–1153

Gill SS, Tuteja N (2010) Reactive oxygen species and antioxidant machinery in abiotic stress tolerance in crop plants. Plant Physiol Biochem 48:909–930

Gill SS, Khan NA, Anjum NA, Tuteja N (2011) Amelioration of cadmium stress in crop plants by nutrients management: morphological, physiological and biochemical aspects. Plant Stress 5:1–23

Glatz A, Vass I, Los DA, Vigh L (1999) The Synechocystis model of stress: from molecular chaperones to membranes. Plant Physiol Biochem 37:1–12

Guan Q, Wang Z, Wang X, Takano T, Liu S (2015) A peroxisomal APX from *Puccinellia tenuiflora* improves the abiotic stress tolerance of transgenic *Arabidopsis thaliana* through decreasing of H_2O_2 accumulation. J Plant Physiol 175:183–191

Halliwell B (2006) Reactive species and antioxidants. Redox biology is a fundamental theme of aerobic life. Plant Physiol 141:312–322

Halliwell B, Gutteridge JMC (1999) Free radicals in biology and medicine, 3rd edn. Oxford University Press, Oxford

Hancock J, Desikan R, Harrison J, Bright J, Hooley R, Neill S (2006) Doing the unexpected: proteins involved in hydrogen peroxide perception. J Exp Bot 57:1711–1718

Hernández JA, Ferrer MA, Jiménez A, Barceló AR, Sevilla F (2001) Antioxidant systems and $O_2^{\cdot -}$ /H_2O_2 production in the apoplast of pea leaves. Its relation with salt-induced necrotic lesions in minor veins. Plant Physiol 127:817–831

Hien DT, Jacobs M, Angenon G, Hermans C, Thu TT, Son LV, Roosens NH (2003) Proline accumulation and 1-pyrroline-5-carboxylate synthetase gene properties in three rice cultivars differing in salinity and drought tolerance. Plant Sci 165:1059–1068

Hu J, Baker A, Bartel B, Linka N, Mullen RT, Reumann S, Zolman BK (2012) Plant peroxisomes: biogenesis and function. Plant Cell 24:2279–2303

Ismail A, Takeda S, Nick PR (2014) Life and death under salt stress: same players, different timing? J Exp Bot. doi:10.1093/jxb/eru159

Jing X, Hou P, Lu Y, Deng S, Li N, Zhao R, Sun J, Wang Y, Han Y, Lang T, Ding M, Shen X, Chen S (2015) Overexpression of copper/zinc superoxide dismutase from mangrove *Kandelia candel* in tobacco enhances salinity tolerance by the reduction of reactive oxygen species in chloroplast. Front Plant Sci 6:23

Jithesh MN, Prashanth SR, Sivaprakash KR, Parida AK (2006) Antioxidative response mechanisms in halophytes: their role in stress defence. J Genet 85:237–254

Job C, Rajjou L, Lovigny Y, Belghazi M, Job D (2005) Patterns of protein oxidation in Arabidopsis seeds and during germination. Plant Physiol 138:790e802

Khan MIR, Khan NA (2014) Ethylene reverses photosynthetic inhibition by nickel and zinc in mustard through changes in PS II activity, photosynthetic-nitrogen use efficiency and antioxidant metabolism. Protoplasma 251:1007–1019

Khan MH, Panda SK (2008) Alterations in root lipid peroxidation and antioxidative responses in two rice cultivars under NaCl-salinity stress. Acta Physiol Plant 30:81–89

Khan MIR, Asgher M, Khan NA (2014) Alleviation of salt-induced photosynthesis and growth inhibition by salicylic acid involves glycinebetaine and ethylene in mungbean (*Vigna radiata* L.) Plant Physiol Biochem 80:67–74

Khan MIR, Nazir F, Asgher M, Per TS, Khan NA (2015) Selenium and sulfur influence ethylene formation and alleviate cadmium-induced oxidative stress by improving proline and glutathione production in wheat. J Plant Physiol 178:9–18

Khan MIR, Iqbal N, Masood A, Mobin M, Anjum NA, Khan NA (2016a) Modulation and significance of nitrogen and sulfur metabolism in cadmium challenged plants. Plant Growth Regul 78:1–11

Khan MIR, Khan NA, Masood A, Per TS, Asgher M (2016b) Hydrogen peroxide alleviates nickel-inhibited photosynthetic responses through increase in use-efficiency of nitrogen and sulfur, and glutathione production in mustard. Front Plant Sci 7:44

Khare T, Kumar V, Kavi Kishor PB (2015) Na^+ and Cl^- ions show additive effects under NaCl stress on induction of oxidative stress and the responsive antioxidative defense in rice. Protoplasma 252:1149–1165

Khokon AR, Okuma E, Hossain MA, Munemasa S, Uraji M, Nakamura Y, Mori IC, Murata Y (2011) Involvement of extracellular oxidative burst in salicylic acid- induced stomatal closure in Arabidopsis. Plant Cell Environ 34:434–443

Kim KH, Alam I, Lee KW, Sharmim SA, Kwak SS, Lee SY, Lee BH (2010) Enhanced tolerance of transgenic tall fescue plants overexpressing 2-Cys peroxiredoxin against methyl viologen and heat stresses. Biotechnol Lett 32:571–576

Kovtun Y, Chiu WL, Tena G, Sheen J (2000) Functional analysis of oxidative stress-activated MAPK cascade in plants. Proc Natl Acad Sci U S A 97:2940–2945

Krasuska U, Gniazdowska A (2012) Nitric oxide and hydrogen cyanide as regulating factors of enzymatic antioxidant system in germinating apple embryos. Acta Physiol Plant 34:683–692

Kukreja S, Nandval AS, Kumar N, Sharma SK, Sharma SK, Unvi V, Sharma PK (2005) Plant water status, H_2O_2 scavenging enzymes, ethylene evolution and membrane integrity of *Cicer arietinum* roots as affected by salinity. Biol Plant 49:305–308

Kumar V, Khare T (2016) Differential growth and yield responses of salt-tolerant and susceptible rice cultivars to individual (Na^+ and Cl^-) and additive stress effects of NaCl. Acta Physiol Plant 38:1–9. doi:10.1007/s11738-016-2191-x

Kumar V, Shriram V, Nikam TD, Jawali N, Shitole MG (2009) Antioxidant enzyme activities and protein profiling under salt stress in indica rice genotypes differing in salt tolerance. Arch Agron Soil Sci 55:379–394

Kumar V, Shriram V, Kavi Kishor PB, Jawali N, Shitole MG (2010) Enhanced proline accumulation and salt stress tolerance of transgenic *indica* rice by over-expressing *P5CSF129A* gene. Plant Biotechnol Rep 4:37–48. doi:10.1007/s11816-009-0118-3

Kumar SR, Sivalingam A, Selvaraj D, Ahmed Z, Ramalingam S (2012) Isolation and characterization of cold inducible genes in carrot by suppression subtractive hybridization. Biol Plant 57:97–104

Kumar V, Sharma M, Lemos M, Shriram V (2013) Efficacy of *Helicteres isora* L. against free radicals, lipid peroxidation, protein oxidation and DNA damage. J Pharm Res 6:620–625. doi:10.1016/j.jopr.2013.05.017

Kumar V, Shriram V, Hussain MA, Kavi Kishor PB (2015) Engineering proline metabolism for enhanced plant salt stress tolerance. In: Wani SH, Hussain MA (eds) Managing salinity tolerance in plants: molecular and genomic perspectives. CRC Press, Taylor & Francis Group, Boca Raton, pp 353–372

Laloi C, Apel K, Danon A (2004) Reactive oxygen signalling: the latest news. Curr Opin Plant Biol 7:323–328

Le Martret B, Poage M, Shiel K, Nugent GD, Dix PJ (2011) Tobacco chloroplast transformants expressing genes encoding dehydroascorbate reductase, glutathione reductase, and glutathione-S-transferase, exhibit altered anti-oxidant metabolism and improved abiotic stress tolerance. Plant Biotechnol J 9:661–673

Lee SH, Ahsan N, Lee KW, Kim DH, Lee DG, Kwak SS, Kwon SY, Kim TH, Lee BH (2007) Simultaneous overexpression of both CuZn superoxide dismutase and ascorbate peroxidase in transgenic tall fescue plants confers increased tolerance to a wide range of abiotic stresses. J Plant Physiol 164:1626–1638

Lim S, Kim YH, Kim SH, Kwon SY, Lee HS, Kim JS, Cho KY, Paek KY, Kwak SS (2007) Enhanced tolerance of transgenic sweet potato plants that express both CuZnSOD and APX in chloroplasts to methyl viologen-mediated oxidative stress and chilling. Mol Breed 19:227–239

Lu ZQ, Liu D, Liu SK (2007) Two rice cytosolic ascorbate peroxidases differentially improve salt tolerance in transgenic Arabidopsis. Plant Cell Rep 26:1909–1917

Ma L, Zhang H, Sun L, Jiao Y, Zhang G, Miao C, Hao F (2012) NADPH oxidase ArbohD and AtrbohF function in ROS-dependent regulation of Na^+/K^+ homeostasis in Arabidopsis under salt stress. J Exp Bot 63:305–317

Mahajan S, Tuteja N (2005) Cold, salinity and drought stresses: an overview. Arch Biochem Biophys 444:139–158

Maksimovic JD, Zhang J, Zeng F, Ziivanovic BD, Shabala L, Zhou M, Shabala S (2013) Linking oxidative and salinity stress tolerance in barley: can root antioxidant enzyme activity be used as a measure of stress tolerance? Plant Soil 365:141–155

Marinho HS, Real C, Cyrne L, Soares H, Antunes F (2014) Hydrogen peroxide sensing, signalling and regulation of transcription factors. Redox Biol 2:35–62

Martı MC, Florez-Sarasa, Camejo D, Ribas-Carbo M, Lázaro JJ, Sevilla F, Jiménez A (2011) Response of mitochondrial thioredoxin *PsTrxo1*, antioxidant enzymes, and respiration to salinity in pea (*Pisum sativum* L.) leaves. J Exp Bot 62:3863–3874

Maruta T, Inoue T, Tamoi M, Yabuta Y, Yoshimura K, Ishikawa T, Shigeoka S (2011) Arabidopsis NADPH oxidases, AtrbohD and AtrbohF, are essential for jasmonic acid-induced expression of genes regulated by MYC2 transcription factor. Plant Sci 180:655–660

McAinsh MR, Pittman JK (2009) Shaping the calcium signature. New Phytol 181:275–294

Mhamdi A, Hager J, Chaouch S et al (2010) Arabidopsis GLUTATHIONEREDUCTASE1 plays a crucial role in leaf responses to intracellular hydrogen peroxide and in ensuring appropriate gene expression through both salicylic acid and jasmonic acid signalling pathways. Plant Physiol 153:1144–1160

Millar AH, Leaver CJ (2000) The cytotoxic lipid peroxidation product, 4-hydroxy- 2-nonenal, specifically inhibits decarboxylating dehydrogenases in the matrix of plant mitochondria. FEBS Lett 481:117–121

Mishra S, Srivastava S, Tripathi RD, Kumar R, Seth CS, Gupta DK (2006) Lead detoxification by coontail (*Ceratophyllum demersum* L.) involves induction of phytochelatins and antioxidant system in response to its accumulation. Chemosphere 65:1027–1039

Mittler R (2002) Oxidative stress, antioxidants and stress tolerance. Trends Plant Sci 9:405–410

Mittler R, Vanderauwera S, Gollery M, Van Breusegem F (2004) Reactive oxygen gene network of plants. Trends Plant Sci 9:490–498

Molassiotis A, Fotopoulos V (2011) Oxidative and nitrosative signalling in plants: two branches in the same tree? Plant Signal Behav 6:210–214. doi:10.4161/psb.6.2.14878

Moller IM, Sweetlove LJ (2010) ROS signalling-specificity is required. Trends Plant Sci 15:370–374

Møller IM, Jensen PE, Hansson A (2007) Oxidative modifications to cellular components in plants. Annu Rev Plant Biol 58:459–481

Montillet JL, Chamnongpol S, Rustérucci C, Dat J, Van de Cotte B, Agnel JP, Battesti C, Inzé D, Van Breusegem F, Triantaphylides C (2005) Fatty acid hydroperoxides and H_2O_2 in the execution of hypersensitive cell death in tobacco leaves. Plant Physiol 138:1516–1526

Mubarakshina MM, Ivanov BN, Naydov IA, Hillier W, Badger MR, Krieger-Liszkay A (2010) Production and diffusion of chloroplastic H_2O_2 and its implication to signalling. J Exp Bot 61:3577–3587

Munne-Bosch S (2005) The role of α-tocopherol in plant stress tolerance. J Plant Physiol 162:743–748

Mur LA, Kenton P, Atzorn R, Miersch O, Wasternack C (2006) The outcomes of concentration-specific interactions between salicylate and jasmonate signalling include synergy, antagonism, andoxidative stress leading to cell death. Plant Physiol 140:249–262

Mylona PV, Polidoros AN (2010) ROS regulation of antioxidant genes. In: Gupta SD (ed) Reactive oxygen species and antioxidants in higher plants. CRC Press, New York

Neill SJ, Desikan R, Clarke A, Hurst RD, Hancock JT (2002) Hydrogen peroxide and nitric oxide as signalling molecules in plants. J Exp Bot 53:1237–1247

Neill SJ, Desikan R, Hancock JT (2003) Nitric oxide signalling in plants. New Phytol 159:11–35

Noctor G (2006) Metabolic signalling in defense and stress: the central roles of soluble redox couples. Plant Cell Environ 29:409–425

Noctor G, Foyer CH (1998) Ascorbate and glutathione: keeping active oxygen under control. Annu Rev Plant Physiol Plant Mol Biol 49:249–279

Ozgur R, Uzilday B, Sekmen AH, Turkan I (2013) Reactive oxygen species regulation and antioxidant defense in halophytes. Funct Plant Biol 40:832–847

Pan Y, Wu LJ, Yu ZL (2006) Effect of salt and drought stress on antioxidant enzymes activities and SOD isoenzymes of liquorice (*Glycyrrhiza uralensis* Fisch). Plant Growth Regul 49:157–165

Passardi F, Longet D, Penel C, Dunand C (2004) The class III peroxidase multigene family in rice and its evolution in land plants. Phytochemistry 65:1879–1893

Paulsen CE, Carroll KS (2010) Orchestrating redox signalling networks through regulatory cysteine switches. ACS Chem Biol 5:47–62

Pesaresi P, Hertle A, Pribil M, Kleine T, Wagner R, Strissel H, Ihnatowicz A, Bonardi V, Scharfenberg M, Schneider A, Pfannschmidt T, Leister D (2009) Arabidopsis STN7 kinase provides a link between short- and long-term photosynthetic acclimation. Plant Cell 21:2402–2423

Petrov VD, Van Breusegem F (2012) Hydrogen peroxide- a central hub for information flow in plant cells. AoB Plants 2012:ls014

Petrov V, Hille J, Mueller-Roeber B, Gechev TS (2015) ROS-mediated abiotic stress-induced programmed cell death in plants. Front Plant Sci 6:69

Pitzschke A, Forzani C, Hirt H (2006) Reactive oxygen species signaling in plants. Antioxid Redox Signal 8:1757–1764

Pitzschke A, Schikora A, Hirt H (2009) MAPK cascade signalling networks in plant defense. Curr Opin Plant Biol 12:421–426

Qureshi MI, Abdin MZ, Ahmad J, Iqbal M (2013) Effect of long-term salinity on cellular antioxidants, compatible solute and fatty acid profile of Sweet Annie (*Artemisia annua* L.) Phytochemistry 95:215–223

Rejeb KB, Abdelly C, Savouré (2014) How reactive oxygen species and proline face stress together. Plant Physiol Biochem 80:278–284

Rhoads DM, Umbach AL, Subbaiah CC, Siedow JN (2006) Mitochondrial reactive oxygen species. Contribution to oxidative stress and interorganellar signaling. Plant Physiol 14:357–366

Rodrigo-Moreno A, Poschenrieder C, Shabala S (2013) Transition metals: a double edge sward in ROS generation and signaling. Plant Signal Behav 8:e23245

Rodriguez MC, Petersen M, Mundy J (2010) Mitogen-activated protein kinase signalling in plants. Annu Rev Plant Biol 61:621–649

Sadanandom A, Poghosyan Z, Fairbairn DJ, Murphy DJ (2000) Differential regulation of plastidial and cytosolic isoforms of peptides methionine sulfoxide reductase in Arabidopsis. Plant Physiol 123:1299–1309

Sagi M, Fluhr R (2006) Production of reactive oxygen species by plant NADPH oxidases. Plant Physiol 141:336–340

Sairam RK, Srivastava GC, Agarwal S, Meena RC (2005) Differences in antioxidant activity in response to salinity stress in tolerant and susceptible wheat genotypes. Biol Plant 49:85–91

Scheibe R, Dietz KJ (2012) Reduction-oxidation network for flexible adjustment of cellular metabolism in photoautotrophic cells. Plant Cell Environ 35:202–216

Sekmen A, Türkan I, Takio S (2007) Differential responses of antioxidative enzymes and lipid peroxidation to salt stress in salt-tolerant *Plantago maritima* and salt-sensitive *Plantago media*. Physiol Plant 131:399–411

Sewelam N, Kazan K, Thomas-Hall S, Kidd BN, Manners JM, Schenk PM (2013) Ethylene response factor 6 is a regulator of reactive oxygen species signalling in *Arabidopsis*. PLoS One 8:e70289

Sewelam N, Jaspert N, Kelen KVD, Schmitz J, Frerigmann H, Tognetti VB, Achmitz J, Frerigmann H, Stahl E, Zeier J, Van Breusegem F, Maurino VG (2014) Spatial H_2O_2 signalling specificity: H_2O_2 from chloroplasts and peroxisomes differentially modulates the plant transcriptome. Mol Plant 7:1191–1210

Sewelam N, Kazan K, Schenk PM (2016) Global plant stress signaling: reactive oxygen species at the cross-road. Front Plant Sci 7:187

Shafi A, Chauhan R, Gill T, Swarnkar MK, Sreenivasulu Y, Kumar S, Kumar N, Shankar R, Ahuja PS, Singh AK (2015) Expression of SOD and APX genes positively regulates secondary cell wall biosynthesis and promotes plant growth and yield in Arabidopsis under salt stress. Plant Mol Biol 87:615–631

Sies H (2014) Role of metabolic H_2O_2 generation redox signaling and oxidative stress. J Biol Chem 289:8735–8741

Singh N, Mishra A, Jha B (2014a) Over-expression of the peroxisomal ascorbate peroxidase (*SbpAPX*) gene cloned from halophyte *Salicornia brachiata* confers salt and drought stress tolerance in transgenic tobacco. Mar Biotechnol 16:321–332

Singh N, Mishra A, Jha B (2014b) Ectopic over-expression of peroxisomal ascorbate peroxidase (*SbpAPX*) gene confers salt stress tolerance in transgenic peanut (*Arachis hypogaea*). Gene 547:119–125

Sinha AK, Jaggi M, Raghuram B, Tuteja N (2011) Mitogen-activated protein kinase signalling in plants under abiotic stress. Plant Signal Behav 6:196–203

Skopelitis DS, Paranychianakis NV, Paschalidis KA, Pliakonis ED, Delis ID, Yakoumakis DI, Kouvarakis A, Papadakis AK, Stephanou EG, Roubelakis-Angelakis KA (2006) Abiotic stress generates ROS that signal expression of anionic glutamate dehydrogenases to form glutamate for proline synthesis in tobacco and grapevine. Plant Cell 18:2767–2781

Stael S, Wurzinger B, Mair A, Mehlme N, Vothknecht UC, Teige M (2012) Plant organellar calcium signalling: an emerging field. J Exp Bot 63:1525–1542

Suzuki N, Koussevitzky S, Mittler R, Miller G (2012) ROS and redox signalling in the response of plants to abiotic stress. Plant Cell Environ 35:259–270

Suzuki N, Miller G, Salazar C, Mondal HA, Shulaev E, Cortes DF, Shuman JL, Luo X, Shah J, Schlauch K, Shulaev V, Mittler R (2013) Temporal-spatial interaction between reactive oxygen species and abscisic acid regulates rapid systemic acclimation in plants. Plant Cell 25:3553–3569

Sweetlove LJ, Foyer CH (2004) Roles for reactive oxygen species and antioxidants in plant mitochondria. In: Day DA, Millar AH, Whelan J (eds) Plant mitochondria: from genome to function, Advances in photosynthesis and respiration, vol 1. Kluwer Academic Press, Dordrecht, pp 307–320

Szechynska-Hebda M, Kruk J, Gorecka M, Karpinska B, Karpinski S (2010) Evidence for light wavelength-specific photoelectrophysiological signaling and memory of excess light episodes in Arabidopsis. Plant Cell 22:2201–2218

Tanou G, Job C, Rajjou L, Arc E, Belghazi M, Diamantidis G, Molassiotis A, Job D (2009) Proteomics reveals the overlapping roles of hydrogen peroxide and nitric oxide in the acclimation of citrus plants to salinity. Plant J 60:795–804

Tanou G, Job C, Belghazi M, Molassiotis A, Diamantidis G, Job D (2010) Proteomic signatures uncover hydrogen peroxide and nitric oxide cross- talk signalling network in citrus plants. J Proteome Res 9:5994–6006

Taylor NL, Heazlewood JL, Day DA, Millar AH (2005) Differential impact of environmental stresses on the pea mitochondrial proteome. Mol Cell Proteomics 4:1122–1133

Torres MA, Dangl JL (2005) Functions of the respiratory burst oxidase in biotic interactions, abiotic stress and development. Curr Opin Plant Biol 8:397–403

Tseng MJ, Liu CW, Yiu JC (2007) Enhanced tolerance to sulfur dioxide and salt stress of transgenic Chinese cabbage plants expressing both superoxide dismutase and catalase in chloroplasts. Plant Physiol Biochem 45:822–833

Turkan I, Demiral T, Sekmen AH (2013) The regulation of antioxidant enzymes in two *Plantago* species differing in salinity tolerance under combination of waterlogging and salinity. Funct Plant Biol 40:484–493

Tuteja N (2010) Cold, salt and drought stress. In: Hirt H (ed) Plant stress biology: from genomics towards system Biology. Wiley-Blackwell, Weinheim, pp 137–159

Tuteja N, Tuteja R (2001) Unravelling DNA repair in human: molecular mechanisms and consequences of repair defect. Crit Rev Biochem Mol Biol 36:261–290

Tuteja N, Singh MB, Misra MK, Bhalla PL, Tuteja R (2001) Molecular mechanisms of DNA damage and repair: progress in plants. Crit Rev Biochem Mol Biol 36:337–397

Tuteja N, Ahmad P, Panda BB, Tuteja R (2009) Genotoxic stress in plants: shedding light on DNA damage, repair and DNA repair helicases. Mutat Res 681:134–149

Ursini F, Maiorin M, Roveri A (1997) Phospholipid hydroperoxide glutathione peroxidase (*PHGPx*): more than an antioxidant enzyme? Biomed Environ Sci 10:327–332

Uzilday B, Ozgur R, Sekmen AH, Yildiztugay E, Turkan I (2014) Changes in the alternative electron sinks and antioxidant defence in chloroplasts of the extreme halophyte *Eutrema parvulum* (*Thellungiella parvula*) under salinity. Ann Bot 115:449–463. doi: 10.1093/aob/mcu1843

Valko M, Rhodes CJ, Moncol J, Izakovic M, Mazur M (2006) Free radicals, metals and antioxidants in oxidative stress-induced cancer. Chem Biol Interact 160:1–40

Verslues PE, Zhu J (2004) Before and beyond ABA: upstream sensing and internal signals that determine ABA accumulation and response under abiotic stress. Biochem Soc Trans 33:375–379

Wang FZ, Wang QB, Kwon SY, Kwak SS, Su WA (2005) Enhanced drought tolerance of transgenic rice plants expressing a pea manganese superoxide dismutase. J Plant Physiol 162:465–472

Wang X, Yang P, Gao Q et al (2008) Proteomic analyses of the response to high-salinity stress in *Physcomitrella patens*. Planta 228:167–177

Wani SH, Kumar V, Shriram V, Sah SK (2016) Phytohormones and their metabolic engineering for abiotic stress tolerance in crop plants. Crop J 4:162–176. doi: 10.1016/j.cj.2016.01.010

Weisany W, Sohrabi Y, Heidari G, Siosemardeh A, Ghassemi-Golezani K (2012) Changes in antioxidant enzymes activity and plant performance by salinity stress and zinc application in soybean (*Glycine max* L.) Plant Omics 5:60

Willekens H, Chamnongpol S, Davey M, Schraudner M, Langebartels C, Van Montagu M, Inzé D, Van Camp W (1997) Catalase is a sink for H_2O_2 and is indispensable for stress defence in C3 plants. EMBO J16:4806–4816

Wiseman H, Halliwell B (1996) Damage to DNA by reactive oxygen and nitrogen species: role in inflammatory disease and progression to cancer. Biochem J 313:7–29

Xu J, Yang J, Duan X, Jiang Y, Zhang P (2014) Increased expression of native cytosolic Cu/Zn superoxide dismutase and ascorbate peroxidase improves tolerance to oxidative and chilling stresses in cassava (*Manihot esculenta* Crantz). BMC Plant Biol 14:208

Yannarelli GG, Fernández-Alvarez AJ, Santa-Cruz DM, Tomaro ML (2007) Glutathione reductase activity and isoforms in leaves and roots of wheat plants subjected to cadmium stress. Phytochemistry 68:505–512

Yao S, Chen S, Xu D, Lan H (2010) Plant growth and responses of antioxidants of *Chenopodium album* to long-term NaCl and KCl stress. Plant Growth Regul 60:15–125

Yildiztugay E, Konakçi CÖ, Küçüködük M, Duran Y (2014) Modulation of osmotic adjustment and enzymatic antioxidant profiling in *Apera intermedia* exposed to salt stress. Turk J Bot 38:99–111

Zhai CZ, Zhao L, Yin LJ, Chen M, Wang QY, Li LC, Xu ZS, Ma YZ (2013) Two wheat glutathione peroxidase genes whose products are located in chloroplasts improve salt and H_2O_2 tolerances in *Arabidopsis*. PLoS One 8:e73989

Zhang Z, Zhang Q, Wu J, Zheng X, Zheng S, Sun X, Qiu Q, Lu T (2013) Gene knockout study reveals that cytosolic ascorbate peroxidase 2 (OsAPX2) plays a critical role in growth and reproduction in rice under drought, salt and cold stresses. PLoS One 8:e57472

ROS Signaling in Plants Under Heavy Metal Stress

8

Sukhmeen Kaur Kohli, Neha Handa, Vandana Gautam, Shagun Bali, Anket Sharma, Kanika Khanna, Saroj Arora, Ashwani Kumar Thukral, Puja Ohri, Yuriy Victorovich Karpets, Yuriy Evgenovich Kolupaev, and Renu Bhardwaj

Abstract

Contamination of soil with toxic heavy metals is a major reason for retarded growth of crops and harmful effects on human health. Cultivation of large number of agricultural crops in contaminated soil is a major concern of environmentalist in the present times. Increased level of heavy metals can enter in to the food chain and may available for human consumption. Metal toxicity-induced oxidative stress eventually leads to refrained enzyme activities due to displacement of essential cofactors with other metal ions and blocking of functional groups such as carboxyl, histidyl and thiol, and proteins. Oxidative burst releases large quantities of reactive oxygen species (ROS) such as superoxide anion, hydrogen peroxide, hydroxyl radical, singlet oxygen, etc., which is one of the primary response of plants to heavy metal stress. Production of ROS is an inherent feature of plant cell and contributes to the process of oxidative damage leading to cell death. Its production is restricted to several cellular compartments such as mitochondria, chloroplast, and peroxisomes etc. ROS production leads to alteration of several physiological processes including degradation of enzymes, proteins, and amino acids and change in structure of cells. ROS are well described as secondary messengers in variety of cellular processes including acclimatization of cells to stress conditions. The signaling of ROS as a

S.K. Kohli • N. Handa • V. Gautam • S. Bali • A. Sharma • K. Khanna • S. Arora • A.K. Thukral
R. Bhardwaj (✉)
Department of Botanical and Environment Sciences, Guru Nanak Dev University, Amritsar, Punjab 143005, India
e-mail: renubhardwaj82@gmail.com

P. Ohri
Department of Zoology, Guru Nanak Dev University, Amritsar, Punjab 143005, India

Y.V. Karpets • Y.E. Kolupaev
Dokuchaev Kharkiv National Agrarian University, Kharkiv 62483, Ukraine

M.I.R. Khan, N.A. Khan (eds.), *Reactive Oxygen Species and Antioxidant Systems in Plants: Role and Regulation under Abiotic Stress*,
DOI 10.1007/978-981-10-5254-5_8

result of oxidative damage is regulated by several other signaling cascades which are interlinked. Their role has been studied under various stress conditions specifically heavy metals which leads to production of NO, H_2O_2, synthetic electrophilic compounds, lipid peroxidation molecules, etc.

Keywords
Oxidative damage • Reactive oxygen species • Hormone signaling • Transcription factors • Hydroxyl radical

Abbreviations

$^{I}O_2$	Superoxide radical
ABA	Abscisic acid
ACC Synthase	Aminocyclopropane-1-carboxylic acid synthase
APX	Ascorbate peroxidase
As	Arsenic
BRs	Brassinosteroids
Ca	Calcium
cADPR	Cyclic ADP ribose
CAT	Catalase
Cd	Cadmium
CDPKs	Calcium-dependent protein kinases
Co	Cobalt
Cr	Chromium
Cu	Copper
DHAR	Dehydroascorbate reductase
DNA	Deoxyribonucleic acid
EBR	24-Epibrassinolide
EGTA	Ethylene Glycol-bis-aminoethyether-N,N,NI,NI tetra acetic acid
Fe	Iron
GPX	Glutathione peroxidase
GR	Glutathione reductase
GSH	Reduced glutathione
GSSG	Oxidized glutathione
H_2O	Water
H_2O_2	Hydrogen peroxide
HSP	Heat shock proteins
I	Iodine
IAA	Indole-3-acetic acid

IP3	Inositol 1,4,5- triphosphate
K	Potassium
LMWOAs	Low molecular weight organic acids
LOX	Lipoxygenase
MAPK	Mitogen-activated protein kinase
MAPKK	Mitogen-activated protein kinase kinase
MAPKKK	Mitogen-activated protein kinase kinase kinase
MDA	Malondialdehyde
MDHAR	Monodehydroascorbate reductase
Mg	Magnesium
miRNA	MicroRNA
NADPH oxidase	Nicotinamide adenine dinucleotide phosphate-oxidase
NO	Nitric oxide
O_2	Oxygen
$O^{2.-}$	Superoxide anion
OH	Hydroxyl ion
$OONO^-$	Peroxynitrite
OXII	Oxylipin
Pb	Lead
PLD	Phospholipase D
POD	Peroxidases
PSI	Photosystem I
PSII	Photosystem II
ROS	Reactive oxygen species
SNAP	S-nitroso-N-acetyl-1,1-pencillamine
SNP	Sodium nitroprusside
SOD	Superoxide dismutase
V	Vanadium
Zn	Zinc

8.1 Introduction

Heavy metals are important biological entities, which play an important role in growth and development of plants. They are implicated in several cellular processes including maintenance of cellular homeostasis and photosynthetic efficiency of plants. Heavy metals iron (Fe), zinc (Zn), and copper (Cu) also act as cofactor of several enzymes (Hansch and Mendel 2009). However increased concentration of heavy metals has drastic effects on plant growth and its development, subsequently resulting in reduced photosynthesis and retardation of growth and other toxicity symptoms (Khan and Khan 2014). Metal toxicity results in production of reactive oxygen species (ROS) and oxidative stress which eventually leads to refrained enzyme activities due to displacement of essential cofactors with other metal ions, blocking of functional groups such as carboxyl, histidyl and thiol, and proteins

(Sharma and Dietz 2009). ROS are collectively considered as free radicals, ions, and reactive molecules that are produced from O_2. About 1% of O_2 is utilized by plants to synthesize ROS. These include superoxide anion, hydrogen peroxide, hydroxyl radical and singlet oxygen, etc. ROS are products of aerobic metabolism (Halliwell 2006). They are distinguished from other signaling molecules due to their high reactivity, half-life, and solubility in lipids (Autreaux and Toledano 2007). Photosynthesizing organisms have an elevated threat of oxidative damage due to large quantities of photosensitizer and polyunsaturated fatty acids (PUFAs) present in chloroplast envelope. The presence of O_2 in the atmosphere which acts as a final electron acceptor leads to formation of ROS such as H_2O_2, OH, and $_1O_2$ in plant cells (Temple et al. 2005; Scandalios 2005). Various reactive oxygen intermediates are formed by oxidation of O_2 to singlet oxygen (Triantaphylides et al. 2008). Hydroxyl ions have a nonselective reactivity toward biological molecules, although H_2O_2 and O_2^- have specific biological targets (Autreaux and Toledano 2007; Fig. 8.1).

In presence of certain transition metals such as Cu or Fe, conversion of O_2 to OH ions takes place via Haber-Weiss or Fenton reaction. Hydroxyl ion is considered as the most reactive species out of all ROS. O_2 also reacts with nitric oxide (NO) another ROS to form peroxynitrite ($OONO^-$) (Gill and Tuteja 2010). Other reactive oxidation products include alkoxy radical, peroxy radical, perhydroxyl radical, organic hydroperoxide, excited carbonyl, etc. (Bhattacharjee 2005). In plant cell, these are considered as harmful as well as advantageous molecules depending on the concentration of ROS production. At low or moderate concentration, ROS operate as second messengers in intracellular signaling mechanism that intervene numerous cellular responses in plants (Fig. 8.2), while at high concentration, these cause deleterious effect on plant system (Fig. 8.3).

Various enzymes involved in the synthesis of ROS are found in diverse subcellular organelles like mitochondria, peroxisomes, chloroplasts, etc. (Groß et al. 2013). Table 8.1 summarizes different organelles of plants in which ROS are generated. Production of ROS in these compartments results in modification of several physiological processes including degradation of enzymes, proteins, and amino acids and change in structure of cells (Choudhary et al. 2013; Gill and Tuteja 2010). In plants, ROS synthesis is regulated by highly specific ROS scavenging pathway, comprising of two components, nonenzymatic and enzymatic. Consequences of production of ROS under heavy metal stress including diminution of antioxidants and lipid peroxidation ultimately result in change in gene expression (Munne-Bosch et al. 2013). ROS have been reported to function as intercellular signaling molecules. Their signaling is modulated by series of chemical reactions with proteinaceous target molecules which eventually results in modification of protein structure and function (Nathan 2003). There are several reports which suggest availability of heavy metal sensors in plants that first perceive the presence of heavy metal in environment. Gene expression studies of these sensors have led to identification of several heavy metals' stress-protective transcriptional profiles.

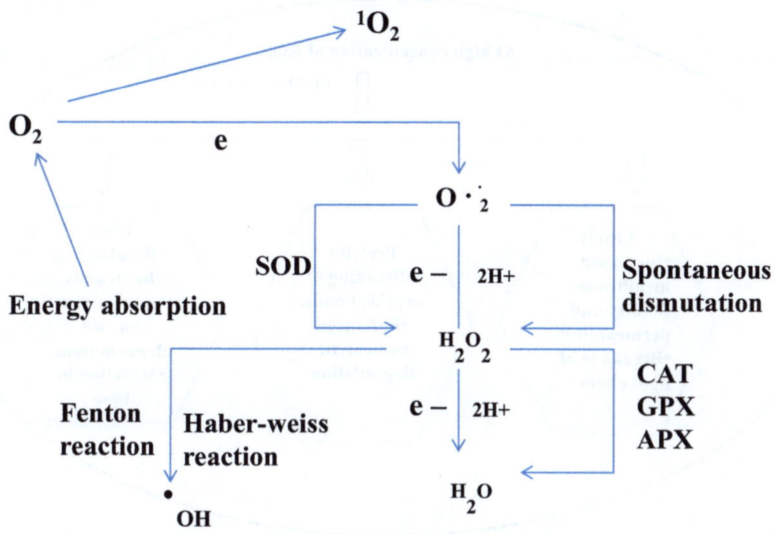

Fig. 8.1 Represents the synthesis of ROS in plant cells

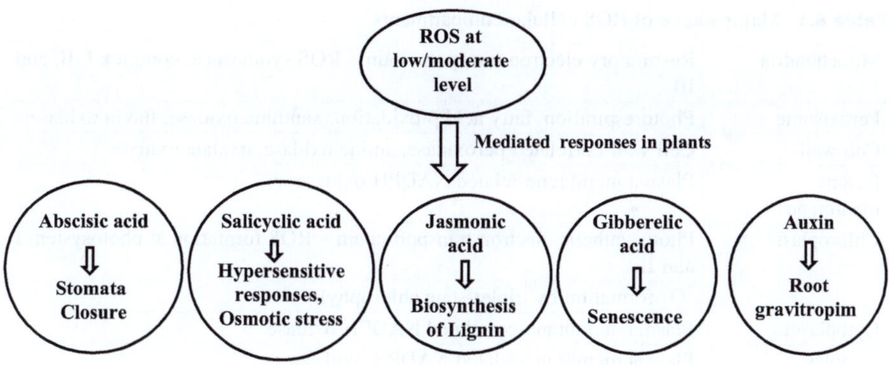

Fig. 8.2 Represents ROS mediated responses in plants

8.2 ROS Signaling Under Heavy Metal Stress

ROS metabolism is closely associated with signaling during variable abiotic stress conditions (Munne-Bosch et al. 2013). Certain ROS such as O_2^- and H_2O_2 act as secondary messengers in intercellular signaling pathways which modulate plant responses to apoptosis, stomatal closure, as well as acclimatization of biotic and abiotic stresses and consequently regulate growth and development of organisms (Foyer and Noctor 2005; Yan et al. 2007). O_2 is considered as least damaging

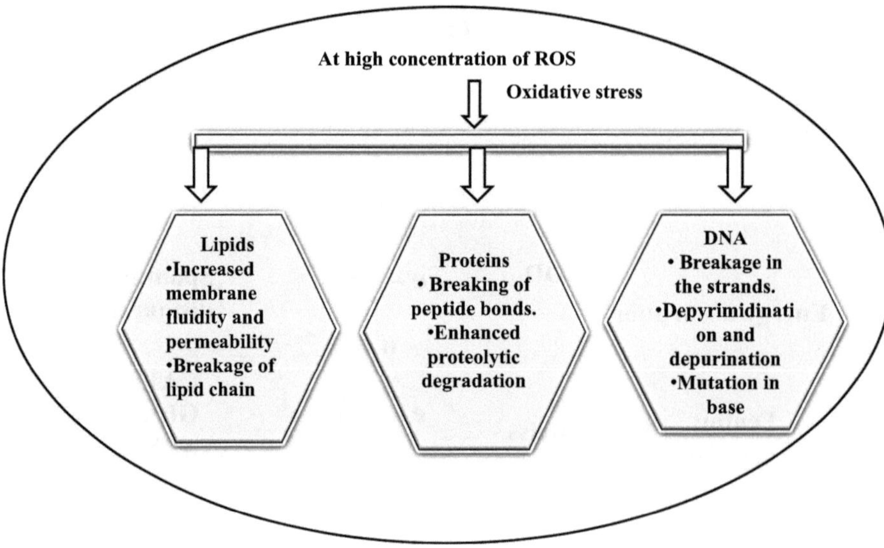

Fig. 8.3 ROS stimulated oxidative stress and caused damage to lipids, proteins and DNA

Table 8.1 Major source of ROS cellular compartments

Mitochondria	Respiratory electron transport chain – ROS synthesis at complex I, II, and III
Peroxisome	Photorespiration, fatty acid β-oxidation, xanthine oxidase, flavin oxidase
Cell wall	Cell wall-correlated peroxidase, amine oxidase, oxalate oxidase
Plasma membrane	Plasma membrane-related NADPH oxidase
Chloroplast	Photosynthetic electron transport chain – ROS formation at photosystem I and II
	$^1O_2$ formation by triplet-state chlorophyll
Cytoplasm	Plasma membrane-correlated NADPH oxidase
Apoplast	Plasma membrane-related NADPH oxidase
	Cell wall-correlated peroxidase, amine oxidase, oxalate oxidase

Adapted from Groß et al. (2013)

species in its ground state due to two unpaired electrons with parallel spin which makes it inactive, and as it becomes activated, it takes part in reactions with organic molecules (Apel and Hirt 2004). The redox state is hastily changed by elevation in ROS and inactivation of antioxidant enzymes during cell signaling under stress conditions which is important to elicit particular cellular responses such as enhanced antioxidative defense system or senescence which is essential to eradicate cells that have been rigorously damaged (del Río 2015; Shi et al. 2014; Yun et al. 2011). Activation of O_2 might take place through two different processes: (a) absorption of an adequate energy to reverse the spin of the unpaired electrons and (b) sequential monovalent reduction. In the first process, $^1O_2$ is formed, while in the second one, stepwise reduction of O_2 to O_2•−, H_2O_2, and •OH takes place.

Stress-protective responses of ROS are regulated by several signaling cascades which are interlinked. For example, KEAP1 and NRF2 complexes are similar to ROS receptors found in mammals and have a property to synchronize various environmental and xenobiotic stress responses (Sharma et al. 2012). Their role have been studied under adverse stress conditions which leads to production of NO, H_2O_2, synthetic electrophilic compounds, lipid peroxidation molecules, etc. (Kobayashi and Yamamoto 2006). ROS are known to regulate a number of signaling cascades including calcium/calmodulin mobilization and phosphorylation of proteins and certain genes. They also modulate functioning of protein kinases, various transcription factors, and protein phosphatases (Sharma et al. 2012). They are directly detected by signaling proteins especially tyrosine phosphate via oxidation of the cysteine residue. Metal responsive transcription factor 1 (MTF-1) plays a significant role in the cellular response and forbearance to heavy metal stress by stimulating certain genes accountable for heavy metal uptake, accumulation, and detoxification (Fusco et al. 2005; Mortel et al. 2008). In response to ROS production, the transcriptional changes are triggered that allow further stimulation of genes involved in several processes like signaling cascade, basic defense, transport, metabolism, enhanced ROS production, and eventually cell death (Palmieri et al. 2008). Certain transcription factors in *Escherichia coli* including OxyR and SoxR have been observed to be redox susceptible and regulate the responses of *E. coli* to ROS (Autreaux and Toledano 2007).

8.2.1 Involvement of MAP Kinases and ROS Signaling Under Heavy Metal Stress

One of the mechanisms which cells adopt to communicate with each other is phosphorylation of proteins via specific protein kinases and dephosphorylation of certain protein phosphatases. For any signal cascade to function properly, a continuous process of activation and deactivation of enzymes through phosphorylation/dephosphorylation in the presence of phosphatases and kinases is imperative (Rodriguez et al. 2010). Protein kinases include protein phosphorylating gene families, which function as regulators of protein activity inside the cell. Protein phosphorylation was primarily studied in eukaryotes. One of the most essential protein kinases important for activation and deactivation of genes is mitogen-activated protein kinases (MAPK) (Joshi et al. 2011). Plants distinguish various stresses through receptors or sensors present on the cell surface and are transduced by MAPKKK (MAP kinase kinase kinase) signaling networks (Taj et al. 2010). One of the largest groups of MAPK is MAPKK kinases found upstream of signaling cascade, which are further divided into two components, i.e., MAPK-type and Raf-like kinases (Rao et al. 2010; Opdenakker et al. 2012). Localization of MAPK is restricted to cytoplasm and nucleus of cells (Joshi et al. 2011). MAPK pathway comprises of the following components: (i) MAP kinase kinase kinase

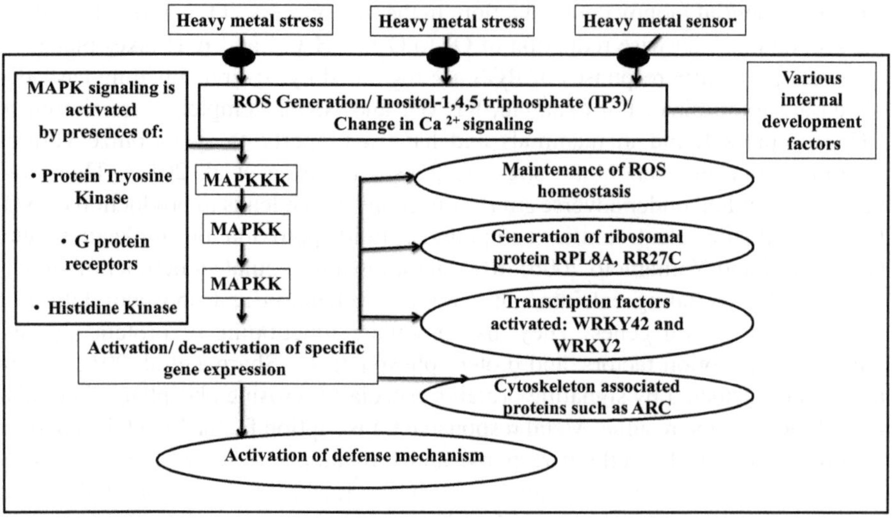

Fig. 8.4 Schematic representation of interplay between MAPK pathway and ROS generation under heavy metal stress

(MAPKKK), (ii) MAP kinase kinase (MAPKK), and (iii) MAP kinase (MAPK) (Sinha et al. 2011; Opdenakker et al. 2012).

Synthesis of ROS and change in Ca^{2+} signaling cascade lead to activation of phosphorylation cascade which eventually results in the interaction of various responsive genes and transcription factors. Figure 8.4 describes the interplay between MAPK and ROS signaling cascade under heavy metal stress. Oxidative stress due to production of ROS in the response to heavy metal stress involves MAPK signaling networks which are activated in the presence of protein tyrosine kinase, G protein receptors, and two-component histidine kinases (Taj et al. 2010). Consequences of MAPK signaling network depend upon the duration of MAPK's activation, which is further modulated by co-localization (scaffolding) and alleviation of stress via phosphatases (Sinha et al. 2011). MAPK signaling networks and its cross talk with heavy metal produced ROS signaling which is widely reported in plants (Rao et al. 2011; Raina et al. 2013; Šamajová et al. 2013) (Table 8.2).

8.2.2 Role of Calcium in Heavy Metal-Induced ROS Signaling

The normal growth and development of plants have direct involvement of Ca as an essential element which is required by apical meristems for mitosis. As calcium pectate in cell walls is essential for cellular binding and acts as a second messenger, its delicately balanced concentrations are responsive to various extracellular signals (Tuteja and Mahajan 2007). Ca has the ability to readily form complexes with several cellular biomolecules involving proteins and organic acids (Kudla et al.

Table 8.2 MAPK and ROS cross talk in several plants under heavy metal stress

S. No.	Heavy metal	Plant studied	MAPK and ROS signaling cross talk under heavy metal stress	References
1.	Cd	*Medicago sativa*	ROS production in response to Cd stress	Jonak et al. (2004)
	Cu		Activation of four distinct MAPKs – SIMK, SAMK, MMK2, and MMK3	
2.	I	*Oryza sativa*	42KDa MAPK activated	Tsai and Huang (2006)
			Protein activity increases	
3.	Cd	*Oryza sativa*	ROS generation	Yeh et al. (2007)
	Cu		Activation of MAPK3 and MPK6	
4.	Pb	*Oryza sativa*	Activation of 42KDa MAPK	Huang and Haung (2008)
			Protein activity is increased	
5.	Cr	*Zea mays*	Zm MPK5 activity is increased	Ding et al. (2009)
			Protein activity is increased	
6.	As	*Brassica juncea*	As induced mediated ROS signaling	Gupta et al. (2009)
			Activation of MAPK cascade	
7.	Several HM	*Zea mays*	Oxidative damage	Wang et al. (2010)
			Activation of ZmMPK3	
8.	Cd	*Arabidopsis thaliana*	Accumulation of ROS leading to oxidative damage	Liu et al. (2010)
9.	Cd	*Arabidopsis thaliana*	Cd-induced production of ROS	Remans et al. (2010)
	Cu		Activation of MAPK cascade	
			NADPH oxidase production enhanced	
			Lipoxygenase activity enhanced	
10.	As	*Oryza sativa*	Activation of OsMPK4, OsMKK4, and OsMPK3	Rao et al. (2011)
11.	Cu	*Arabidopsis thaliana*	Cd-induced production and accumulation of ROS	Smeets et al. (2013)
	Cd		Activation of MEKK, MKK2, and WRKY25 (*Arabidopsis* transcription factor)	
			Increased superoxide dismutase activity	
			Fluctuation in activity of miRNA	
			Single induced kinase (OXII) activated	
12.	Cu	*Arabidopsis thaliana*	Cd-induced production of ROS	Opdenakker et al. (2012)
	Cd		Increase in levels of mRNA	
			Activation of MAPKKK and MAPK	
13.	Fe	*Arabidopsis thaliana*	Upregulation of antioxidative enzymes	Smeets et al. (2013)
			MEKK, MKKK2, and WRKY25 cascade activated	

(continued)

Table 8.2 (continued)

S. No.	Heavy metal	Plant studied	MAPK and ROS signaling cross talk under heavy metal stress	References
14.	Cu	*Lolium perenne*	Production of NO enhanced	Dong et al. (2014)
			Activation of MAPK cascade	
15.	Cd	*Arabidopsis thaliana*	ROS-induced production of MPK3/6	Schellingen et al. (2015)
			Enhanced ethylene signaling and NADPH oxidases	

MAPK mitogen-activated protein kinase, *ROS* reactive oxygen species, *SIMK* salt stress-induced MAPK, SAMK salt stress-activated MAPK, *MMK2* mammalian MAPK2, *MMK3* mammalian MAPK3, *MAPK3* mitogen-activated protein kinase 3, *MPK6* mitogen protein kinase 6, *Zm MPK5 Zea mays* mitogen protein kinase 5, *Zm MPK3 Zea mays* mitogen protein kinase 3, *NADPH oxidase* nicotinamide adenine dinucleotide phosphate-oxidase, *OsMPK4 Oryza sativa* mitogen protein kinase 4, *OsMPK3 Oryza sativa* mitogen protein kinase 3, *MEKK* mitogen-activated protein kinase kinase, *MKK2* mitogen-activated protein kinase kinase 2, *miRNA 398* microRNA 398, *MAPKKK* mitogen-activated protein kinase kinase kinase, *MKKK2* mitogen protein kinase kinase kinase 2, *NO* nitric oxide

2010). This property of Ca can be very harmful for the cellular environments, as Ca at higher concentrations can form insoluble complexes. While at appropriate levels, these are involved in signal transduction pathways (Kudla et al. 2010). These appropriate cellular levels are maintained by storing Ca^{2+} mainly in vacuole, endoplasmic reticulum, and apoplast, and any increase in the cytosolic level can be considered as a response to biotic or abiotic stress (McAinsh and Pittman 2009; Dodd et al. 2010; Stael et al. 2012).

When plant cells come in contact with any type of stress, an elevation in the cytosolic Ca^{2+} level is considered to be a response to stress (Tuteja and Mahajan 2007). This increase is brought about by influx of Ca^{2+} ions from the apoplast via plasma membrane or other cellular organelles. It was postulated by Trewavas (1999) that such a localized increase of cytosolic Ca^{2+} can also cause the production of inositol 1,4,5-trisphosphate (IP_3) or cyclic ADP ribose (cADPR), which are soluble second messengers, and they have the ability to activate the Ca^{2+} channels which are spatially separated. The stress signals that caused an elevation of Ca^{2+} ions also require Ca^{2+} sensors which can further aid in signal transduction. These sensors are small proteins that can bind to Ca^{2+} and undergo a conformational change. Major Ca^{2+} sensors include calmodulin, calmodulin-like proteins, calcium-dependent protein kinases (CDPKs), and calcineurin B-like proteins (Bouche et al. 2005; McCormack et al. 2005; Das and Pandey 2010; Asai et al. 2013; Sewelam et al. 2016).

The defense system of the plants is activated in response to the imbalance caused by ROS overproduction. H_2O_2, in particular, has been reported to interact with Ca^{2+} and affect the signal transduction pathways. H_2O_2 and Ca^{2+} both are vital messengers that are responsible for stress signaling, and it has been suggested by several workers that these two have a reciprocal relationship (Cuypers et al. 2016).

Initial reports by Rentel et al. (2004) provided the evidence of responses of H_2O_2-responsive *GST1* gene that showed changes in its expression in response to changed Ca^{2+} and hence suggested the probable cross talk between the two. It has been postulated that exposure to heavy metals and subsequent increase of ROS increase the influx of Ca^{2+} ions by modifying the Ca channels in plasma membrane (Mazars et al. 2010; Manara 2012). On the other hand, Ca^{2+} also controls the production of H_2O_2 through NADPH oxidases (Mazars et al. 2010). It was suggested that H_2O_2 overproduction in oxidative burst requires influx of Ca^{2+} so as to activate NADPH oxidases. The enhanced Ca^{2+} levels in the cytosol further lead to activation of calmodulin that passes the signal to catalase (CAT) (Lamb and Dixon 1997; Yang and Poovaiah 2002; Khan et al. 2012). CAT is one of the important components of enzymatic antioxidative defense system that converts H_2O_2 to H_2O. Therefore, Ca $^{2+}$ controls the H_2O_2 homeostasis in response to oxidative burst. The evidence was also provided by a study on bright yellow two *Nicotiana tabacum* subjected to Cd stress which showed that increased contents of Ca^{2+} are required to activate NADPH oxidases (Garnier et al. 2006). The activation of NADPH oxidases is brought about by direct binding of Ca^{2+} to N-terminus of the enzyme and activating CDPKs that cause the phosphorylation of N-terminus (Sagi and Fluhr 2006; Kobayashi et al. 2007; Ogasawara et al. 2008). In a study on *Arabidopsis thaliana* subjected to Cd stress, the possible role of CDPKs in metal stress responses was indicated by *CDPK1* gene that was transcriptionally induced in response to Cd stress plants (Smeets et al. 2013). The study was also supported by reports on H_2O_2-treated *Triticum aestivum* in which isoforms of CDPKs were responsive to H_2O_2 (Li et al. 2008; Schulz et al. 2013). The H_2O_2 scavenging ability of peroxisomes under heavy metal stress is also considered to be enhanced due to cytosolic Ca^{2+} that can increase the activity of CAT3 (Costa et al. 2010). It was also reported by Baliardini et al. (2015) in *Arabidopsis* that the gene *CAX1*, which codes for Ca^{2+}/H^+ exchanger, was also positively correlated with Cd stress tolerance. The study also showed that *CAX1* gene was instrumental in sustaining the levels of Ca^{2+} in cytosol which further resulted in low accumulation of ROS under Cd exposure.

8.2.3 Hormone Signaling and ROS Signaling Under Heavy Metal Stress

In plants under stress conditions, cross talk among plant hormones helps in ameliorating the negative effects of abiotic stress. Cross talk between brassinosteroids (BRs) and other plant hormones like abscisic acid (ABA), auxins, ethylene, cytokinins, jasmonic acid (JA), salicylic acid, and gibberellins (GA) have been reported to play crucial role in growth and development of plants under abiotic stress conditions (Choudhary et al. 2012a; Kurepin et al. 2015). Cross talk of hormones such as ABA, JA, and ethylene with BRs specifies their significant role in modulation of stress responses. Interplay between BRs and ABA is found to be antagonistic in accelerated seed germination (Zhang et al. 2009). Their co-application is found to be effective in increasing tolerance of plant to various abiotic stresses. ABA is most effective in altering plant responses to drought stress. Other than ABA, an interaction between JA, cytokinins, auxins, BRs, and GA is found to modulate the expression of stress-responsive genes (Huang et al. 2008). Another report by Kurepin et al. (2015) suggested involvement of endogenous levels of hormones like SA, ABA, and ET in synthesis of GB. Synergistic effects of BRs along with ethylene, GA, and auxins have been found to increase plants' tolerance to flooding stress (Gommers et al. 2013; Ayano et al. 2014). It has also been reported that JA induced the synthesis of ethylene by modulating ACC synthase and oxidase activities (Kruzmane et al. 2002; Saniewski et al. 2003). Upregulation of gene responsible for ABA biosynthesis (*vp14)* has been reported, after BRs stimulated NO production which resulted in heavy metal stress tolerance in *Zea mays* (Zhang et al. 2011). Copper homeostasis has also been reported to modulate the signaling pathway of salicylic acid, ABA, and nitric oxide (Wimalasekera et al. 2011; Yan and Dong 2014). Under copper metal stress in *Raphanus sativus*, content of indole-3-acetic acid (IAA) was observed to reduce, but application of epibrassinolide (EBR) resulted in increase of IAA content due to the enhanced expression of *RsCYP79B3* gene. However, ABA content was reduced after the downregulation of *RsABA3* and upregulation of *RsCYP707A3* gene in response to EBR application under copper toxicity (Choudhary et al. 2012b). In *R. sativus*, EBR application resulted in increase of IAA content and decrease of ABA content under chromium metal stress (Choudhary et al. 2010). Reduction in stress caused by heavy metals like lead, Zn, and cadmium has also been observed in *Triticum aestivum* seedlings after the interaction of ABA and gibberellins (Erguuml and Oncel 2012). It has been reported that Cu metal modulates the biosynthesis of various plant hormones suggesting the role of Cu in hormonal cross talk. It has also been suggested that regulation of transcripts under environmental stress conditions that leads to the hormonal signaling in plants could result in increased micronutrient uptake (Penarrubia et al. 2015). Several reports of cross talk of ethylene with other phytohormones or with signaling cascade that is an imperative plant adaptation to heavy metal-induced oxidative burst can be cited (Thapa et al. 2012; Montero-Palmero et al. 2014a; Montero-Palmero et al. 2014b). *Arabidopsis* plant given Al treatment led to the upregulation expression of *AtACS* (*AtACS2*, *AtACS6*, and *AtACS8*) and *AtACO* (*AtACO1* and *AtACO2*) genes which are ethylene biosynthesis-related genes (Sun et al. 2010). Treatment with Al also elevated the transcript of AUXIN RESISTANT1 (AtAUX1) and PINFORMED2 (AtPIN2) indicating cross talk between auxin and ethylene.

8.3 Superoxide Radical Signaling Under Heavy Metal Stress

Superoxide radical (O_2^-) is the first ROS formed as a result of oxidative stress (Das and Roychoudhury 2014). Increase in O_2^- ions was observed by Choudhary et al. (2012a) in *R. sativus* subjected to Cu metal toxicity which would have resulted in the upregulation of *RsCOPT1* and *RsCOPT2* (involved in Cu uptake) and *RsMT1C*

Table 8.3 O_2^--induced signaling response in various plants under heavy metal stress

S. No.	Plant	Heavy metal	O_2^- signaling response	References
1.	*Brassica juncea*	Cd	$O_2^{\cdot -}$ signaling induced degradation of photosynthetic efficiency	Hayat et al. (2007)
			Enhanced hormonal signaling	
2.	*Pisum sativum*	Cd	Methyl jasmonate and ethylene concentrations were increased	Rodriguez-Serrano et al. (2009)
			Enhanced expression of *PrP4A*, *HSP71.2*, and *chitinase*	
3.	*Raphanus sativus*	Cu	*RsCOPT1* and *RsCOPT2* genes were upregulated resulting in enhanced Cu uptake	Choudhary et al. (2012a)
			Upregulations of *RsMT1C* and *RsCCH1* involved in metallothioneins and Cu chaperones	
4.	*Raphanus sativus*	Cr	Downregulation of *RsADC* involved in biosynthesis of putrescine	Choudhary et al. (2012b)
			Upregulation of *RsSPDS* involved in spermidine biosynthesis	
			SOD, CAT, GR, and GPOX activities were enhanced along with increased contents of proline and glycine betaine	
5.	*Brassica juncea*	Co	SOD, POD, DHAR, and MDHAR activities were increased	Karuppanapandian and Kim (2013)
			DNA ligase activity increased	
			MAPK-like protein kinases were activated	
6.	*Brassica napus*	Cr	POD and GR activities were increased in leaves	Ali et al. (2014)
			GSH and GSSG contents were enhanced in leaves	
7.	*Lolium perenne*	Cu	K, Fe, Mg, and Zn contents were decreased in roots	Dong et al. (Dong et al. 2014)
			Photosynthesis rate was decreased	
8.	*Oryza sativa*	Cu	GSH and proline content were increased	Mostofa et al. (2015)
9.	*Brassica napus*	Pb	HSP90-1 and MT-1 transcript levels increased	Gill et al. (2015)
			Expression of *SOD*, *POD*, *CAT*, *APOX*, and *GR* was increased	
			GSH and GSSG contents were enhanced	

(continued)

Table 8.3 (continued)

S. No.	Plant	Heavy metal	O_2^- signaling response	References
10.	*Phaseolus coccineus*	Cu	Activities of SOD, CAT, APOX, and POD were increased	Hanaka et al. (2016)
			hGSH and proline content increased	
			LMWOAs were decreased	

HSP heat shock protein, *SOD* superoxide dismutase, *CAT* catalase, *GR* glutathione reductase, *GPOX* glutathione peroxidase, *POD* guaiacol peroxidase, *DHAR* dehydroascorbate reductase, *MDHAR* monodehydroascorbate reductase, *MAPK* mitogen-activated protein kinase, *GSH* glutathione, *GSSG* oxidized glutathione, *DNA* deoxyribonucleic acid, *MT-1* mouse metallothionein-1 gene, *hGSH* homoglutathione, *LMWOAs* low molecular weight organic acids

and *RsCCH1* (involved in biosynthesis of metallothioneins and Cu chaperones). Under chromium metal stress in *R. sativus*, O_2^- ion produced as a result of oxidative stress has resulted in the downregulation of *RsADC* involved in biosynthesis of putrescine, upregulation of *RsSPDS* involved in spermidine biosynthesis, enhanced activities of SOD, CAT, GR, and GPOX, and increased contents of proline and glycine betaine (Choudhary et al. 2012b). In the leaves of *Brassica juncea* plants grown under cobalt stress, O_2^- ions were observed to increase as a consequence of oxidative stress. Furthermore, it has been observed that SOD (superoxide dismutase), POD (guaiacol peroxidase), DHAR (dehydroascorbate reductase), and MDHAR (monodehydroascorbate reductase) activities were increased along with enhanced DNA ligase activity and activation of MAPK-like protein kinases (Karuppanapandian and Kim 2013). Chromium metal-induced production of O_2^- ions in *B. napus* has resulted in enhanced contents of GSH and GSSG along with increased activities of POD and GR (glutathione reductase) (Ali et al. 2014). In *B. napus*, lead metal toxicity induced the production of O_2^- ions resulting in enhanced *HSP90-1* and *MT-1* transcript levels and expression of *SOD*, *POD*, *CAT*, *APOX*, and *GR* genes (Gill et al. 2015). In *Phaseolus coccineus*, O_2^- content was noticed to enhance under copper metal stress which resulted in increase in contents of hGSH and proline, activities of SOD, CAT (catalase), APOX (ascorbate peroxidase), and POD, along with reduction in contents of low molecular weight organic acids (LMWOAs) (Hanaka et al. 2016) (Table 8.3).

8.4 Singlet Oxygen Signaling

Singlet oxygen ($^1O_2$) is the first excited state of molecular oxygen (O_2). Photosystem II (PSII) is associated with the production of $^1O_2$. During photosynthesis, deficient indulgence of energy results in the development of triplet-state chlorophyll which reacts with three molecules of molecular oxygen to produce singlet oxygen (Kochevar 2004; Jajic et al. 2015). There is narrow range of information about signaling character of singlet oxygen as it is very reactive and has got a very short half-life of ~100 ns. Because of this, it can travel merely a little space of <100 nm in the cells (Moan 1990; Niedre et al. 2002). Because of its highly excited state and short life, singlet oxygen quickly reacts with the molecules present in immediate neighborhood to its production site, i.e., in the chloroplast. This indicates that singlet oxygen may convey definite signals from end-to-end spatial aspect of ROS production (Triantaphylidès and Havaux 2009). During the conditions of heavy metal stress, singlet oxygen production is enhanced which leads to breakage of double bonds in unsaturated fatty acids that ultimately causes damage to thylakoid membrane. These events cause imbalance in complexes of the photosynthetic apparatus, thereby generating response signals for modulation of gene expression via the pathway that detects the physical condition of the membrane (Glatz et al. 1999). This study indicates that under stress conditions, the receptors on the cell membranes or cell surface do not recognize the stimulus straight in the form of stress but through the chloroplast stress signals. Oxidative damage engages specific signals that are used to formulate suitable modifications of cell structure or gene expression in reaction to metal stress. There are some other reports that prove that primarily, singlet oxygen does not function as a toxin, but to a certain extent, it plays a role of signal that triggers a number of stress response corridor. The *flu* mutant of *Arabidopsis* is used to study the specific activity of singlet oxygen, because it allocates the stimulation of $^1O_2$ within plastids in a controlled manner. The singlet oxygen stress responses originate either by establishment of a genetically controlled stress-response series or by the physicochemical injury caused by $^1O_2$ (Kim et al. 2008). Singlet oxygen signaling in *flu* mutants of *Arabidopsis* has been found to be coupled with programmed cell death. The progression of programmed cell death stimulates definite unambiguous features in terms of gene stimulation as compared with other ROS (Triantaphylidès and Havaux 2009; Roel et al. 2003). Executors 1 and 2 are the signaling proteins which hold back the singlet oxygen-stimulated cell death (Lee et al. 2007). After the generation of singlet oxygen in plastids, about 300 nuclear genes get synchronized within 30 min time (Danon et al. 2005; Ochsenbein et al. 2006). Additionally, around 50 other genes that are responsible for expression of transcription factors (numerous DNA-binding proteins, *WRKY* transcription factors, and zinc-finger proteins) have also been reported to within 30 min of its discharge; the expression of those genes increases three times (Gadjev et al. 2006; Laloi et al. 2006; Op den Camp 2003). Singlet oxygen stimulates the expression of the nuclear-encoded glutathione peroxidase homologous gene *GPXH* in *Chlamydomonas reinhardtii* (Fischer et al. 2011). The study showed that $^1O_2$ aroused *GPXH* expression by triggering a signaling mechanism on the outer thylakoid membrane. This is the sign that $^1O_2$ can initiate a reaction at other sites apart from where its production was triggered.

Table 8.4 Hydroxyl radical signaling under heavy metal stress

S. No.	Heavy metal	Plant	Physiological effects	Hydroxyl radical signaling responses	References
1.	Cu	*Phaseolus vulgaris* L.	Decreased photosynthetic competence, low quantum efficiency of PSII, injuries to DNA, RNA, lipids, carbohydrates, and proteins	Stimulation of increase in quantum yield of photoinhibition	Patsikka et al. (1998)
2.	Cu	*Prunus cerasifera*	Depleted PSII action centers and led to photoinhibition and disruption of its repair cycle	Modulation of transcription of *Sod* and *Cat* gene expression	Lombardi and Sebastian (2015)
3.	Cd	*Lycopersicon esculentum* Ibiza F1	Decreased biomass production	Induction of phytochelatin synthesis and cadmium sequestration, which might decrease its cytosolic accessibility and thus limit its toxic effects	Mediouni et al. (2006)
4.	Cu	*Spinacia oleracea* L.	Depression in seedling fresh weight	Stimulation of root elongation	Sharma et al. (2010)
5.	Mn	*Cucumis sativus* L.	Inhibitory effect of Si on guaiacol-POD activity decreases OH radical accumulation in leaves	Regulation of Fenton reaction and direct effect on guaiacol-POD activity	Maksimovic´ et al. (2012)
6.	Cd	*Phaseolus vulgaris* L.	Cellular oxidative damage, lipid peroxidation	Enhancement of antioxidant enzymes	Foroozesh et al. (2012)
7.	Cu	*Brassica napus* L.	Drop in chlorophylls (Chl*a* and Chl*b*) and carotenoids content	Enhancement in bioaccumulation	Pesko and Král'ová (2013)
8.	Co, Cd	*Brassica oleracea* var. botrytis L.	Chlorosis of young emerging leaf	Stimulation of antioxidant defense system	Sinha et al. (2015)

(continued)

Table 8.4 (continued)

S. No.	Heavy metal	Plant	Physiological effects	Hydroxyl radical signaling responses	References
9.	Pb	*Gossypium hirsutum* L.	Oxidative damage, stunted growth, wilting, and leaf abscission	Increase in protein content and antioxidants	Bharwana et al. (2016)
10.	Cd	*Vigna aconitifolia* L.	Cell membrane damage, lipid peroxidation	Increase in nonenzymatic antioxidants, protein carbonyls, and nonprotein thiols	Vijendra et al. (2016)

PSII photosystem II, *DNA* deoxyribonucleic acid, *RNA* ribonucleic acid, *Sod* superoxide dismutase, *Cat* catalase, POD peroxidase, *OH* hydroxyl radical, *Chla* chlorophyll a, *Chlb* chlorophyll b

8.5 Hydroxyl Radical Signaling

The most reactive and highly toxic ROS are hydroxyl radicals (HO·), which are generated by reaction of H_2O_2 with Fe^{2+} and Cu^{+} at normal temperature and neutral pH (Quan 2008). Metal stress leads to the formation of excess of ROS including OH. Arsenic metal leads to the production of OH˙ radicals by O_2•– dismutation and Fenton reaction and Haber-Weiss reaction (Bongiovanni et al. 2007; Hughes 2002; Flora et al. 2008). HO· can make a way into the membranes and leave the mitochondria, resulting in peroxidation of mitochondrial membrane and formation of lipid aldehydes (Rhoads et al. 2006). Hydroxyl radicals produced in response to Cd and Hg stress lead to the blockage of functional groups in biomolecules, and the damage caused cannot be restricted by the antioxidants (Schutzendubel and Polle 2002). However the condition of increased oxidation is a harmful event, but sometimes, it may act as a signal for regulating plant metabolism to the changing circumstances of growth (Miller et al. 2010). Therefore, the oxidative stress caused by the OH˙ radicals as a result of heavy metal stress must not be considered as indication of cellular damage, but it may be considered as a signal for plant, to stimulate appropriate acclimation to stressed condition (Jaspers and Kandasjarvi 2010). Table 8.4 discusses hydroxyl radical signaling under heavy metal stress.

Table 8.5 H_2O_2-induced signaling response in various plants under heavy metal stress

S. No.	Plant	Heavy metal	H_2O_2-induced signaling response	References
1.	*Pisum sativum*	Cd	NADPH oxidase leads to H_2O_2 accumulation in peroxisomes and mitochondria	Romero-Puertas et al. (2004)
			Cd induces ROS, H_2O_2, O_2^- production, and lower ROS level function as signal molecules, and these signal molecules induce defense genes	
			Regulation of ROS through protein phosphatases, Ca^{2+}channels, cGMP	
2.	*Oryza sativa*	Cd	H_2O_2 signals in increasing enzymatic activities of GPX, CAT, SOD, APX	Hu et al. (2009)
			Rise in glutathione and ascorbic acid levels	
			Cd sequestration takes place in roots through H_2O_2-Cd signaling	
3.	*Oryza sativa*	Ni	Ni-induced H_2O_2 results in oxidative stress, enhanced lipid peroxidation, and decreased protein thiols	Maheshwari and Dubey (2009)
			SOD, GPX, APX activities increase due to H_2O_2-Ni signaling	
4.	*Scytosiphon lomentaria*	Cu	Production of lipoperoxides by cascade of reactions	Contreras et al. (2009)
			Lipoperoxides accumulation due to LOX	
			Modulation of activities of antioxidant enzyme such as CAT, GPOX, APOX, DHAR effected	
5.	*Oryza sativa*	As	Oxidative stress leads to elevation of enzymatic and nonenzymatic antioxidants	Mishra et al. (2011)
			Enhancement in thiols and phytochelatin synthesis	
6.	*Ulva compressa*	Cu	Kreb cycle activity effected	Gonzalez et al. (2012)
			Increase in activities of pyruvate dehydrogenases, 2-oxoglutarate dehydrogenases, and isocitrate dehydrogenases	
			These activities were hindered due to the presence of Ca chelating agent EGTA (ethylene glycol-bis(β-aminoethylether)-N,N,N',N'-tetraacetic acid)	
7.	*Zea mays* (sweet corn)	Cu	Signaling pathways affected, due to declined expression of ProDH	Wen et al. (2013)
			Enhancement in proline content and arginase activity	
			Increased enzymatic expression of ornithine and glutamate pathway	

(continued)

Table 8.5 (continued)

S. No.	Plant	Heavy metal	H_2O_2-induced signaling response	References
8.	*Brassica juncea*	Ni Zn	Ethylene application prevents photosynthesis by maintaining PSII pathway through ribulose-1-P carboxylase and photosynthesis N-use efficiency (PNUE)	Khan and Khan et al. (2014)
9.	*Arabidopsis thaliana*	Cd	Signaling pathway showed increased expression of gene encoding Ca^{2+}/H^{+} exchanger (CAX1)	Baliardini et al. (2015)
			CAX1 maintain Ca^{2+} balance so as to avoid H_2O_2 accumulation	
10.	*Triticum aestivum*	Zn	H_2O_2 signaling results in the regulation of antioxidative enzymatic activities through ROS generation from plasma membrane NADPH oxidase	Duan et al. (2015)
			Reduction in the levels of diamine oxidase, cell wall soluble POD, and polyamine oxidase activity	
11.	*Arabidopsis thaliana*	Cd	Ion exchange channels are disturbed due to enhancement of K, Ca, Zn, Fe accumulation	Gupta et al. (2016)
			Redox homeostasis is differentially regulated	
12.	*Trapa bicornis* (water caltrop)	Pb	Pb along with H_2O_2 mediates upregulation of transcripts of CAT, APX, and lignin biosynthesis-related gene encoding F-5-H	Li and Yuan (2016)
			Production of F-5-H signals leads to modification of cell wall	

cGMP cyclic guanosine monophosphate, *ROS* reactive oxygen species, *GPX* guaiacol peroxidase, *CAT* catalase, *SOD* superoxide dismutase, *APX* ascorbate peroxidase, *GPOX* glutathione peroxidases, *APOX* ascorbate peroxidases, *DHAR* dehydroascorbate reductase, *LOX* lipoxygenases, *PDH* pyruvate dehydrogenases, *ODH* 2-oxoglutarate dehydrogenases, *IDH* isocitrate dehydrogenases, *ProDH* proline dehydrogenase 1, *RuBisCO* ribulose-1-P carboxylase, CAX1 calcium exchanger 1, *POD* peroxidases, *F-5-H* ferulate-5-hydroxylase

8.6 H_2O_2 Signaling Under Heavy Metal Stress

Oxidative stresses lead to generation of ROS like H_2O_2 (Tanou et al. 2009; Qiaoa et al. 2013). Initially, ROS were considered as lethal products produced by aerobic metabolism, but now they are known to play vital role in cellular signaling (Moller and Sweetlove 2010). There are three ways in which heavy metals, such as Cd and Cu, may influence ROS accumulation: (i) miR398 expression induced by excess of Cd^{2+}, then inhibition of Cu/Zn/SOD (CSD) function, and ultimately ROS accumulation takes place; (ii) regulatory action of Ca^{2+} gets inhibited by excess of Cd^{2+},

Table 8.6 NO-induced signaling response in various plants under heavy metal stress

S. No.	Plant	Heavy metal	Signaling response of NO	References
1.	*Cassia tora L.*	Al	NO signaling induces oxidative stress by action of reduction in epoxygenases and antioxidant enzyme activity	Wang and Yang (2005)
			Increase in lipid peroxidation and reactive oxygen species	
2.	*Glycine max*	Cd	Antioxidative activities of NO result in scavenging of ROS directly	Kopyra et al. (2006)
3.	*Triticum aestivum*	Cu	NO acts as gaseous signal, enhances activities of SOD, CAT	Hu et al. (2007)
			Reduction in H_2O_2, ultimately increasing antioxidative capacity	
			Reduced levels of MDA, lipoxygenases	
4.	*Triticum aestivum*	Cd	NO acts as signaling molecule in metal-induced root growth inhibition	Groppa et al. (2008)
			Induces lipid peroxidation in roots	
5.	*Triticum aestivum*	Al	Enhanced antioxidant system	Zhang et al. (2008)
			Rise in the proline content, catalase, ascorbate peroxidase, superoxide dismutase activity	
			Reduced H_2O_2 and malondialdehyde levels	
6.	*Arabidopsis thaliana*	Cd	Signaling events lead to the expression of marker senescence- associated gene 12 (SAG12) which increase NO and phytochelatin synthesis	Michele et al. (2009)
			Modulation of phytochelatin content through NO, by S-nitrosylation	
7.	*Oryza sativa*	As	NO added in the form of SNP decreases As stress through scavenging	Singh et al. (2009)
			NO signals with As to enhance antioxidant activities of enzymes such as CAT, POD, APOX, SOD	
8.	*Brassica napus*	Zn	Lipid peroxidation activity increases biochemically	Wang et al. (2009)
			Roots and leaves show increase in NADH oxidase and peroxidase (POD) activities and decrease in SOD, CAT, and APOX	
9.	*Arabidopsis thaliana*	Cd	Pathway involving upregulation of IRT1 genes related to Fe uptake contributes to Cd accumulation	Besson-Bard et al. (2009)
			Cellular pathway equivalent to Fe deprivation is involved	

(continued)

Table 8.6 (continued)

S. No.	Plant	Heavy metal	Signaling response of NO	References
10.	*Arabidopsis thaliana*	As	ROS and RNS interactive pathways lead to lipid peroxidation	Leterrier et al. (2012)
			GSH produced by S-nitrosylation process reacts with NO to produce GNSO acting as signal molecule	
			$ONOO^-$ produced through NO signaling causes nitro-oxidative stress	
11.	*Lupinus luteus* L.	Cd	Due to TUNEL positive reaction	Arasimowicz-Jelonek et al. (2012)
			NADPH oxidase-dependent superoxide anion is formed through NO signaling	
			NO and O_2^- signaling results in PCD in roots	
			Post-stress signals in leaves increased	
12.	*Arabidopsis thaliana*	Cd	Metabolic pathways in peroxisomes lead to formation of peroxynitrite ($ONOO^-$) formation in excess	Corpas and Barroso (2014)
			Nitro-oxidative stress initiated	
13.	*Typha angustifolia*	Cd	SNP alleviate electrolyte leakage and MDA in leaves and roots	Zhao et al. (2016)
			Regulation of antioxidant metabolism and Cd accumulation in cell wall of roots	
14.	*Oryza sativa*	Cd	SNAP treatment leads to Cd tolerance	Yang et al. (2016)
			Ion homeostasis regulated by H^+ATPase is reduced	
			PLD activity increases and accumulation of glutathione and other antioxidative enzymes	

ROS reactive oxygen species, *SOD* superoxide dismutase, *CAT* catalase, *MDA* malondialdehyde, *LOX* lipoxygenases, *APOX* ascorbate peroxidase, *SOD* superoxide dismutase, *SAG12* senescence-associated gene 12, *SNP* sodium nitroprusside, *GPX* guaiacol peroxidase, *NADH oxidase* nicotinamide adenine dinucleotide phosphate-*oxidase*, *POD* peroxidase, *IRT1* iron-regulated transporter1, *RNS* reactive nitrogen species, *GSH* glutathione, *GNSO* S-nitroso-L-glutathione, *PCD* programmed cell death, *TUNEL* terminal deoxynucleotidyl transferase dUTP nick end labeling, *SNAP* S-nitroso-N-acetyl-l,l-penicillamine, *PLD* phospholipase D

which enhances ROS accumulation by CSD activity inhibition; and (iii) NADPH oxidases produced by excess of Cd^{2+} lead to H_2O_2 production (Lin and Aarts 2012). A *Rbohs* gene produced in *Gossypium hirsutum* due to overexpression of ROS through Ca^{2+}-dependent pathway determines elongation of fiber because of loosening of cell wall and turgidity. On the other hand, H_2O_2 leads to inhibition of cotton fiber. ROS together with Ca^{2+} regulates elongation of fiber cell. Thus, these studies provide us with the information regarding signaling pathways of *GhFAnnxA* in fiber development (Zhang et al. 2016). Another report in *Arabidopsis thaliana* plant suggests role of aquaporin *AtPIP1;4* gene in leaves, induced by bacteria and H_2O_2

accumulation in the cytoplasm. *AtPIP1;4* gene acts as facilitator of H_2O_2 transport across plasma membrane and leads to the H_2O_2 translocation from the apoplast to the cytoplasm. Furthermore, if any mutation occurs in *AtPIP1;4* gene locus, it stops apocytoplastic signal transduction pathways showing its importance in defense mechanisms (Tian et al. 2016). Important example of H_2O_2 signaling is accumulation of p53 tumor suppressor genes which induces production of H_2O_2 which is further modulated by GPX1. High production of H_2O_2 triggers p53-dependent cell apoptosis (Sablina et al. 2005). Table 8.5 discusses H_2O_2-induced signaling response in various plants under heavy metal stress.

Thiol peroxidase has a key role in regulating several H_2O_2 signaling pathways. Thiol peroxidase regulates the flux and levels of H_2O_2 which are produced in response to cytokine-induced cellular signaling and growth factor in subcellular compartments, e.g., PRX 3, a mitochondrial peroxidase that is reported to modulate apoptosis signaling by scavenging of H_2O_2 (Chang et al. 2004).

8.7 NO Signaling Under Heavy Metal Stress

NO plays an important physiological role in plants, by acting as a signaling molecule within them. It behaves as a bioactive signaling entity only at low concentration and acts as stress molecule when present in high concentrations. Further, they may also be linked to damaging of macromolecules through protein nitration (Corpas and Barroso 2015). Variety of NO species include NO_2, N_2O_3, NO_2^-, and NO_3^- (Arasimowicz and Floryszak-Wieczorek 2007). NO is synthesized in the similar way as H_2O_2 during stress responses and has been reported as secondary messenger under heavy metal stress (Khan et al. 2014). Table 8.6 discusses NO-induced signaling response in various plants under heavy metal stress. There are a number of reactive nitrogen species (RNS) produced under heavy metal stresses, through different mechanisms such as peroxynitrite (ONOO−), resultant of the NO reaction with $O_2^{\bullet-}$ (Radi 2013), and *S*-nitrosothiols (SNOs), formed through the reaction between NO and thiol groups (Broniowska and Hogg 2012). Further, the communications between tripeptide (GSH) and NO produced *S*-nitrosoglutathione (Corpas et al. 2013; Begara-Morales et al. 2014). The signaling of NO is well studied in mammals via cGMP-dependent or cGMP-independent pathways. Although, presence of cGMP pathway in plants is also widely reported (Arasimowicz and Floryszak-Wieczorek 2007).

8.8 Conclusion and Future Prospects

Coordinated and suitable signaling cascade in response to heavy metal stress has been targeted by researchers in recent times. High concentration of ROS in cells as a result of heavy metal stress limits growth and development of plants. Plants have developed a multifaceted signaling pathway to acclimatize to erratic stress conditions. Identification of specific heavy metal sensors and their signaling under stressed conditions is well elucidated. ROS rather than being just lethal constituents of cells have imperative roles as signaling molecules which regulate stress acclimatization, cell death, and development. Though major advances have been achieved to comprehend ROS signaling under heavy metal stress, it is still far from precision that ROS and its transcriptional reprogramming have a critical role in stress regulation and metabolism.

References

Ali B, Mwamba TM, Gill RA, Yang C, Ali S, Daud MK, Wu Y, Zhou W (2014) Improvement of element uptake and antioxidative defense in *Brassica napus* under lead stress by application of hydrogen sulfide. Plant Growth Regul 74:261–273

Apel K, Hirt H (2004) Reactive oxygen species: metabolism, oxidative stress, and signal transduction. Annu Rev Plant Biol 55:373–399

Arasimowicz M, Floryszak-Wieczorek J (2007) Nitric oxide as a bioactive signaling molecule in plants stress responses. Plant Sci 172:876–887

Arasimowicz-Jelonek M, Floryszak-Wieczorek J, Deckert J, Rucińska-Sobkowiak R, Gzyl J, Pawlak-Sprada S, Abramowski D, Jelonek T, Gwóźdź EA (2012) Nitric oxide implication in cadmium-induced programmed cell death in roots and signaling response of yellow lupine plants. Plant Physiol Biochem 58:124–134

Asai S, Ichikawa T, Nomura H, Kobayashi M, Kamiyoshihara Y, Mori H et al (2013) The variable domain of a plant calcium-dependent protein kinase (CDPK) confers sub cellular localization and substrate recognition for NADPH oxidase. J Biol Chem 288:14332–14340

Autreaux BD, Toledano MB (2007) ROS as signaling molecules: mechanisms that generate specificity in ROS homeostasis. Mol Cell Biol 8:813–824

Ayano M, Kani T, Kojima M, Sakakibara H, Kitaoka T, Kuroha T, Angeles-Shim RB, Kitano H, Nagai K, Ashikari M (2014) Gibberellin biosynthesis and signal transduction is essential for internode elongation in deepwater rice. Plant Cell Environ 37:2313–2324

Baliardini C, Meyer CL, Salis P, Saumitou-Laprade P, Verbruggen N (2015) CATIONEXCHANGER1 cosegregates with cadmium tolerance in the metal hyperaccumulator *Arabidopsis halleri* and plays a role in limiting oxidative stress in *Arabidopsis* Spp. Plant Physiol 169:549–559

Begara-Morales JC, Sánchez-Calvo B, Luque F, Leyva-Pérez MO, Leterrier M, Corpas FJ, Barroso JB (2014) Differential transcriptomic analysis by RNA-Seq of GSNO-responsive genes between *Arabidopsis* roots and leaves. Plant Cell Physiol 55:1080–1095

Besson-Bard A, Gravot A, Richaud P, Auroy P, Duc C, Gaymard F, Taconnat L, Renou JP, Pugin A, Wendehenne D (2009) Nitric oxide contributes to cadmium toxicity in Arabidopsis by promoting cadmium accumulation in roots and by up-regulating genes related to iron uptake. Plant Physiol 149:1302–1315

Bharwana SA, Ali S, Farooq MA, Farid M, Bashir N, Ahmad R (2016) Physiological and biochemical changes induced by lead stress in cotton (*Gossypium hirsutum* L.) seedlings. Acad J Agri Res 4:160–167

Bhattacharjee S (2005) Reactive oxygen species and oxidative burst: roles in stress, senescence and signal transduction in plants. Curr Sci 89:1113–1121

Bongiovanni GA, Soria FA, Eynard AR (2007) Effects of the plant flavonoids silymarin and quercetin on arsenite-induced oxidative stress in CHO-K1 cells. Food Chem Toxicol 45:971–976

Bouche N, Yellin A, Snedden WA, Fromm H (2005) Plant-specific calmodulin-binding proteins. Annu Rev Plant Biol 56:435–466

Broniowska KA, Hogg N (2012) The chemical biology of *S*-nitrosothiols. Antioxid Redox Signal 17:969–980

Choudhary SP, Kanwar M, Bhardwaj R, Yu JQ, Tran LSP (2012a) Chromium stress mitigation by polyamine-brassinosteroid application involves phytohormonal and physiological strategies in *Raphanus sativus* L. PLoS One 7(3):e33210

Choudhary SP, Oral HV, Bhardwaj R, Yu JQ, Tran LSP (2012b) Interaction of brassinosteroids and polyamines enhances copper stress tolerance in *Raphanus sativus*. J Exp Bot 63:5659–5675. doi:10.1093/jxb/ers219

Contreras L, Mella D, Moenne A, Correa JA (2009) Differential responses to copper-induced oxidative stress in the marine macroalgae *Lessonia nigrescens* and *Scytosiphon lomentaria* (Phaeophyceae). Aquat Toxicol 94:94–102

Corpas FJ, Alché JD, Barroso JB (2013) Current overview of S-nitrosoglutathione (GSNO) in higher plants. Front Plant Sci 4:126. doi:10.3389/fpls.2013.00126

Corpas FJ, Barroso JB (2014) Peroxynitrite (ONOO−) is endogenously produced in *Arabidopsis* peroxisomes and is overproduced under cadmium stress. Ann Bot 113:87–96

Corpas FJ, Barroso JB (2015) Functions of Nitric Oxide (NO) in roots during development and under adverse stress conditions. Plants 4:240–252. doi:10.3390/plants4020240

Costa A, Drago I, Behera S, Zottini M, Pizzo P, Schroeder JI et al (2010) H_2O_2 in plant peroxisomes: an in vivo analysis uncovers a Ca^{2+}-dependent scavenging system. Plant J 62:760–772

Cuypers A, Hendrix S, dos Reis RA, Smet SD, Deckers J, Giel H, Jozefczak M, Loix C, Vercampt H, Vangronsveld J, Keunen E (2016) Hydrogen peroxide, signaling in disguise during metal phytotoxicity. Front Plant Sci 7:1–25

Danon A, Miersch O, Felix G, Camp RG, Apel K (2005) Concurrent activation of cell death-regulating signalling pathways by singlet oxygen in *Arabidopsis thaliana*. Plant J 41:68–80

Das R, Pandey GK (2010) Expressional analysis and role of calcium regulated kinases in abiotic stress signaling. Curr Genomics 11:2–13

Das K, Roychoudhury A (2014) Reactive oxygen species (ROS) and response of antioxidants as ROS-scavengers during environmental stress in plants. Front Environ Sci 2:53. doi:10.3389/fenvs.2014.00053

De Michele R, Vurro E, Rigo C, Costa A, Elviri L, Di Valentin M, Careri M, Zottini M, Sanità di Toppi L, Lo Schiavo F (2009) Nitric oxide is involved in cadmium-induced programmed cell death in Arabidopsis suspension cultures. Plant Physiol 150:217–228

del Río LA (2015) ROS and RNS in plant physiology: an overview. J Exp Bot 66:2827–2837

Ding HD, Zhang XH, Xu SC, Sun LL, Jiang MY, Zhang AY, Jin YG (2009) Induction of protection against paraquat-induced oxidative damage by abscisic acidin maize leaves is mediated through mitogen-activated protein kinase. J Integr Plant Biol 51:961–972

Dodd AN, Kudla J, Sanders D (2010) The language of calcium signaling. Annu Rev Plant Biol 61:593–620

Dong Y, Xu L, Wang Q, Fan Z, Kong J, Bai X (2014) Effects of exogenous nitric oxide on photosynthesis, antioxidative ability, and mineral element contents of perennial ryegrass under copper stress. J Plant Interact 9:402–411

Duan XLX, Ding F, Zhao J, Guo A, Zhang L, Yao J, Yang Y (2015) Interaction of nitric oxide and reactive oxygen species and associated regulation of root growth in wheat seedlings under zinc stress. Ecotoxicol Environ Saf 113:95–102

Erguuml N, Oncel I (2012) Effects of some heavy metals and heavy metal hormone interactions on wheat (*Triticum aestivum* L. cv. Gun 91) seedlings. Afri J Agri Res 7:1518–1523

Fischer BB, Krieger-Liszkay A, Hideg E, Šnyrychová I, Wiesendanger M, Eggen RIL (2011) Role of singlet oxygen in chloroplast to nucleus retrograde signaling in *Chlamydomonas reinhardtii*. FEBS Let 581:5555–5560

Flora SJS, Mittal M, Mehta A (2008) Heavy metal induced oxidative stress & its possible reversal by chelation therapy. Indian J Med Res 128:501–523

Foroozesh P, Bahmani R, Pazouki A, Asgharzadeh A, Rahimdabbagh S, Ahmadvand S (2012) Effect of cadmium stress on antioxidant enzymes activity in different bean genotypes. J Agri Biol Sci 7:351–356

Foyer CH, Noctor G (2005) Oxidant and antioxidant signaling in plants: are-evaluation of concept of oxidative stress in a physiological context. Plant Cell Environ 28:1056–1071

Fusco N, Micheletto L, Dal Corso G, Borgato L, Furini A (2005) Identification of cadmium-regulated genes by cDNA-AFLP in the heavy metal accumulator *Brassica juncea* L. J Exp Bot 56:3017–3027

Gadjev I, Vanderauwera S, Gechev TS, Laloi C, Minkov IN, Shulaev V, Apel K, Inze D, Mittler R, Van Breusegem F (2006) Transcriptomic footprints disclose specificity of reactive oxygen species signalling in *Arabidopsis*. Plant Physiol 141:436–445

Garnier L, Simon-Plas F, Thuleau P, Agnel JP, Blein JP, Ranjeva R et al (2006) Cadmium affects tobacco cells by a series of three waves of reactive oxygen species that contribute to cytotoxicity. Plant Cell Environ 29:1956–1969

Gill SS, Tuteja N (2010) Reactive oxygen species and antioxidant machinery in abiotic stress tolerance in crop plants. Plant Physiol Biochem 48:909–930

Gill RA, Ali B, Islam F, Farooq MA, Gill MB, Mwamba TM, Zhou W (2015) Physiological and molecular analyses of black and yellow seeded *Brassica napus* regulated by 5-aminolivulinic acid under chromium stress. Plant Physiol Biochem 94:130–143

Glatz A, Vass I, Los DA, Vigh L (1999) The synechocystis model of stress: from molecular chaperones to membranes. Plant Physiol Biochem 37:1–12

Gommers CMM, Visser EJW, St Onge KR, Voesenek LACJ, Pierik R (2013) Shade tolerance: when growing tall is not an option. Trends Plant Sci 18:65–71. doi:10.1016/j.tplants.2012.09.008

González A, Cabrera M, Henríquez MJ, Contreras RA, Morales B, Moenne A (2012) Crosstalk among calcium, hydrogen peroxide and nitric oxide and activation of gene expression involving calmodulins and calcium-dependent protein kinases in *Ulva compressa* exposed to copper excess. Plant Physiol 158:1451–1462

Gropa MD, Rosales EP, Lannone MF, Benavides MP (2008) Nitric oxide, polyamines and Cd induced phytotoxicity in wheat roots. Phytochemistry 69:2609–2615

Groß F, Dumer J, Gaupels F (2013) Nitricoxide, antioxidants and prooxidants in plant defence responses. Front Plant Sci 29:419. doi:10.3389/fpls.2013.00419

Gupta M, Sharma P, Sarin N, Sinha AK (2009) Differential response of arsenic stress in two varieties of *Brassicajuncea* L. Chemosphere 74:1201–1208

Gupta DK, Pena LB, Romero-Puertas MC, Hernández A, Inouhe M, Sandalio LM (2016) NADPH oxidases differentially regulate ROS metabolism and nutrient uptake under cadmium toxicity. Plant Cell Environ. doi:10.1111/pce.12711

Halliwell B (2006) Reactive species and antioxidants. Redox biology is a fundamental theme of aerobic life. Plant Physiol 141:312–322

Hanaka A, Wójcik M, Dresler S, Mroczek-Zdyrska M, Maksymiec W (2016) Does methyl jasmonate modify the oxidative stress response in *Phaseolus coccineus* treated with Cu? Ecotoxicol Environ Saf 124:480–488

Hansch R, Mendel RR (2009) Physiological functions mineral micronutrients (Cu, Zn, Mn, Fe, Ni, Mo, B, Cl). Curr Opin Plant Biol 12:259–266

Hayat S, Ali B, Aiman Hasan S, Ahmad A (2007) Brassinosteroid enhanced the level of antioxidants under cadmium stress in *Brassica juncea*. Environ Exp Bot 60(1):33–41

Hu KD, Hu LY, Li YH, Zhang FQ, Zhang H (2007) Protective roles of nitric oxide on germination and antioxidant metabolism in wheat seeds under copper stress. Plant Growth Regul 53:173–183

Hu YL, Ge Y, Zhang CH, Ju T, Cheng WD (2009) Cadmium toxicity and translocation in rice seedlings are reduced by hydrogen peroxide pre-treatment. Plant Growth Regul 59:51–61

Huang TL, Huang HJ (2008) ROS and CDPK-like kinase-mediated activation of MAP kinase in rice roots exposed to lead. Chemosphere 71:1377–1385

Huang D, Wu W, Abrams SR, Cutler AJ (2008) The relationship of drought related gene expression in Arabidopsis thaliana to hormonal and environmental factors. J Exp Bot 59:2991–3007

Hughes MF (2002) Arsenic toxicity and potential mechanisms of action. Toxicol Lett 133:1–16
Jajic I, Sarna T, Strzalka K (2015) Senescence, stress, and reactive oxygen species. Plants 4:393–411
Jaspers P, Kandasjarvi J (2010) Reactive oxygen species in abiotic stress signalling. Physiol Plant 138:405–413
Jonak C, Nakagami H, Hirt H (2004) Heavy metal stress. Activation of distinct mitogen-activated protein kinase pathways by copper and cadmium. Plant Physiol 136(2):3276–3283
Joshi RK, Kar B, Nayak S (2011) Characterization of mitogenactivated protein kinases (MAPKs) in the *Curcuma longa* expressed sequence tag database. Bioinformation 7:180–183
Karuppanapandian T, Kim W (2013) Cobalt-induced oxidative stress causes growth inhibition associated with enhanced lipid peroxidation and activates antioxidant responses in Indian mustard (*Brassica juncea* L.) leaves. Acta Physiol Plant 35:2429–2443
Khan MIR, Khan NA (2014) Ethylene reverses photosynthetic inhibition by nickel and zinc in mustard through changes in PSII activity, photosynthetic nitrogen use efficiency, and antioxidant metabolism. Protoplasma 251:1007–1019. doi:10.1007/s00709-014-0610-7
Khan TA, Mazid M, da Silva JAT, Mohammad F, Khan MN (2014) Role of NO-mediated signaling under abiotic stress (heavy metal)-induced oxidative stress in plants: An overview. Fun Plant Sci Biotech 6:91–107
Kim C, Meskauskiene R, Apel K, Laloi C (2008) No single way to understand singlet oxygen signaling in plants. EMBO Rep 9:435–439
Kobayashi M, Yamamoto M (2006) Nrf2-keap1 regulation of cellular defense mechanism against electrophiles and reactive oxygen species. Adv Enzym Regul 46:113–140
Kobayashi M, Ohura I, Kawakita K, Yokota N, Fujiwara M, Shimamoto K et al (2007) Calcium-dependent protein kinases regulate the production of reactive oxygen species by potato NADPH oxidase. Plant Cell 19:1065–1080
Kochevar IE (2004) Singlet oxygen signaling: from intimate to global. Sci STKE 2004:e7
Kopyra M, Stachon-Wilk M, Gwó'zd'z EA (2006) Effects of exogenous nitric oxide on the antioxidant capacity of cadmium-treated soybean cell suspension. Acta Physiol Plant 28:525–536
Kruzmane D, Jankevica L, Ievinsh G (2002) Effect of regurgitant from *Leptinotarsadecemlineata* on wound responses in *Solanum tuberosum* and *Phaseolus vulgaris*. Physiol Plant 115:577–584
Kudla J, Batistic O, Hashimoto K (2010) Calcium signals: the lead currency of plant information processing. Plant Cell 22:541–563
Kurepin LV, Ivanov AG, Zaman M, Pharis RP, Allakhverdiev SI, Hurry V, Hüner NPA (2015) Stress-related hormones and glycinebetaine interplay in protection of photosynthesis under abiotic stress conditions. Photosynth Res. doi:10.1007/s11120-015-0125-x
Laloi C, Przybyla D, Apel K (2006) A genetic approach towards elucidating the biological activity of different reactive oxygen species in *Arabidopsis thaliana*. J Exp Bot 57:1719–1724
Lamb C, Dixon RA (1997) The oxidative burst in plant disease resistance. Annu Rev Plant Physiol Plant Mol Biol 48:251–275
Lee KP, Kim C, Landgraf F, Apel K (2007) EXECUTER-1 and EXECUTER-2 dependent transfer of stress related signals from the plastid to the nucleus of *Arabidopsis thaliana*. Proc Nat Acad Sci USA 104:10270–10275
Leterrier M, Airaki M, Palma JM, Chaki M, Barroso JB, Corpas FJ (2012) Arsenic triggers the nitric oxide (NO) and *S*-nitrosoglutathione (GSNO) metabolism in *Arabidopsis*. Environ Pollut 166:136–143
Li DH, Yuan Y (2016) Hydrogen peroxide enhances antioxidative defense in the leaves of water caltrop (*Trapa bicornis*) seedlings treated with lead. Biologia 71:100–108
Li A, Wang X, Leseberg CH, Jia J, Mao L (2008) Biotic and abiotic stress responses through calcium-dependent protein kinase (CDPK) signaling in wheat (*Triticum aestivum*L.) Plant Signal Behav 3:654–656
Lin YF, Aarts MG (2012) The molecular mechanism of zinc and cadmium stress response in plants. Cell Mol Life Sci 69:3187–3206

Liu XM, Kim KE, Kim KC, Nguyen XC, Han HJ, Jung MS (2010) Cadmium activates Arabidopsis MPK3 and MPK6 via accumulation of reactive oxygen species. Phytochemistry 71:614–618

Lombardi L, Sebastian L (2015) Copper toxicity in Prunus cerasifera: growth and antioxidant 715 enzymes responses of in vitro grown plants. Plant Sci 168:797–802

Maheshwari R, Dubey RS (2009) Nickel-induced oxidative stress and the role of antioxidant defence in rice seedlings. Plant Growth Regul 59:37–49

Maksimovic' JD, Mojovic' M, Maksimovic' V, Römheld V, Nikolic M (2012) Silicon ameliorates manganese toxicity in cucumber by decreasing hydroxyl radical accumulation in the leaf apoplast. J Exp Bot 63:2411–2420

Manara A (2012) Plant responses to heavy metal toxicity. In: Furini A (ed) Plants and heavy 722 metals. Springer briefs in biometals. Springer, New York, pp 27–53

Mazars C, Thuleau P, Lamotte O, Bourque S (2010) Cross-talk between ROS and calcium in regulation of nuclear activities. Mol Plant 3:706–718

McAinsh MR, Pittman JK (2009) Shaping the calcium signature. New Phytol 181:275–294

McCormack E, Tsai YC, Braam J (2005) Handling calcium signaling: *Arabidopsis* CaMs and CMLs. Trends Plant Sci 10:383–389

Mediouni C, Benzarti O, Tray B, Ghorbel MH, Jemal F (2006) Cadmium and copper toxicity for tomato seedlings. Agr Sustan Devel 26:227–232

Miller G, Suzuki N, Ciftci-Yilmaz S, Mittler R (2010) Reactive oxygen species homeostasis and signalling during drought and salinity stresses. Plant Cell Environ 33:453–467

Mishra S, Jha AB, Dubey RS (2011) Arsenite treatment induces oxidative stress, upregulates antioxidant system, and causes phytochelatin synthesis in rice seedlings. Protoplasma 248:565–577

Moan J (1990) On the diffusion length of singlet oxygen in cells and tissues. J Photochem Photobiol 6:343–344

Moller IM, Sweetlove LJ (2010) ROS signaling-specificity is required. Trends Plant Sci 7:370–374

Montero-Palmero MB, Martín-Barranco A, Escobar C, Hernández LE (2014a) Early transcriptional responses to mercury: a role for ethylene in mercury-induced stress. New Phytol 201:116–130

Montero-Palmero MB, Ortega-Villasante C, Escobar C, Hernández LE (2014b) Are plant endogenous factors like ethylene modulators of the early oxidative stress induced by mercury? Environ Toxicol 2:34

Mostofa MG, Hossain MA, Fujita M, Tran LSP (2015) Physiological and biochemical mechanisms associated with trehalose-induced copper-stress tolerance in rice. Sci Rep 5:11433. doi:10.1038/srep11433

Munne-Bosch S, Queval G, Foyer CH (2013) The impact of global change factors on redox signaling under pinning stress tolerance. Plant Physiol 161:5–19

Nathan C (2003) Specificity of a third kind: reactive oxygen and nitrogen intermediates in cell signaling. J Clin Invest 111:769–778

Niedre M, Patterson MS, Wilson BC (2002) Direct near-infrared luminescence detection of singlet oxygen generated by photodynamic therapy in cells in vitro and tissues in vivo. J Photochem Photobiol 75:382–391

Ochsenbein C, Przybyla D, Danon A, Landgraf F, Gobel C, Imboden A, Feussner I, Apel K (2006) The role of EDS1 (enhanced disease susceptibility) during singlet oxygen-mediated stress responses of *Arabidopsis*. Plant J 47:445–456

Ogasawara Y, Kaya H, Hiraoka G, Yumoto F, Kimura S, Kadota Y et al (2008) Synergistic activation of the *Arabidopsis* NADPH oxidase AtrbohD by Ca2+ and phosphorylation. J Biol Chem 283:8885–8892

Op den Camp RG (2003) Rapid induction of distinct stress responses after the release of singlet oxygen in *Arabidopsis*. Plant Cell 15:2320–2332

Opdenakker K, Remans T, Vangronsveld J, Cuypers A (2012) Mitogen – Activated Protein (MAP) kinases in plant metal stress: regulation and responses in comparison to other biotic and abiotic stresses. Int J Mol Sci 13:7828–7853

Palmieri MC, Sell S, Huang X et al (2008) Nitric oxide-responsive genes and promoters in Arabidopsis thaliana: a bioinformatics approach. J Exp Bot 59:177–186

Patsikka E, Aro EM, Tyystjärvi E (1998) Increase in the quantum yield of photo inhibition contributes to copper toxicity in vivo. Plant Physiol 117:619–627

Penarrubia L, Romero P, Carrio-Segui A, Andres-Bordería A, Moreno J, Sanz A (2015) Temporal aspects of copper homeostasis and its crosstalk with hormones. Front Plant Sci 6:255. doi:10.3389/fpls.2015.00255

Pesko M, Král'ová A (2013) Physiological response of *Brassica napus* L. plants to Cu (II) treatment. Proceedings of ECOpole 7:155–161

Qiaoa W, Li C, Fan LM (2013) Cross-talk between nitric oxide and hydrogen peroxide in plant responses to abiotic stresses. Environ Exp Bot 100:84–93

Radi R (2013) Peroxynitrite, a stealthy biological oxidant. J Biol Chem 288:26464–26472

Raina SK, Wankhede DP, Sinha AK (2013) *Catharanthus roseus* mitogen-activated protein kinase 3 confers UV and heat tolerance to *Saccharomyces cerevisiae*. Plant Signal Behav 8:1

Rao KP, Richa T, Kumar K, Raghuram B, Sinha AK (2010) In silico analysis reveals 75 members of mitogen-activated protein kinase kinase kinase gene family in rice. DNA Res 17:139–153

Rao KP, Vani G, Kumar K, Wankhede DP, Misra M, Gupta M, Sinha AK (2011) Arsenic stress activates MAP kinase in rice roots and leaves. Arch Biochem Biophys 506:73–82

Remans T, Opdenakker K, Smeets K, Mathijsen D, Vangronsveld J, Cuypers A (2010) Metal-specific and NADPH oxidase dependent changes in lipoxygenase and NADPH oxidase gene expression in *Arabidopsis thaliana* exposed to cadmium or excess copper. Funct Plant Biol 37:532–544

Rentel MC, Lecourieux D, Ouaked F, Usher SL, Petersen L, Okamoto H et al (2004) OXI1 kinase is necessary for oxidative burst-mediated signaling in *Arabidopsis*. Nature 427:858–861

Rodriguez MC, Petersen M, Mundy J (2010) Mitogen activated protein kinase signaling in plants. Annu Rev Plant Biol 6:621–649

Rodríguez-Serrano M, Romero-Puertas MC, Pazmino DM, Testillano PS, Risueno MC, Luis A, Sandalio LM (2009) Cellular response of pea plants to cadmium toxicity: cross talk between reactive oxygen species, nitric oxide, and calcium. Plant Physiol 150:229–243

Roel GL, Przybyla D, Ochsenbein C, Laloi C, Kim C, Danon A, Wagner D, Hideg E, Göbel C, Feussner I, Nater M, Apel K (2003) Rapid induction of distinct stress responses after the release of singlet oxygen in *Arabidopsis*. Plant Cell 15:2320–2332

Romero-Puertas MC, Rodriguez-Serrano M, Corpas FJ, Gomez M, Del Rio LA, Sandalio LM (2004) Cadmium-induced subcellular accumulation of $O_2^{\cdot -}$ and H_2O_2 in pea leaves. Plant Cell Environ 27:1122–1134

Sablina AA, Budanov AV, Llyinskaya GV, Aqapova LS, Kravchenko JE, Chumakov PM (2005) The antioxidant function of p53 tumor suppressor. Nat Med 11:1306–1313

Sagi M, Fluhr R (2006) Production of reactive oxygen species by plant NADPH oxidases. Plant Physiol 141:336–340

Šamajová O, Plíhal O, Al-Yousif M, Hirt H, Šamaj J (2013) Improvement of stress tolerance in plants by genetic manipulation of mitogen-activated protein kinases. Bioteh Adv 31:118–128

Saniewski M, Ueda J, Miyamoto K, Urbanek H (2003) Interactions between ethylene and other plant hormones in regulation of plant growth and development in natural conditions and under abiotic and biotic stresses. In: Vendrell M, Klee H, Pech JC, Romojaro F (eds) Biology and biotechnology of the plant hormone ethylene III. IOS Press, Amsterdam, pp 263–270

Scandalios JG (2005) Oxidative stress: molecular perception and transduction of signals triggering antioxidant gene defenses. Braz J Med Biol Res 38:995–1014

Schellingen K, Van Der Straeten D, Remans T, Loix C, Vangronsveld J, Cuypers A (2015) Ethylene biosynthesis is involved in the early oxidative challenge induced by moderate Cd exposure in *Arabidopsis thaliana*. Environ Exp Bot 117:1–11

Schulz P, Herde M, Romeis T (2013) Calcium-dependent protein kinases: hubs in plant stress signaling and development. Plant Physiol 163:523–530

Schutzendubel A, Polle A (2002) Plant responses to abiotic stresses: heavy metal-induced oxidative stress and protection by mycorrhization. J Exp Bot 53:1351–1365

Sewelam N, Kazan K, Schenk PM (2016) Global plant stress signaling: reactive oxygen species at the cross-road. Front Plant Sci 7:1–21

Sharma SS, Dietz KJ (2009) The relationship between metal toxicity and cellular redox imbalance. Trends Plant Sci:144–150

Sharma RK, Devi S, dan Dhyani PP (2010) Comparative assessment of the toxic effects of copper and cypermethrin using seeds of *Spinacia Oleracea* L. plants. Trop Ecol 51:375–387

Sharma P, Jha AB, Dubey RS, Pessarakli M (2012) Reactive oxygen species, oxidative damage and antioxidative defense mechanism in plants under stressful conditions. J Bot Article ID 217037 1–26

Shi H, Ye T, Zhu JK, Chan Z (2014) Constitutive production of nitric oxide leads to enhanced drought stress resistance and extensive transcriptional reprogramming in *Arabidopsis*. J Exp Bot 65:4119–4131

Singh HP, Kaur S, Batish DR, Sharma VP, Sharma N, Kohli RK (2009) Nitric oxide alleviates arsenic toxicity by reducing oxidative damage in the roots of *Oryza sativa* (rice). Nitric Oxide 20:289–297

Sinha AK, Jaggi M, Raghuram B, Tuteja N (2011) Mitogen activated protein kinase signaling in plants under abiotic stress. Plant Signal Behav 6:196–203

Sinha P, Shukla AK, Sharma YK (2015) Amelioration of heavy-metal toxicity in cauliflower by application of salicylic acid. Commun Soil Sci Plant Anal 46:1309–1319

Smeets K, Opdenakker K, Remans T (2013) The role of the kinase OXI1 in cadmium- and copper-induced molecular responses in *Arabidopsis thaliana*. Plant Cell Environ 36:1228–1238

Stael S, Wurzinger B, Mair A, Mehlmer N, Vothknecht UC, Teige M (2012) Plant organellar calcium signaling: an emerging field. J Exp Bot 63:1525–1542

Sun P, Tian QY, Chen J, Zhang WH (2010) Aluminium-induced inhibition of root elongation in Arabidopsis is mediated by ethylene and auxin. J Exp Bot 61:347–356

Taj G, Agarwal P, Grant M, Kumar A (2010) MAPK machinery in plants: recognition and response to different stresses through multiple signal transduction pathways. Plant Signal Behav 5:1370–1378

Tanou G, Molassiotis A, Diamantidis G (2009) Induction of reactive oxygen species and necrotic death-like destruction in strawberry leaves by salinity. Environ Exp Bot 65:270–281

Temple MD, Perrone GG, Dawes IW (2005) Complex cellular responses to recative oxygen species. Trends Cell Biol 15:319–326

Thapa G, Sadhukhan A, Panda SK, Sahoo L (2012) Molecular mechanistic model of plant heavy metal tolerance. Biometals 25:489–505

Tian S, Wang X, Li P, Wang H, Ji H, Xie J, Qiu Q, Shen D, Dong H (2016) Plant aquaporin AtPIP1;4 links apoplastic H2O2 induction to disease immunity pathways. Plant Physiol 171:1635–1650

Trewavas A (1999) Le calcium, c'est la vie: calcium makes waves. Plant Physiol 120:1–6

Triantaphylidès C, Havaux M (2009) Singlet oxygen in plants: production, detoxification and signaling. Trends Plant Sci 14:219–228

Triantaphylides C, Krischke M, Hoeberichts FA, Ksas B, Gresser G, Havaux M, Van-Breusegem FV, Mueller MJ (2008) Singlet oxygen is major reactive oxygen species involved in photo-oxidative damage to plants. Plant Physiol 148:960–968

Tsai TM, Huang HJ (2006) Effects of iron excess on cell viability and mitogen-activated protein kinase activation in rice roots. Physiol Plant 127:583–592

Tuteja N, Mahajan S (2007) Calcium signaling network in plants. Plant Signal Behav 2:79–85

Vijendra PD, Huchappa KM, Lingappa R, Basappa G, Jayanna SG, Kumar V (2016) Physiological and biochemical changes in moth bean (Vigna aconitifolia L.) under cadmium stress. J Bot 1–13

Wang YS, Yang ZM (2005) Nitric oxide reduces aluminum toxicity by preventing oxidative stress in the roots of Cassia tora L. Plant Cell Physiol 46:1915–1923

Wang C, Zhang SH, Wang PF, Hou J, Zhang WJ, Li W, Lin ZP (2009) The effect of excess Zn on mineral nutrition and antioxidative response in rapeseed seedlings. Chemosphere 75:1468–1476

Wang J, Ding H, Zhang A, Ma F, Cao J, Jiang M (2010) A novel mitogen-activated protein kinase gene in maize (*Zea mays*), ZmMPK3, is involved in response to diverse environmental cues. J Integr Plant Biol 52:442–452

Wen JF, Gong M, Liu Y, Hu JL, Deng MH (2013) Effect of hydrogen peroxide on growth and activity of some enzymes involved in proline metabolism of sweet corn seedlings under copper stress. Sci Hortic 164:366–371

Wimalasekera R, Villar C, Begum T, Scherer GF (2011) Copper amine oxidase1 (CuAO1) of Arabidopsis thaliana contributes to abscisic acid-and polyamine-induced nitric oxide biosynthesis and abscisic acid signal transduction. Mol Plant 4:663–678. doi:10.1093/mp/ssr023

Yan S, Dong X (2014) Perception of the plant immune signal salicylic acid. Curr Opin Plant Biol 20:64–68. doi:10.1016/j.pbi.2014.04.006

Yan J, Tsuichihara N, Etoh T, Iwai S (2007) Reactive oxygen species and nitric oxide are involved in ABA inhibition of stomatal opening. Plant Cell Environ 30:1320–1325

Yang T, Poovaiah BW (2002) Hydrogen peroxide homeostasis: activation of plant catalase by calcium/calmodulin. Proc Nat Acad Sci USA 99:4097–4102

Yang L, Ji J, Harris-Shultz KR, Wang H, Wang H, Abd-Allah EF, Luoand Y, Hu X (2016) The dynamic changes of the plasma membrane proteins and the protective roles of nitric oxide in Rice subjected to heavy metal cadmium stress. Front Plant Sc 7:190. doi:10.3389/fpls.2016.00190

Yeh CM, Chien PS, Huang HJ (2007) Distinct signaling pathways for induction of MAP kinase activities by cadmium and copper in rice roots. J Exp Bot 58:659–671

Yun BW, Feechan A, Yin M, Saidi NB, Le Bihan T, Yu M, Moore JW, Kang JG, Kwon E, Spoel SH, Pallas JA, Loake GJ (2011) S-nitrosylation of NADPH oxidase regulates cell death in plant immunity. Nature 478:264–268

Zhang H, Li YH, Hu LY, Wang SH, Zhang FQ, Hu KD (2008) Effects of exoge-nous nitric oxide donor on antioxidant metabolism in wheat leaves under aluminum stress. Russ J Plant Physiol 55:469–474

Zhang S, Cai Z, Wang X (2009) The primary signaling outputs of brassionosteroid are regulated by abscisic acid signaling. Proc Natl Acad Sci 106:4543–4548

Zhang A, Zhang J, Zhang J, Ye N, Zhang H, Tan M, Jiang M (2011) Nitric oxide mediates brassinosteroid-induced ABA biosynthesis involved in oxidative stress tolerance in maize leaves. Plant Cell Physiol 52:181–192. doi:10.1093/pcp/pcq187

Zhang F, Jin X, Wang L, Li S, Wu S, Cheng C, Zhang T, Guo W (2016) GhFAnnxA affects fiber elongation and secondary cell wall biosynthesis associated with Ca21 influx, ROS homeostasis, and actin filament reorganization. Plant Physiol 171:1750–1770

Zhao H, Jin Q, Wang Y, Chu L, Li X, Xu Y (2016) Effects of nitric oxide on alleviating cadmium stress in *Typha angustifolia*. Plant Growth Regul 78:243–251

Effects of Different Metal Stresses on the Antioxidant Defense Systems of Medicinal Plants

9

Iffat Zareen Ahmad, Asad Ahmad, Abdul Mabood, and Heena Tabassum

Abstract

Traditional herbal plant-based medicines were characterized by having good amount of antioxidants. But due to the rapid industrialization and urbanization, addition of heavy metal accumulation in soil-plant affected the growth of plants including medicinal plants. Under metal stress, both the antioxidant enzymes and phytochemicals constituting nonenzymatic components coordinately work to encounter abiotic stresses. Antioxidant enzymes such as polyphenol oxidase (PPO), superoxide dismutase (SOD), catalase (CAT), peroxidase (POD), phenylalanine ammonia-lyase (PAL), ascorbate peroxidase (APX), lipoxygenase (LOX), glutathione reductase (GR), glutathione peroxidase (GPX), guaiacol peroxidase (GPO), glutathione-S-transferase (GST), monodehydroascorbate reductase (MDHAR), and dehydroascorbate reductase (DHAR) are altered in stressed conditions which leads to the changes in free radical content. In addition, different components of nonenzymatic defense system such as glycine betaine (GB), proline, glutathione (GSH), ascorbic acid (AsA), tocopherols, carotenoids, flavonoids, and phenolic compounds also play crucial role at cellular level. Almost every type of stress leads to the overproduction of reactive oxygen species (ROS). They are vigorously produced in response to most abiotic stresses and used as signaling molecules by cells. As they are highly reactive species, a balance between their production and detoxification must be ensured. By transformation of plants in which activity of single antioxidant enzyme is overexpressed, it is possible to confer a degree of tolerance to stress. The success of plant transformation has allowed the scientists to confer greater stress tolerance in plants by improvements in the antioxidative defense system.

I.Z. Ahmad (✉) • A. Ahmad • A. Mabood • H. Tabassum
Department of Bioengineering, Integral University, Dasauli, Kursi Road, Lucknow 226026, India
e-mail: iffat@iul.ac.in

M.I.R. Khan, N.A. Khan (eds.), *Reactive Oxygen Species and Antioxidant Systems in Plants: Role and Regulation under Abiotic Stress*,
DOI 10.1007/978-981-10-5254-5_9

Keywords
Metal stress • Medicinal plants • Antioxidants • ROS

Abbreviations

APX	Ascorbate peroxidase
AsA	Ascorbic acid
ASO	Ascorbate oxidase
CA	Citric acid
CAT	Catalase
d-ALA-D	Delta-aminolevulinic acid dehydratase
DHAR	Dehydroascorbate reductase
FRAP	Ferric reducing ability of plasma
GB	Glycine betaine
GPO	Guaiacol peroxidase
GPX	Glutathione peroxidase
GR	Glutathione reductase
GSH	Glutathione
GSSG	Glutathione disulfide
GST	Glutathione-S-transferase
LOX	Lipoxygenase
MDA	Malondialdehyde
MDHAR	Monodehydroascorbate reductase
NM	Nonmetalliferous
NPSH	Nonprotein thiol
PAL	Phenylalanine ammonia-lyase
PC	Phosphatidylcholine
PCD	Programmed cell death
PE	Phosphatidylethanolamine
PG	Phophatidylglycerol
PI	Phosphatidylinositol
PLNH	Paecilomyces lilacinus NH
PM	Plasma membrane
POD	Peroxidase
PPO	Polyphenol oxidase
PS	Phosphatidylserine
PSH	Protein thiol
PSII	Photosystem II
ROS	Reactive oxygen species
Rtn	Rutin

SOD	Superoxide dismutase
TAST	Total acid-soluble thiol
TBARS	Thiobarbituric acid reactive substances
TEM	Transmission electron microscopy
TSS	Total soluble sugars
TWC	Tissue water content

9.1 Introduction

Natural compounds derived from medicinal plants play a central part in the health care and drug development in classical as well as advanced systems of medicine. Further, herbal drugs have advantages as therapeutic agents because of their easy availability, cost-effectiveness, and negligible side effects when compared to modern medicine (Agbor and Ngogang 2005). In nature excess, ROS production is highly reactive toward biological molecules which are lipids, proteins, and deoxyribonucleic acids. This leads to imbalance between oxidants and antioxidants in the plant cell. Although human body is protected by innate antioxidant defense system, a proper supply of antioxidants from natural sources is required (Rimbach et al. 2005).

Abiotic stresses results to the excessive production of ROS in plants which are extremely reactive and toxic cause damage to important biomolecules which eventually leads to oxidative stress. The ROS consist of both free radical forms such as superoxide hydroxyl, per hydroxy, and alkoxy radicals and non-radical forms including hydrogen peroxide and singlet oxygen. Stress leads to ROS buildup which is scavenged by enzymatic antioxidants such as SOD, APX, GPX, and CAT and nonenzymatic low molecular metabolites, such as ascorbate (ASH), GSH, α-tocopherol, carotenoids, and flavonoids (Gill et al. 2011; Mittler et al. 2004). The antioxidants establish in approximately all the organelles, representing the significance of ROS detoxification for the survival of the cell (Gill et al. 2011). In stress condition, chloroplasts and mitochondria endure carbon metabolism and energy balance. For maintaining elevated stage of metabolic coordination without overproduction of ROS and oxidative damage, it is mandatory to sustain the flow of energy via organelles under all physiological conditions during growth (Baxter et al. 2007; Takahashi and Murata 2008). Plants comprise antioxidants supplied exogenously through soil. It is stated that two-thirds of the plant population of the world are medicinally important, and almost all of them have tremendous antioxidant potential (Krishnaiah et al. 2011).

In the present chapter, we laid emphasis on the recent findings related to varied medicinal plants' responses to various metal stresses. The aim of this chapter is to discuss the potential role of metal stresses on antioxidant status of medicinal plants. We concentrated on the metal stresses which led to oxidative damage and its encounter by antioxidant defenses that are activated by plants for its survival. We begin with a brief introduction of ROS and their role in oxidative stress and their combat by antioxidant defense system in plants. We have also presented data on different medicinal plants and their responses to counteract oxidative stress caused by different metals. We end the chapter with a discussion on the potential role of medicinal plants and the phytochemicals produced by them as an important source of edible antioxidants. The knowledge of genetic transformation has placed within our grasp the possibility of engineering greater stress tolerance in plants by overexpression of the enzymatic and nonenzymatic components of the defense system.

9.2 ROS and Antioxidant System

ROS are highly reactive molecules which are produced by living cells during normal cellular metabolism and in response to environmental stresses. ROS have damaging effects on biomolecules, lipids, and proteins and nucleic acids being the most important. The imbalance between oxidants and antioxidants wherein more oxidants are produced is termed "oxidative stress," and its regulation is important for cell viability, division, growth, and organ function. Plants comprise of different strategies to suppress oxidation process which include enzymatic and nonenzymatic components that play role in protecting them from harmful effects of ROS.

9.2.1 Antioxidant System in Plants

The antioxidant defense mechanism of the plant is composed of an array of antioxidant phytochemicals and enzymes (Fig. 9.1). All types of abiotic stresses increase the production of ROS in plant cells which leads to the disruption of cellular homeostasis (Shah et al. 2001; Mittler 2002; Sharma and Dubey 2005, 2007; Hu et al. 2008; Han et al. 2009; Maheshwari and Dubey 2009; Tanou et al. 2009; Mishra et al. 2011; Srivastava and Dubey 2011; Khan and Khan 2014; Khan et al. 2014, 2015, 2016a, b). Additionally, ROS are generated in plant cells during normal metabolic processes also (Foyer and Harbinson 1994; Foyer 1997; del Rio et al. 2006; Blokhina and Fagerstedt 2010; Heyno et al. 2011). The photosynthetic electron transport chain provides active oxygen to plant tissues, and it generates singlet oxygen, $^1O_2$, and superoxide, $O^{\cdot -2}$(Asada 1994). Further, certain photosystem II constituents also have the mechanisms of converting molecular O_2 to high-energy singlet oxygen ($^1O_2$). A metabolic pathway in chloroplasts which is insensitive to cyanide competes with photosynthetic electron transport for electrons (Bennoun 1994). Additionally, the reaction center protein of PSII (DI) and the apoplastic space seems to have less protection against oxidative damage (Castillo and Greppin 1988; Lwe et al. 1993). The increased production of ROS during abiotic stresses can pose a threat to cells by causing lipid peroxidation, oxidation of proteins and nucleic acids, inhibition of enzyme activity, and initiation of programmed cell death (PCD) pathway leading to cell death (Shah et al. 2001; Mittler 2002; Verma and Dubey 2003; Meriga et al. 2004; Sharma and Dubey 2005;

Maheshwari and Dubey 2009; Mishra et al. 2011; Srivastava and Dubey 2011). Despite their harmful activity, ROS are well established as secondary messenger in numerous cellular processes including environmental stress tolerance (Desikan et al. 2001; Neill et al. 2002; Yan et al. 2007).

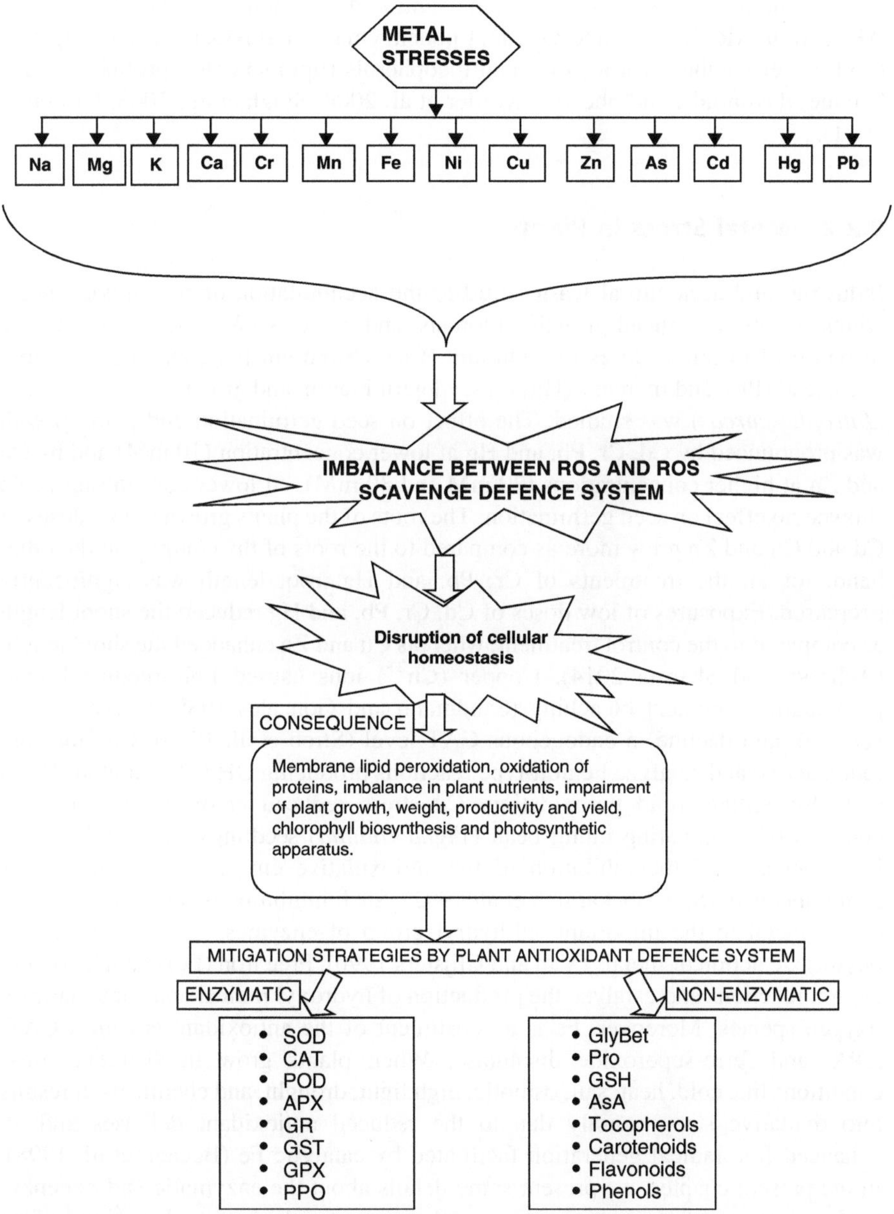

Fig. 9.1 Different types of metal stresses and their enzymatic and nonenzymatic responses

To cope up with the excess ROS production, plant cell's organelles like chloroplast, mitochondria, and peroxisomes depute antioxidant defense systems. An extensive research has shown the significance of the cellular antioxidant machinery for defense against various stresses (Tuteja 2007; Singh et al. 2008; Gill et al. 2011). Enzymatic and nonenzymatic antioxidants are the two components of antioxidant defense system. Enzymatic antioxidants include SOD, CAT, POD, APX, MDHAR, DHAR, and GR, and nonenzymatic antioxidants are GSH, AsA (both water soluble), carotenoids and tocopherols (lipid soluble), proline, glycine betaine, flavonoids, and phenols (Mittler et al. 2004; Singh et al.; 2008; Gill et al. 2011).

9.2.2 Metal Stress in Plants

Industrial and agricultural wastes lead to the accumulation of phytotoxic metals which results in reduced growth, chlorosis, and necrosis (Woolhouse 1983). The response of different doses of cadmium (Cd), chromium (Cr), zinc (Zn), copper (Cu), lead (Pb), and mercury (Hg) on seed germination and growth of barbados nut (*Jatropha curcas*) was studied. The effect on seed germination and plant growth was pronounced by Cd, Cr, Pb, and Hg at lower concentration (10 mM) and by Cu and Zn at higher concentrations (20 mM and 30 mM). At lower concentration, Zn showed no effect on seed germination. The roots of the plants grown in low doses of Cd and Cu and Zn grew more as compared to the roots of the control; on the other hand, for all the treatments of Cr, Pb, and Hg, root length was significantly decreased. Exposures of low doses of Cd, Cr, Pb, and Hg reduced the shoot length as compared to the control treatment, whereas Cu and Zn enhanced the shoot length (Acharya and Sharma 2014). Copper (Cu^{2+}) ions caused light-mediated lipid peroxidation, pigment bleaching (Sandmann and Gonzales 1989) in rye (*Secale cereale*), and decline in endogenous CAT level (Streb et al. 1993). Cu^{2+}ions are redox active and catalyze Fenton-type reactions producing OH (Elstner et al. 1988) and also initiate lipid peroxidation. Cd stress reduced chlorophyll and heme contents of germinating mung bean (*Vigna radiata*) seedlings by stimulation of lipoxygenase and the inhibition of the antioxidative enzymes, SOD and CAT simultaneously (Somashekaraiah et al. 1992). Such inhibition is due to the binding of the metal to the important sulfhydryl group of enzymes, which increase the phytotoxic action of metals (Van and Clijsters 1990). Free iron (Fe) can take part in Fenton reactions and catalyze the production of hydroxyl radical and some harmful oxygen species. Moreover, Fe is a constituent of the antioxidant enzymes CAT, APX, and ferro-superoxide dismutase. When plants grow in different stress conditions like cold, heat, salt, osmotic, high light, drought, and chemicals, it results into oxidative stress mainly due to the reduced antioxidant defenses and an enhanced free radical generation facilitated by catalytic Fe (Becana et al. 1998). In the present chapter, we present some details about the enzymatic and nonenzymatic responses of plants to various metal stresses for the better adaptation to face the environmental constraints. Around 80% of the plants that grew on this planet have medicinal properties and are classified as medicinal plants. Most of the medicinal properties are due to the antioxidant phytochemicals present in them. Since these plants are rich source of natural antioxidants, therefore their formulations are used as both preventive medicines and drugs to cure diseases. Due to pollution, most of the soil in which these plants grow contains metals above threshold levels, and there could be an effect of their toxicity on the antioxidant status of the plants. The present chapter explores the effects of the metals on the growth and antioxidant phytochemicals present in plants. The edible medicinal plants are rich source of antioxidants in the diet, and it will be very interesting to study the alterations in the antioxidant system of the plant under the influence of metals.

9.3 Effects of Metal Stress on Antioxidant Status of Medicinal Plants (Table 9.1)

9.3.1 Sweet Wormwood (*Artemisia annua*)

Effect of As(III), Cr(VI), and NaCl on plant growth; antioxidant enzymes, SOD, thiobarbituric acid reactive substances (TBARS); protein; cDNA augmentation of important genes of artemisinin pathway; and its biosynthesis was studied to find out the alterations in plant characteristics and artemisinin yield in sweet wormwood. Enhanced TBARS and SOD activity, decreased CAT activity, and total chlorophyll content were seen in metal and NaCl stress. Accumulation of As was higher in roots than shoot, while Cr accumulated more in shoots. cDNA template of metal-treated plants as well as control was amplified with 3-hydroxy-3methylglutaryl-coenzyme A reductase (HMGR), ADS, and CYP71AV1 genes of artemisinin biosynthetic pathway and displayed less expression with Cr(VI), while As(III) showed higher expression than control (Pauln and Shakya 2013).

9.3.2 Brahmi (*Bacopa monnieri*)

Brahmi plants were exposed to Cd stress and analyzed for the metal accumulation and its impact on different enzymatic and nonenzymatic antioxidants, TBARS, photosynthetic pigments, and protein content. The Cd buildup was found to be enhanced with increasing metal concentration and time period, and more Cd was collected in the root. TBARS content of the roots and shoots increased with increased concentration of Cd and its time of exposure. Enhancement in the activities of SOD, APX, and GPX was documented in stressed roots and shoots of brahmi. A considerable decrease in CAT activity in Cd-treated brahmi was seen. There was enhancement in the cysteine and nonprotein thiols (NPSH) in the roots of brahmi followed by a decrease. In case of shoots, cysteine and NPSH contents were found to be increased at all the Cd concentrations and exposure periods. The levels of ascorbic acid, the total chlorophyll, and protein contents were reduced markedly.

Table 9.1 Enzymatic and nonenzymatic response of some medicinal plants toward metal stress

S. No.	Plant	Stress	Site	Enzymatic effects	Nonenzymatic effects	References
1.	*Artemisia annua*	As, Cr, Na	Root and shoot	SOD↑	TBARS↑	Pauln and Shakya (2013)
				CAT↓		
2.	*Bacopa monnieri*	Cd	Root and shoot	SOD↑	TBARS↑	Singh et al. (2006)
				CAT↓	Cysteine$_{L}$↑	
				APX↑	NPSH$_{L}$↑	
				GPX↑	Cysteine$_{R}$↓	
					NPSH$_{R}$↓	
					AsA↓	
		Cd	Root, shoot, and leaves	SOD↑	Cysteine$_{L}$↓	Srivastava et al. (2006)
				CAT↓	NPSH$_{L}$↓	
				APX↑	GSH↓	
				GR↑		
				GPO↑		
				LPO↑		
		Fe	Leaves and root	SOD$_{L}$↓	AsA$_{L,R}$↑	Sinha and Saxena (2006)
				SOD$_{R}$↑	Cysteine$_{L}$ –	
				POD$_{L}$↓	Cysteine$_{R}$↑	
				POD$_{R}$↑	NPSH↓	
				APX$_{L,R}$↑	Proline↑	
				LPO↑		
		K	Shoot	SOD↓	MDA↑	Ahire et al. (2014)
				CAT↓	Proline↑	
				APX↓	Glycine-betaine↑	
				GPX↓		

		Ca		SOD↑	Bacoside-A↑	
				CAT↑		
				APX↑		
				GPX↑		
3.	*Camellia sinensis*	Cu (TS-462)	Leaves	SOD↑	TBARS↓	Ghanati et al. (2005) and Saha et al. (2012)
				CAT↑		
				APX↑		
				LPO↑		
		Cu (TS-520)		SOD↓	TBARS↓	
				CAT↑		
				LPO↑		
		Al		SOD↑	–	
		Cu		POD↑	Phenol↑	
				LPO↑		
		Al	Leaves	SOD↑	MDA↑	Li et al. (2011)
				CAT↓	GSH -	
				APX↑		
				GPO↓		
		Al	Root and cultured cell	SOD↑	Phenol↓	Ghanati et al. (2005)
				POD↓		
				CAT↑		
				APX↑		
				LPO↓		
		Cu (S3A3)	Root and leaves	SOD↑	MDA↑	Dey et al. (2015)
				POD↑	Phenol↑	
				CAT↑		
				APX↑		

(continued)

Table 9.1 (continued)

S. No.	Plant	Stress	Site	Enzymatic effects	Nonenzymatic effects	References
		Cu (Ts491)		Tolerant		
		Al	Shoot and root	SOD↑	Phenol↓	Mukhopadyay et al. (2012)
				CAT↑		
				APX↑		
				LPO↑		
				GR↑		
4.	*Cardaminopsis arenosa*	Cd, Zn	Leaves	–	GSH↑	Nadgorska-Socha et al. (2013)
		Fe, Mn, Cu		–	GSH↓	
					NPSH↑	
		Cd, Fe		–	NPSH↓	
		Fe, Mn, Cu		–	Proline↓	
		Cd, Zn, Pb, Mn		POD↑	–	
		Cu		SOD↑	–	
	Plantago lanceolata	Pb		–	GSH↑	
		Cu, Pb		–	GSH↑	
					NPSH↑	
		Cd, Zn, Pb, Mn		SOD↑	Proline↓	
				POD↑		
		Fe, Mn		SOD↑	–	
		Cu		SOD↓	–	
5.	*Cleome gynandra*	Cu, Cd	Plant	SOD↑	GSH↑	Haribabu and Sudha (2011)
				CAT↑	Proline↑	
					Phenolics↓	

6.	Egeria *densa*	Ni	Shoot and leaves	SOD↑	Carotenoids↑	Maleva et al. (2012)
				CAT -	NPSH↑	
					PSH↑	
		Cd		SOD↓	Carotenoids↑	
				CAT↑	NPSH↑	
					PSH↑	
		Cu		SOD↓	Carotenoids↓	
				CAT↓	NPSH↑	
					PSH↑	
		Zn		SOD↑	Carotenoids↑	
				CAT -	NPSH↑	
					PSH↑	
		Mn		SOD↑	Carotenoids↓	
				CAT↑	NPSH↑	
					PSH -	
		Ni+Cd		SOD↑	Carotenoids↑	
				CAT↑	NPSH -	
					PSH -	
		Ni+Cu		SOD↑	Carotenoids↓	
				CAT↑	NPSH -	
					PSH -	
		Ni+Zn		SOD↑	Carotenoids↑	
				CAT↑	NPSH↑	
					PSH -	
		Mn+Ni		SOD↑	Carotenoids↑	
				CAT↑	NPSH↑	
					PSH -	

(continued)

Table 9.1 (continued)

S. No.	Plant	Stress	Site	Enzymatic effects	Nonenzymatic effects	References
		Mn+Cd		SOD↑	Carotenoids↑	
				CAT↓	NPSH↑	
					PSH↓	
		Mn+Cu		SOD↓	Carotenoids↓	
				CAT↓	NPSH↑	
					PSH↑	
		Mn+Zn		SOD↑	Carotenoids↓	
				CAT↑	NPSH↑	
					PSH↑	
7.	*Glycyrrhiza uralensis*	Cd	Seedling	SOD↑	–	Zheng et al. (2010)
				POD↑		
				CAT↑		
				PPO↑		
				PAL↑		
		Mn	Leaves, shoot, and root	SOD↓	–	Shengjun et al. (2015)
				POD↓		
				CAT↓		
8.	*Helianthus annus*	Cr	Plant	SOD↓	Proline↑	Vijayalakshmi et al. (2010)
				CAT↓	Vitamins A, B, C↓	
				GSH↓	β-carotene↓	
	Solanum nigrum			SOD↑		
				CAT -	Phenolics↓	
				GSH↑		

9.	*Jatropha curcas*	Al	Seedling	SOD↑	–	Ou-yang et al. (2014)
				POD↑		
				CAT↑		
				$CAT_{Cotyledon}$↓		
				PAL↓		
		Pb	Embryo	SOD↓	–	Gao et al. (2009)
				POD↓		
				CAT↑		
				PAL↓		
		Pb	Seedling	SOD↑	–	Shu et al. (2014)
				POD↑		
				CAT↓		
		Cd, Pb	Seedling	POD↓	MDA↑	Liang et al. (2012)
				CAT↓	Proline↑	
		Zn, Ca, Mg, Pb, Cu, Fe, Mn, Ni, Cd	Culture	–	Proline↑	Raj and Mohan (2016)
		Pb, Cd, Cr	Leaves, shoot, and root	SOD↑	Proline↑	Devi et al. (2014)
					Carotenoid↓	
				POD↑	Phenol -	
				$CAT_{L,S\&R}$↑		
				PPO↑		
				$PAL_{S\&R}$↑		
				PAL_{L}↓		
		Ni	Embryo	SOD↓		Yan et al. (2008)
				POD↓		
				CAT↓		
				PAL↑		

(continued)

Table 9.1 (continued)

S. No.	Plant	Stress	Site	Enzymatic effects	Nonenzymatic effects	References
		Cd, Cr, Zn	Seedling	SOD↑		
		Zn	Embryo and seedling	SOD↑	–	Luo et al. (2010)
				POD↑		
				CAT↑		
				PAL↑		
10.	*Kandelia candel*	Na, Cd, Pb, Hg	Leaves and root	CAT_L↑	–	Zhang et al. (2007)
				CAT_R↓		
				SOD_R↓		
				POD_R↓		
	Bruguiera gymnorrhiza			CAT_L –		
				SOD_R↓		
				CAT_R↓		
				POD_R↓		
				LPO_L↑		
11.	*Mentha spicata*	Cr, Cd, Co, Ni, Pb	Root, stem, and leaves	SOD↑	MDA↑	Patel et al. (2016)
				POD↑	Proline↑	
				CAT↑		
12.	*Ocimum basilicum*	As	Seedling	CAT↑	$Proline_L$↑	Dehabadi et al. (2013)
				SOD↑	$Phenolic_L$↑	
				POD↑		
				LPO↑		
		Cd, Pb, Zn	Root and shoot	GST↑	Phenol↑	Stancheva et al. (2014)
				GR↑	Flavonoids↑	
				GPO↑	MDHAR↓	

				APX↓	DHAR↓	
		Cd	Plant	–	Phenol↑	Khavari-Nejad et al. (2013)
					Falavonoid↑	
					Proline↑	
	Origanum vulgare	Cd, Pb, Zn	Root and shoot	GPX↑	Ascorbate↑	Stancheva et al. (2014)
					GSH↑	
					Phenol↑	
					MDHAR↓	
					DHAR↓	
13.	*Olea europaea*	Cd	Root and leaves	SOD↑	Proline↑	Zouari et al. (2016)
				CAT↑	TBARS↑	
				GPX↑		
14.	*Pfaffia glomerata*	Cd	Root and leaves	SOD↑	–	Marques and Soares (2011)
				CAT_L↓		
				APX_L↑		
				APX_R↓		
				GPX_L↑		
				GPX_R↑		
				GR_L↑		
				GR_R↓		
		Hg	Root and shoot	SOD↑	AsA_S↑	Calgaroto et al. (2010)
				CAT_S↑	$NPSH_S$↑	
				APX_S↑	$Proline_S$↑	
				APX_R↓	Carotenoid –	
				$LPO_{R\&S}$↑	AsA_R↑	
					$NPSH_R$↑	
					$Proline_R$↓	

(continued)

Table 9.1 (continued)

S. No.	Plant	Stress	Site	Enzymatic effects	Nonenzymatic effects	References
		Cd	Root and shoot	SOD_R↑	–	Gomes et al. (2012)
				CAT_L↑		
				GPO_R↑		
				GPO_L↑		
		Zn, Hg	Root and shoot	SOD↑	–	Calgaroto et al. (2011)
				CAT_S↑		
				APX_S↑		
				LPO↓		
15.	*Salicornia brachiate*	Cd, Ni, Na, As	Flower	SOD↑	Proline↑	Sharma et al. (2010)
				CAT↑		
16.	*Solanum nigrum*	Cu	Root and shoot	$SOD_{S\&R}$	Carotenoid –	Fidalgo et al. (2013)
				$CAT_{S\&R}$	$Proline_R$↑	
				APX_S↓	MDA_S↑	
				LPO↑	MDA_R↑	
		Cd	Shoot	SOD↑	GSH↑	Xu et al. (2009)
				CAT↑		
		Cd	Culture	SOD↓	MDA↓	Gao et al. (2010)
				POD↑	GSH↓	
		Cr	Root and leaves	SOD_L↑	$Proline_L$↑	Uddin et al. (2015)
				POD_L↑		
	Parthenium hysterophorus	Cr	Root	–	Citric $acid_R$↑	
17.	*Silybum marianum*	Cd	Seedling	–	Proline↑	Khatamipour et al. (2011)
18.	*Trifolium resupinatum*	Cd	Root and shoot	POD↑	–	Ghasemi et al. (2014)
				CAT↓		

19.	*Urtica dioica*	Cd	Root, stem, and leaves	GR↑	GSH↑	Tarhan and Kavakcioglu (2016)
				GST↑		
				LPO_L↑		
20.	*Withania somnifera*	Fe	Root and leaves	SOD↑	–	Rout et al. (2015)
				CAT↑		
				GPO↑		
		Cu	Seedling	–	Carotenoid↓	Singh et al. (2014)
					Phenol↓	
					MDA↑	
		Cu	Plant	SOD↓	–	Khatun et al. (2008)
				CAT↓		
				APX↑		
				GPX↑		
				GPO↑		
				GST↑		
				GR↓		
				MDHAR↑		
				DHAR↑		
21.	*Zygophyllum album* and *Zygophyllum coccineum*	Al, Cu, Zn, Fe	Shoot	SOD↑	–	Morsy et al. (2012)
				CAT↓		
				APX↑		
				GPO↑		
				LPO↑		
				ASO↑		

(continued)

Table 9.1 (continued)

S. No.	Plant	Stress	Site	Enzymatic effects	Nonenzymatic effects	References
22.	*Zygophyllum fabago*	Cd	Seedling	SOD↑	Proline↑	Yildiztugay and Ozfidan-Konakci (2015)
				POD↑	TBARS↑	
				APX↑		
				GR↑		
23.	*Zygophyllum xanthoxylon*	Ni	Seedling	SOD↑	$MDA_{S\&R}$↑	Lu et al. (2010)
				POD↑	$Proline_{S\&R}$↑	
				APX↑		

Brahmi was able to counteract metal-induced oxidative damage by employing a mechanism of activation of various enzymatic and nonenzymatic antioxidants (Singh et al. 2006).

The phytotoxicity of Cd and its detoxification by brahmi plants have been studied. Plants accumulate considerable quantity of Cd in various plant parts (root, stem, and leaf), the maximum being in roots. Cd led to oxidative stress as is evident by enhanced lipid peroxidation and electrical conductivity in dose- and time-dependent manner. Photosynthetic pigments decreased continually, while protein content was increased slightly at lower concentrations. Antioxidant enzymes, SOD, GPO, APX, and GR, increased while CAT was reduced under stress. Cysteine, GSH, and NPSH increased initially but reduced later on as metal concentration and time increased. Phytochelatins were produced notably at lower concentration in roots and at higher concentration of Cd in leaves. At higher concentration, the phytochelatins decreased in roots; GSH, probably due to reduced GR activity, was also decreased (Srivastava et al. 2006).

The response of Fe was studied in brahmi, and increased lipid peroxidation, being more pronounced in leaves than roots, was observed. In roots, SOD activity was seen to increase under stress, whereas it decreased in leaves when compared to their respective control group. POD activity was enhanced in roots, but it decreased in shoots when compared to control. APX activity and ascorbic acid content were found to increase in both roots and leaves of the metal-treated plants when compared with their respective controls. The level of cysteine content in the roots increased initially when exposed to Fe, but no significant change was observed in leaves. In both roots and leaves, NPSH content was increased but decreased at higher metal concentrations. The amount of proline showed significant ($p < 0.01$) increase at higher metal concentrations in both parts of the plant. The level of bacoside-A also increased in metal-stressed plants (Sinha and Saxena 2006).

In vitro-grown shoots of brahmi were studied for growth, osmotic regulation, antioxidant enzymes, and bacoside-A contents in the presence of different concentrations of KCl and $CaCl_2$. Considerable decline was observed in number of shoot per culture, length of shoot, fresh and dry weight, and tissue water content (TWC) under increasing concentrations of KCl and $CaCl_2$ when compared with control. Less harm to the membrane as analyzed by malondialdehyde (MDA) content was observed in control with respect to steep increase in KCl- and $CaCl_2$-treated shoots. Increased contents of total soluble sugars (TSS), proline, and glycine betaine were observed in KCl- and $CaCl_2$-treated shoots as compared to the controls. An increasing concentration of $CaCl_2$ showed increase in osmolyte accumulation. The amount of Na+ was reduced with increasing concentrations of KCl and $CaCl_2$. K+ concentration buildup and its accumulation increased markedly in KCl-treated shoots as compared to control, while it was reduced in $CaCl_2$-treated shoots indicating that it prevents the uptake of K+ ions. Ca^{2+} accumulation was considerably enhanced with increasing concentrations of $CaCl_2$ but decreased at higher concentrations. Shoots treated with KCl and $CaCl_2$ showed enhanced activities of antioxidant enzymes (SOD, CAT, APX, and GPX), but KCl inhibited their activities as concentration increased. The synthesis of bacoside-A was enhanced with an increase in KCl and $CaCl_2$ concentration (Ahire et al. 2014).

9.3.3 Tea Plant (*Camellia sinensis*)

Tea plants treated with heavy metals (e.g., Cd, Cu, Al) exhibited reduction in growth and antioxidative responses (Upadhyaya and Panda 2013). The antioxidant responses to increasing concentrations of copper were investigated in the leaves of two species (TS-462 and TS-520) of tea commonly grown in the Darjeeling hills. Treatment with increased levels of Cu resulted in increased lipid peroxidation, reduced chlorophyll content, elevated levels of phenolic compounds, and an enhancement in peroxidase enzyme contents. Two new peroxidase isoenzymes (POD1 and POD2) were seen in plants exposed to Cu. TS-520 was found to be more susceptible toward increasing concentrations of Cu. SOD activity increased in TS-462 but declined in TS-520 when exposed to higher Cu concentrations. A sharp increase in the activity of APX was seen at the 10 days of treatment in the more resistant species. Moreover, CAT levels increased only slightly in both the species. Cu treatment caused accumulation of lipid peroxidation products in the shoots. The amount of TBARS improved gradually with dose of Cu and duration of treatment in both the species up to 7 days beyond which the levels declined. Al stress also caused an increase in SOD activity in cultured tea cells (Ghanati et al. 2005; Saha et al. 2012).

Seedlings of tea plants were studied for the effect of Al on antioxidant defense mechanism of leaf and the ultrastructure of cell. It was seen that MDA content declined at low Al concentration but increased at high Al concentration. Hydrogen peroxide content was enhanced at high Al dose, and no differences were observed at low Al dose. SOD activity was constant at low Al concentration but enhanced suddenly at high Al concentration. CAT and GPO activities were reduced followed by an increase, reaching maximum at low Al dose. APX activity increased and GSH level fluctuated with increasing Al stress. TEM analysis of leaves under Al stress showed that although cell ultrastructural integrity was maintained at low concentration of Al, substantial membrane damage was detected at high concentration (Li et al. 2011).

A probable link between the effects of Al on the growth of tea plants and the reactive oxygen species scavenging method in roots of complete tea plants and suspension-cultured tea cells was investigated. The enzymatic activities (SOD, CAT, and APX) were enhanced by Al both in roots of intact plants and cultured cells when compared to control. The amount of membrane lipid peroxidation, the activity of wall-bound peroxidase, and the lignin and wall-bound phenolic contents were reduced by the Al stress in cultured tea cells (Ghanati et al. 2005). The effect of increased concentration of Cu stress was studied on physiological and biochemical parameters of two tea cultivars (S3A3 and TS-491). The buildup of Cu in the diverse parts of the tea plants was positively correlated to the Cu treatment. The level of Cu was more in roots than in leaves and also the new stems. The activities of SOD, POD, and CAT increased with increasing Cu concentration but in different fashion in both the cultivars. The activities of antioxidant enzymes, APX, CAT, SOD, and POD, in cultivar S3A3 showed an increase, whereas TS-491was tolerant to Cu stress. MDA and phenol contents have also direct correlation with increasing Cu concentration in both the cultivars (Dey et al. 2015).

The effect of Al was studied on tea plants taking various factors into consideration in hydroponic system. Al ions enhanced the growth of shoots and roots, but at greater concentrations, it became injurious and caused oxidative damage. The accumulation pattern of Al showed dose-dependent manner. The parameters such as pigment content, photosynthetic rate, transpiration, and stomatal conductance were declined under Al stress showing decreased growth and activity. An increase in lipid peroxidation, cell membrane damage, release of superoxide radicals, and increase in H_2O_2 content were observed. Contents of phenols increased initially but declined at higher levels. As there was elevation in the concentration of Al, the activities of antioxidant enzymes showed enhancement. Expression of citrate synthase gene was upregulated in the mature leaves, both young and mature roots simultaneously as the concentration of Al was increased (Mukhopadyay et al. 2012).

9.3.4 Sand Rock Cress (*Cardaminopsis arenosa*) and English Plantain (*Plantago lanceolata*)

Sand rock cress and English plantain were studied for the effects of different doses of Cd, Pb, Zn, Cu, Fe, and Mn. The parameters taken into account were the bioavailability and bioaccumulation of these metals in the leaves of plants. The antioxidant response of these plants was studied, and it was found that the activities of GPO and SOD and the amounts of antioxidant phytochemicals like GSH, proline, and NPSH were determined. The contents of Zn, Cd, and Pb in the shoots of sand rock cress properly showed the amount of metal in the metalliferous and nonmetalliferous (NM) soil. The measured metal concentrations in English plantain showed that the bioaccumulated Zn, Cd, and Pb concentrations were greater or in the ranges mentioned as toxic for plant tissues (Nadgorska-Socha et al. 2013).

An enhancement in the total GSH concentration was seen in leaves of sand rock cress and English plantain collected from the polluted locality in comparison to its concentration in plants from the NM area. In the leaves of English plantain collected from site containing metals and in the leaves of sand rock cress from metalliferous site, the total GSH level in the leaves was comparable to the NM area (Nadgorska-Socha et al. 2013). The highest total GSH concentration was recorded in the shoots of sand rock cress from one of the calamine areas which was twice than in the plant samples from the NM area. The total glutathione content showed positive correlation with Zn and Cd content in sand rock cress shoots and with Pb content in English plantain shoots. Further, an indirect correlation was found between Fe, Mn, and Cu content and GSH level in sand rock cress leaves. In contrast, an opposite trend was observed in case of NPSH. Their concentrations were mostly lesser in the leaves of plants that were exposed to metal contamination when compared with the NM area. They showed positive correlation with Cu and Pb contents in English plantain leaves, in contrast to a negative relationship between Cd or Fe contents and NPSH levels in sand rock cress leaves. A decrease in proline content was seen in the leaves of English plantain in polluted areas in contrast to the nonpolluted areas. The proline content showed negative correlation with Fe, Mn, and Cu concentrations in the leaves of sand rock cress plants and with Cd, Zn, Pb, and Mn in the leaves of English plantain (Nadgorska-Socha et al. 2013).

The activities of POD were more in plants grown in metalliferous soil than in those from the NM area. The POD activity in the leaves of sand rock cress and English plantain from both the calamine sites was comparable with the activity in plant leaves from the NM area (Nadgorska-Socha et al. 2013). SOD activity was higher in the case of sand rock cress and English plantain leaves collected at the polluted calamine sites than SOD activity in plant leaves at the NM location. SOD activity was enhanced in the leaves of both species as the concentrations of Fe and Mn were increased, whereas its activity was enhanced with increase in Cu content in sand rock cress leaves. An indirect relation was found between the Cu concentration and SOD activity in English plantain leaves (Nadgorska-Socha et al. 2013).

9.3.5 Shona Cabbage (*Cleome gynandra*)

The effects of heavy metals on antioxidant defense system were studied in shona cabbage plants. The reduced value of phenolics with an increase in concentration of Cu and Cd showed that phenols form chelation with metal and thereby reduce the toxicity of these metals in the plant during their bioaccumulation. The antioxidant activity was found to be the maximum in the plant grown in soil without any treatment (control), and the free radical scavenging activity was decreased in the plant grown in the heavy metal-contaminated soils. The proline value was considerably greater in plants exposed to increased concentrations of Cu and Cd. SOD, CAT, and GSH were enhanced markedly in the plant sample grown in heavy metal-contaminated soils. Among the two metals studied, cadmium was more toxic to the plant than Cu (Haribabu and Sudha 2011).

9.3.6 Brazilian Waterweed (*Egeria densa*)

Brazilian waterweeds were exposed to sulfate salts of Ni, Cd, Cu, Zn, and Mn individually and in mixtures to study the effects of these metals on shoot growth, structural and functional factors of the photosynthetic apparatus, lipid peroxidation, antioxidant enzymatic activities (SOD and CAT), and the content of NPSH and protein thiols (PSH) in leaves.

The accumulation of heavy metals in shoots was reduced in the order: Mn> Cu > Cd > Zn > Ni. The maximum reduction in chlorophyll content was caused by Mn and Cu, whereas the maximum reduction in carotenoid content was induced by

Cu, and it also decreased the quantum efficiency of photosystem II (PSII) (Fv/Fm) to a great extent. The presence of Cd increased the contents of chlorophyll and carotenoids without changing the photochemical efficiency of PSII; Cd retarded the shoot growth but had no appreciable effect on leaf mesostructure. The addition of another metal to the growth medium alleviated the detrimental action of individual ions in most treatments owing to the enhanced activities of SOD and CAT and because of the significant increase in the content of NPSH. It is supposed that the observed antagonism of metal ions is related to their competitive interactions restricting the entry of heavy metals into the cell.

The chloroplast dimensions in Brazilian waterweed cells showed no uniform change under the action of heavy metals. The addition of Ni caused a significant reduction of chloroplast volume. The lesser reduction was noted under combined application of all metals examined. On the other hand, the prolonged exposure of plants to Cd or Cd + Ni induced the marked increase in the chloroplast volume, which might be responsible for chloroplast swelling. It is not excluded that the presence of Cd resulted in partial destruction of chlorophyll, which was evident from the pale leaf color. The supplementation of the growth medium with Ni, Zn, and Mn elevated SOD activity by 20% on the average compared to control values. The combined application of most metals increased the SOD activity substantially, especially in the presence of Ni + Cd, Ni + Zn, and Ni + Mn combinations. The heavy metals examined had fewer effects on CAT activity when applied individually. However, the combined application of two metals enhanced CAT activity by 1.5–2.0 times, with an exception of Mn + Cd and Mn + Cu combinations.

Synthesis of NPSH and PSH is a known means of plant protection against deleterious action of heavy metals. The content of NPSH was enhanced markedly in the presence of individual heavy metals as well as their combinations. The largest increase was observed in the presence of metal pair Mn + Cu. NPSH, GSH in particular, play an important role in the protection of membrane against free radical damage (Maleva et al. 2012).

9.3.7 Chinese Liquorice (*Glycyrrhiza uralensis*)

Chinese liquorice seeds were sown and grown in different concentrations of cadmium acetate, in order to investigate the effects of Cd on the growth, uptake studies, and the activities of SOD, POD, CAT, PPO, and PAL in Chinese liquorice seedlings. Results showed that increased Cd adversely affected shoot elongation and seedling biomass. SOD activity in the cotyledons, hypocotyls, and radicles increased gradually. POD activity was increased in all the parts as the concentration of Cd was increased, whereas CAT activity increased gradually with increasing Cd concentrations. PPO activity was enhanced markedly in the cotyledons, hypocotyls, and radicles. There was marked alterations in PAL activity in all the parts with increasing Cd concentrations (Zheng et al. 2010).

The effects of Mn metal were investigated on the physiological and growth characteristics of Chinese liquorice. All of the physiological and growth characteristics of Chinese liquorice showed increase with increased concentration of Mn. SOD and POD increased in Mn stress initially and showed significant difference afterward, while CAT was the highest at higher concentration. Then all of them began to decline. Dry weights of the shoots and roots at lower concentration were the highest and showed significant difference at higher concentration (Shengjun et al. 2015).

9.3.8 Liquorice (*Helianthus annuus*)

Two plants, sunflower and wonder berry (*Solanum nigrum*), were allowed to grow in soils contaminated with chromium. The parameters such as the contents of biochemical compounds, vitamins and minerals, and antioxidant status were studied. From the results, it was shown that the heavy metal, chromium, has significant effects on the biochemical factors, vitamin and mineral contents, and antioxidant status in both the plants. When the antioxidant property of the two plants was compared, it was seen that wonder berry has higher antioxidant property than sunflower (Vijayalakshmi et al. 2010).

The total carbohydrates, total proteins, fats, chlorophyll, and total amino acid content except proline were markedly decreased in the plants sample of both sunflower and wonder berry grown in heavy metal Cr-contaminated soil. The vitamin and mineral contents were decreased significantly in the heavy metal-contaminated soil. The vitamins A, B, and C, β-carotene, total phenolics, and other minerals were considerably more in the control plant showing that the plant is rich in antioxidant phytochemicals. The activities of antioxidant enzymes, SOD, GSH, and CAT, were significantly reduced in metal-exposed plants as compared to control in sunflower, whereas the metal stress enhanced the activity of SOD and GSH in wonder berry and no effect was seen on CAT (Vijayalakshmi et al. 2010).

9.3.9 Barbados Nut (*Jatropha curcas*)

The effect of different concentrations of Al metal was studied on growth and antioxidant enzyme (SOD, POD, CAT, and PAL) activities in barbados nut seedlings. It was seen that as the concentration of Al increased, the cotyledon biomass showed an initial increase followed by a decrease, but the biomass of hypocotyls and radicles decreased gradually. The Al stress enhanced SOD activity in the cotyledons, hypocotyls, and radicles as compared to the control. SOD activity in the hypocotyls increased markedly with increasing Al concentrations. The study of SOD isoforms was done by native polyacrylamide gel electrophoresis (PAGE) and analyzed. The activity staining showed that cotyledons, hypocotyls, and radicles expressed at least four SOD isoenzyme bands. Al stress has a significant effect on the POD activity in the cotyledons which increased considerably. On the activity gels, minimum six bands POD isoforms in the cotyledons, hypocotyls, and radicles were observed. POD isoenzymes (II and III) in the cotyledons showed an increased staining intensity with the increasing of Al concentration. In the hypocotyls and radicles, the staining intensities of isoenzymes IV and III were more, respectively. CAT activities increased in hypocotyls and radicles as compared to control, while in cotyledons, CAT activities increased first and then decreased as the concentration of Al increased. Compared to the control, the PAL activities showed increase in all the samples. In the cotyledons and radicles, PAL activities were increased first and then decreased with the increasing Al concentration (Ou-yang et al. 2014).

Barbados nut embryos were grown in vitro to examine the effects of Pb on cotyledon responses. The biomass of cotyledon showed enhancement initially and then showed reduction as Pb concentration was increased. The enzymatic activity of SOD increased in the beginning after which it declined. The POD activity showed a similar trend, and the CAT activity showed enhancement at all Pb concentrations. However, the PAL activity was inhibited significantly under the influence of Pb except for 100 μM concentration. Isozyme analysis by electrophoresis suggested a significant correlation between the concentration of Pb and patterns of SOD, POD, and CAT isoenzymes formation, and these results were in accordance with changes of the antioxidant enzyme activities as assayed in solution (Gao et al. 2009). In another study, effects of Pb treatment on growth and micronutrient uptake in barbados nut seedlings were evaluated by means of microcosm experiments. The outcome showed that SOD activity increased as the concentration of lead increased. There was significant direct correlation between lead concentration and the activities of SOD and POD. CAT activity was seen under lower lead stress but was inhibited under higher Pb exposure. Pb had a stimulating effect on the height of seedlings and leaf area at lower concentrations. The barbados nut seedlings can accumulate greater amounts of lead available in the soil but can translocate only low amounts to the shoots. Results indicated that in barbados nut seedlings, SOD and POD enzymes played an important role in combating the oxidative stress induced by Pb. The addition of Pb significantly increased Zn content in plant tissues and enhanced the transport of Fe from roots to shoots but contributed to a decrease in measured Cu, Fe, and Mn content (Shu et al. 2014).

The effects of Cd and Pb separately and in combination were studied with respect to performance, physiological and biochemical responses, and cellular ultrastructure of the barbados nut seedling exposed to individual and their combined stress. In response to different doses of Cd and Pb, plant biomass, gas exchange rate, and photosynthetic pigments decreased, while other parameters showed significant increase. The activities of antioxidant enzymes (SOD, POD) significantly increased when the seedlings were treated with low concentrations of Cd and Pb but drastically decreased under high doses of Pb and Cd. The level of CAT was seen to be increased irrespective of concentration of metal. This showed that low Cd and Pb levels stimulated the antioxidant system, but higher Cd and Pb levels surpassed the defensive capability of barbados nut. The cellular ultrastructure of seedling fronds was evaluated by transmission electron microscopy. The deleterious effects stress on mitochondria and chloroplast were most obvious. The cell walls and plasma membranes also showed distortion under intracellular observation (Liang et al. 2012).

A pot culture experiment was performed to explore parameters such as growth performance and biochemical and physiological responses of the barbados nut in fly ash amended soil. The elemental composition (Zn, Ca, Mg, Pb, Cu, Fe, Mn, Ni, and Cd) was studied by atomic absorption spectrophotometer in base material at the beginning and at the end of the study. The 3-year response was reported and observed that the proline content in barbados nut leaves increased as the fly ash concentration increased. After 3 years of complete plant growth, the heavy metals uptake increased with respect to the availability. According to the results, it has been proved that when heavy metal uptake by barbados nut increases through fly ash, the proline content increases according to its capability to defense itself in stress conditions (Raj and Mohan 2016).

Barbados nut was analyzed for its accumulation capacity of Pb, Cd, and Cr. The results showed that the plants treated with metal plants accumulated huge quantity of Pb, Cd, and Cr as compared to the control. The activity of enzymatic antioxidant such as SOD, CAT, PPO, POD, and PAL was evaluated, and the data showed significant variations in response to heavy metal stress as compared to the control plants. The enhanced activity of SOD, CAT, PPO, POD, and PAL were seen in root and shoot. PAL activity was reduced in leaf in the presence of stress. The concentration of nonenzymatic antioxidants such as proline increased with increasing heavy metal concentration, but phenol concentration showed fluctuation. Carotenoids and chlorophyll contents were found to be reduced with increasing concentrations of heavy metal. Barbados nut was capable of self-protection against multi-metal stress through activation of various enzymatic and nonenzymatic antioxidants (Devi et al. 2014).

Barbados nut seeds were germinated in vitro, and the toxicity of high Ni concentrations on seedling growth was studied. The fresh weight of cotyledons in all the tested Ni concentrations was less than control, whereas dry weight was increased with increasing stress. SOD activity increased significantly at low concentration and then showed reduction at high concentration. POD activities were induced remarkably at low concentration of Ni, but the activity decreased with increasing concentrations. Similarly, an indirect correlation between CAT activity and Ni concentrations was also observed. PAL activity had a direct correlation with nickel concentrations, and the highest activity was found at moderate Ni stress. The results of electrophoresis showed a significant correlation between Ni concentrations and isoenzyme patterns of SOD and POD, and these analyses were in accordance with the alterations in the activity assayed in solutions. This study investigated morphological, physiological, biochemical, and metal uptake patterns in seedlings exposed to Zn, Cr, and Cd in a water-based liquid system. The buildup of metals in all parts of the seedlings was found to be highest for Zn, followed by Cd and Cr. The activities of antioxidant enzymes, CAT and SOD, showed enhancement on treatment with heavy metals, but low concentration of Cd and Cr significantly reduced SOD activity as compared to Zn which is indicative of damage to the plant due to lipid peroxidation as is evident by a reduction in the % membrane stability index at higher metal levels. The increase in the levels of all the heavy metals tested showed a marked decrease in chlorophyll content; however stimulatory effect was observed at low Zn concentrations. Root length and surface area were both significantly reduced with increasing metal concentration (Yan et al. 2008).

Barbados nut seeds were geminated, and embryos were grown in vitro in the presence of different Zn concentrations to explore its effects on the growth and the alterations in SOD, POD, CAT, and PAL activities in barbados nut seedlings. The biomass of the cotyledons, hypocotyls, and radicles was increased with increasing zinc concentration. SOD activity in all the parts was increased gradually as the concentration of Zn increased. While POD activity in the cotyledons, hypocotyls, and radicles was enhanced with increasing Zn concentrations, CAT activity in the cotyledons, hypocotyls, and radicles reached maximum under Zn stress. PAL activity showed a similar trend as compared to the changes in CAT activity. Electrophoresis study suggested a substantial correlation between Zn concentrations and isoenzyme patterns of SOD, POD, and CAT, and these results were in accordance with the alterations in the activities assayed in solutions. These results suggest that the antioxidant enzymes (SOD, POD, CAT, and PAL) may play a significant role in the defensive mechanisms of barbados nut seedlings treated with excessive Zn (Luo et al. 2010).

9.3.10 Mangrove (*Kandelia candel*) and Black Mangrove (*Bruguiera gymnorrhiza*)

Mangrove and black mangrove, two mangrove plants, were grown in multiple heavy metal stress (Na, Ca, Pb, and Hg), and their effects on the activity of antioxidant enzymes and lipid peroxidation were studied in leaves and roots. In leaves of heavy metal-treated plants, SOD and POD activities showed fluctuation when treated with different concentrations of metals as compared to control, while CAT activity was enhanced as the with stress in mangrove increased but remained unchanged in leaves of black mangrove. When compared with control, there was an increment in the activities of SOD, CAT, and POD activities in roots of heavy metal-treated plants but showed declination later on. The increase in enzyme activities demonstrated that mangrove is more tolerant to heavy metals than black mangrove. Lipid peroxidation was increased only in leaves of heavy metal-stressed plants of black mangrove (Zhang et al. 2007).

9.3.11 Spearmint (*Mentha spicata*)

The effect of heavy metals including Cr, Cd, Co, Ni, and Pb was studied in spearmint with respect to the antioxidant system and oil quality in leaf, root, and shoot. For this, different ratios of sludge and soil was taken, and it was observed that carvone, limonene, dihydrocarvone, and other oil constituents along with biomass were maximum in 75:25 sludge and soil ratio. The activity of the antioxidant enzymes (SOD, CAT, and POD), MDA, and proline contents were enhanced as concentration of metals increased in the experiments (Patel et al. 2016).

9.3.12 Basil (*Ocimum basilicum*)

Arsenic toxicity and the preventive role of coronatine (COR) which is a functional analogue of jasmonates were studied in sweet basil seedlings. Arsenic toxicity caused reduction in growth and biosynthesis of photosynthetic pigments and induction of lipid peroxidation together and enhancement of hydrogen peroxide and electrolyte leakage in leaves. As-treated seedlings accumulated phenolic compounds and proline in the leaf and showed enhanced activities of CAT, SOD, and POD when COR was added (Dehabadi et al. 2013).

The effects of Cd, Pb, and Zn metals were studied on growth and antioxidant status of basil and oregano (*Oregano vulgare*). The pot experiment was conducted on plants grown in severely polluted soil with Cd and Pb and on unpolluted soil. Cd was accumulated in both the plants in the roots, while Pb was seen in the oregano shoots only. The thickness of leaf blade in both plants increased when grown in soil polluted with heavy metals. Heavy metal stress increased gas exchange, stomatal conductivity, and transpiration rate, but water-use efficiency was reduced, whereas gas exchange and transpiration were declined in treated oregano plants, but an increase was observed in stomatal conductivity and water-use efficiency. In basil, heavy metal stress increased the contents of low molecular weight antioxidants such as phenols and flavonoids, whereas in oregano, the levels of ascorbate, glutathione, and phenols were increased. The heavy metal stress increased the activities of three antioxidant enzymes of the glutathione metabolism, namely, GPX, GST, and GR, as well as GPO. The study further revealed that the activities of GST, GR, and GPO were better expressed in basil than in oregano shoots, whereas GPX activity was seen to be higher in the shoots of oregano grown in polluted soil. In contrast, reduced activities of three enzymes of ascorbate metabolism, namely, MDHAR, DHAR, and APX, were observed. The activities of MDHAR and DHAR were decreased in both plant species grown in polluted soil, while APX activity was decreased in basil shoots. The reduced ratios for glutathione/oxidized glutathione, ascorbate/dehydroascorbate, and ascorbate/H_2O_2 increased in the aboveground parts of treated oregano plants only, which indicated that in oregano the functioning of the ascorbate-glutathione cycle is effective. High levels of Cd, Pb, and Zn in plant organs resulted in reduced essential oil yield in basil only (Stancheva et al. 2014).

The effect of Cd on the phenol, flavonoid, and proline contents in Ardestan and Isfahan cultivars of basil plants was studied. The content of phenol, flavonoid, and proline increased in both of the cultivars with increasing concentration of cadmium chloride. The increase in Ardestan cultivar was more pronounced than Isfahan cultivar (Khavari-Nejad et al. 2013).

9.3.13 Olive (*Olea europaea*)

The alleviation of Cd toxicity was studied by exogenous proline in young olive plants. The Cd metal stress led to the buildup of Cd in both root and shoot with a reduction in the exchange of gases, photosynthetic pigment contents, uptake of essential elements (Ca, Mg, and K), and plant biomass. In addition, an increase in the activities of antioxidant enzymes (SOD, CAT, GPX) and proline content was observed. Additionally, high amounts of hydrogen peroxide, TBARS, and electrolyte leakage were seen. It was shown that olive plants grown in Cd stress and proline treatment showed enhancement in the activities of antioxidant enzymes, photosynthetic activity, nutritional status, plant growth, and oil content of fruit. In general, it appears that proline supplementation alleviated the harmful effects of young olive plants exposed to Cd stress (Zouari et al. 2016).

9.3.14 Brazilian Ginseng (*Pfaffia glomerata*)

The role of the antioxidant enzymes in adaptive responses of the accumulator Brazilian ginseng species under Cd stress was studied. The rate of lipid peroxidation in leaves and roots was lesser during initial phase of the experiment at all Cd levels. SOD activity increased in leaves and in roots as the concentration of Cd was increased. Cd stress induced an increase in APX activity in leaves, whereas it was reduced in roots at high concentration of Cd. At the end of the experiment, CAT activity in leaves was reduced as the concentration of Cd increased, whereas the GR and GPX activities were increased, and in roots, GR activity was reduced (Marques and Soares 2011).

Oxidative damage caused by Hg was studied in Brazilian ginseng plantlets. Accumulation of Hg in tissue increased with increasing concentration of Hg. The fresh weight of root and shoot and delta-aminolevulinic acid dehydratase (d-ALA-D) activity were considerably reduced, and there was no effect on chlorophyll and carotenoid contents at high levels of Hg. H_2O_2 concentration in shoot increased curvilinearly with increased Hg levels, whereas lipid peroxidation was enhanced in roots and shoots. SOD activity showed a direct correlation with H_2O_2 concentration, whereas CAT activity was increased in shoots. Shoot APX activity was decreased at low Hg concentration and increased at high concentration. Conversely, root APX activity was only increased at low Hg concentration. In general, AsA, NPSH, and proline contents were enhanced upon addition of Hg, with the exception of proline in roots, which showed decline (Calgaroto et al. 2010).

The effect of Cd stress was evaluated, and alterations in physiological and anatomical characters in Brazilian ginseng were observed. The levels of Cd were found to be higher in roots than in shoots. Cd enhanced the xeromorphic characteristics in leaves and increased thickness of root endodermis. The activities of enzymatic antioxidant systems of root and leaves showed differential effects of Cd. Root endodermis thickness, stomatal size and numbers, root SOD, and GPO, as well as leaf GPO and CAT activities, can all be considered Cd-tolerance markers in

Brazilian ginseng (Gomes et al. 2012).The probable role of Zn to recover the oxidative damage caused by Hg was studied in Brazilian ginseng plantlets. In combined Zn and Hg treatment, shoot and root were smaller than that plants exposed to Hg only. An enhancement in the Zn content in the shoot of plants exposed to Zn-Hg occurred, although in the roots Zn content was not changed when compared to the control. Fresh and dry weights, as well as the activity of d-ALA-D in Hg-exposed plants, were markedly decreased. Percentage survival, fresh and dry weights, and d-ALA-D activity of plants treated by Zn-Hg were more than of Hg treatment alone. Moreover, Zn treatment decreased lipid peroxidation caused by Hg; this effect might be responsible for increased root SOD activity and shoot CAT and APX activities (Calgaroto et al. 2011).

9.3.15 Pickleweed (*Salicornia brachiate*)

Pickleweed plants were treated with salts of three different heavy metals, $CdCl_2$. H_2O, $NiCl_2$.6H2O, and $NaAsO_2$, in the presence of NaCl. There was no effect on fresh weight of plants. All the doses severely affected the chlorophyll content of the treated plants. Accumulation of Cd^{2-}, Ni^{2-}, and As^{2-} in foliar tissue was found to be considerably higher under stress. Proline concentration was markedly increased at all concentrations of metals. H_2O_2 content showed gradual increase with the increasing metal concentrations. The CAT activity was notably increased when exposed to all three metals, whereas SOD activity was considerably higher at maximum concentration of Cd^{2-} and Ni^{2-}stresses (Sharma et al. 2010).

9.3.16 Wonder Berry (*Solanum nigrum*)

The effect of Cu stress was studied on wonder berry plants. A dose-dependent reduction in growth with increasing concentration of Cu was observed, whereas Cu stress had no effect on total chlorophyll and carotenoid content. An enhanced MDA content which is a marker for lipid peroxidation was quantified in shoots, while in roots MDA levels showed a dose-dependent increase with increasing Cu concentrations. An increase in proline content in roots of plants treated with Cu was observed. The activities of antioxidant enzymes, SOD and CAT, did not show effects with respect to control, in both roots and shoots. APX was negatively affected in shoots by the highest Cu level (Fidalgo et al. 2013).

The defensive role of proline was studied in callus and in vitro-grown shoots of wonder berry against cadmium toxicity. The pretreatment with proline decreased the generation of ROS and showed protection of the plasma membrane integrity of callus grown in Cd stress. Also, exogenous proline enhanced the accumulation of Cd in callus and regenerated shoots. There was also an increase in SOD and CAT activity and intracellular total glutathione content (Xu et al. 2009).The improvement in the phytoextraction of Cd and the antioxidative defense response of wonder berry by application of a newly isolated strain (*Paecilomyceslilacinus*NH1)

(PLNH1) and citric acid (CA) was studied. The data was evaluated, and the results showed that application of CA or PLNH1 considerably stimulated the growth of wonder berry grown in Cd stress, but the synergistic effect of CA and PLNH1 on wonder berry growth was more obvious. The coexistence of CA and PLNH1 could enhance about 30% of Cd accumulation in different organs of wonder berry as compared to the treatment without the addition of CA and PLNH1, whereas CA or PLNH1 added separately enhanced about 10–15% of Cd accumulation in different organs of wonder berry. The results indicated that the antioxidative defense in wonder berry under Cd stress was significantly improved as a result of synergistic effects of CA and PLNH1. The activities of antioxidative enzymes in response to Cd stress were considerably reduced following application of CA and PLNH1, and the stress in the plant due to Cd in the soil was significantly alleviated (Gao et al. 2010).

SOD activity in plants in soil without Cd stress (control) increased with culture time, whereas the presence of Cd in soil increased SOD activity in plants that gradually decreased with culture time. The synergistic effect of CA and PLNH1 increased SOD enzyme activity during the initial phase of culture, after which it was slowly reduced and recovered to normal level. SOD activity was lesser in the presence of PLNH1 treatment than under unpolluted soil treatment, while it did not considerably change with culture time. The POD activity was enhanced with an increase in culture time, except in case of CA and PLNH1 synergy effects. The alteration in the activity of POD was higher in polluted soil as compared to control. The POD activity at initial stage of culture was considerably higher under the coexistence of CA and PLNH1 treatment as compared to other treatments. MDA content was initially more in control soil and single treatment than under combined treatment of CA and PLNH1 after which it was slowly reduced and then recovered to normal level. The MDA content in plant was higher in case of PLNH1 treatment as compared to CA treatment and the CA and PLNH1 synergistic effect. Total acid-soluble thiol (TAST) was reduced considerably with culture time except in soil without CA or PLNH1. The total GSH content was higher in control soil as compared to treated soil. In addition, the total GSH was more in plants in the presence of PLNH1 than for those in the presence of CA treatment and their combined treatments. The phosphatidylcholine (PC) content increased when plants were grown in soil without CA and PLNH1 with culture time, whereas its synthesis in plants grown in control soil or single treatment markedly decreased with increasing culture time (Gao et al. 2010).

Plants of wonder berry were planted in pots in the presence of variable concentrations of Cr for 3 weeks in pots and evaluated for various parameters of growth, antioxidants, and ion. Moreover, after the treatment the plants were grown in rhizobox-like system for 48 h, and organic acids were checked in the rhizospheric solution. It was observed that the Cr treatment enhanced the fresh and dry weights of shoot and root in wonder berry, whereas the reverse effect was seen in Santa-Maria (*Parthenium hysterophorus*) in lower levels of Cr or in control. The buildup of Cr and Cl was increased in both the plants, whereas that of Ca, Mg, and K content in root, shoot, and root exudates was reduced with increasing doses of Cr. Higher concentrations of Cr showed enhanced activities of SOD, POD, and proline content in leaves of wonder berry, whereas lower doses of Cr were found to have stimulatory effects in Santa-Maria. Santa-Maria showed greatest exudation of organic acid contents. With increasing concentrations of Cr treatments, citric acid concentration in root secretions was increased in both the plants (Uddin et al. 2015).

9.3.17 Milk Thistle (*Silybum marianum*)

The influence of cadmium was studied on seed germination percentage, growth of seedling, and proline content in milk thistle by growing the seedlings at variable doses of cadmium chloride. The outcome of the experiment was a considerable inhibitory effect under Cd treatment as compared to control. As concentration increased, a considerable decrease in germination of seed, length of shoot and root, and fresh weight of seedlings was observed. In contrast, an increase in Cd concentration showed an enhancement in shoot/root length ratio and proline content (Khatamipour et al. 2011).

9.3.18 Persian Clover (*Trifolium resupinatum*)

The study was conducted to study the alterations in the activity of CAT and POD as well as protein concentration in Persian clover explored in the presence of Cd based on a randomized block design with three repetitions. Four concentrations of $Cd(NO_3)_2$ were used, and the plants were under treatment for 10 days. The data showed that CAT activity was reduced as the concentration of Cd increased. In contrast, POD activity was enhanced by an increase in the utilization of Cd. The evaluation of protein content showed that the amount of plantlet protein was decreased at high Cd concentrations. The findings also showed that Cd accumulation in roots was higher than in shoots (Ghasemi et al. 2014).

9.3.19 Stinging Nettle (*Urtica dioica*)

Lead stress was given to stinging nettle, and it was observed that the concentration of Pb in stems changed synchronously with those in leaves of the plant, and nonessential metals disrupted the balance of mineral nutrient elements. DNA damage was studied, and it was observed that the samples treated with metals yielded a large number of new DNA fragments (total 12) in comparison with the control sample. Also, very high accumulation of Zn in stinging nettle leaves from the Veles region was observed. Cd content in stinging nettle leaves and stems, respectively, was also in the phytotoxic range. Copper content in all plant species did not exceed the threshold limit. The abundance of Mg, Ca, and Na in the result of this analysis was in agreement with previous findings that these three elements represent the most abundant metal constituents in plants (Maiga et al. 2005; Ozcan and Akbulut 2008). The treatment with high concentrations of redox-reactive metals resulted in decreased and not in increased activities of antioxidative enzymes. The total antioxidant activity ferric reducing ability of plasma (FRAP) in samples exposed to metals was lower than total antioxidant level in control sample. It could be concluded that the heavy metal stress decreased total antioxidant level and induced DNA damage in stinging nettle (Gjorgieva et al. 2013).

The glutathione metabolism was studied in stinging nettle with respect to the antioxidative response under Cd stress. The Cd stress led to increase in GR and GST activities in all organs; but, GSH-Px activity was Cd dependent and varies in different plant organ. The GSH/glutathione disulfide (GSSG) ratio was seen to be above control in the stems in Cd stress, and the lipid peroxidation was close to the control values in the roots and stems but was enhanced in the leaves. The buildup of Fe, Zn, Mn, and Cu besides Cd was also studied in treated plants, and it was seen that the Cd concentration was increased in all organs, whereas the Fe, Zn, Mn, and Cu content were reduced in concentration and time-dependent manner (Tarhan and Kavakcioglu 2016).

9.3.20 Ashwagandha (*Withania somnifera*)

Iron stress was induced by adding higher quantity of $FeSO_4$ in ashwagandha plants. It caused disorder in nutrient balance and induced oxidative stress in plant. Root and leaf tissues were studied for CAT, SOD, and GPO activities. The results showed an increase in content with respect to exposure time (Rout et al. 2015).In vitro-grown ashwagandha seedlings were grown in variable concentrations of Cu. Decrease in stem length, root elongation, and leaf area was observed on increasing the concentration of Cu^{+2}. The stress also showed reduction in chlorophyll and carotenoid contents in leaves and stems as compared to controls, and the effect was more severe in leaves than in stems. Lipid peroxidation (MDA content) was highest in roots followed by leaves and stems. The phenol content of leaves, stems, and roots was reduced at higher concentration of Cu^{+2} (Singh et al. 2014).

The LOX activity was enhanced along with increased H_2O_2 and superoxide anions ($O_2^{-\bullet}$) generation in Cu-treated plants. APX, MDHAR, DHAR, GST, and GPO activities in leaves were increased in Cu stress as compared to control. Six APX and four GPO isoforms were identified on native polyacrylamide gels in treated plants. There was remarkable decrease in SOD, CAT, GR, and GPX activities (Khatun et al. 2008).

9.3.21 *Zygophyllum album* L. and *Zygophyllum coccineum* L.

Two species of *Zygophyllum* (*Z. album* and *Z. coccineum*) were grown to study the effects of soil heavy metal (Al, Cu, Zn, Fe) in the accumulation of heavy metals in shoot, LPO, enzyme activities, and the root plasma membrane (PM) lipid composition. Heavy metal accumulation and LPO were enhanced in the shoot of both species grown in the polluted area. The activities of ascorbate oxidase (ASO), GPO, APX, and SOD were increased, whereas that of CAT was decreased in both species under the stress conditions. PM total lipids, phospholipids, glycolipids, and sterols were reduced in *Z. album* and *Z. coccineum* as a result of the polluted soil. Heavy metal stress increased phosphatidylethanolamine (PE) and decreased phosphatidylinositol (PI) and phosphatidylglycerol (PG), with no significant change in PC in the root PM of both species. Phosphatidylserine (PS) was decreased in the PM of *Z. album*, whereas it was increased in the PM of *Z. coccineum* under the stress conditions. The composition and concentration of fatty acids of the root PM were altered in heavy metal stress, resulting in increased sat/unsat ratio of both species (Morsy et al. 2012).

9.3.22 Syrian Bean Caper (*Zygophyllum fabago*)

Syrian bean caper was studied to see the effects of the flavonoid rutin (Rtn) on stress responses of some heavy metal such as antioxidant defense systems and water status. The seedlings were exposed to $CdCl_2$ stress without or with Rtn. In Cd stress decreased growth, water content, leaf osmotic potential, and chlorophyll fluorescence were observed. All of these responses could be partially alleviated by addition of Rtn. Activities of SOD, POD, APX, and GR were enhanced after initial exposure to Cd. An abrupt rise in superoxide anion radical and hydrogen peroxide content and decline in hydroxyl radical scavenging activity resulted in enhancement of lipid peroxidation which is a marker oxidative stress. However, exogenously supplemented Rtn greatly improved the stress tolerance of plants via a reduction in Cd accumulation, modulation of POD activity, increase in proline content, reduction in TBARS and ROS contents, and subsequent reduction of oxidative damage of membrane (Yildiztugay and Ozfidan-Konakci 2015).

9.3.23 *Zygophyllum xanthoxylon*

Z. xanthoxylon seedlings were grown in five different concentrations of Ni. Ni stress resulted in significant reduction in fresh biomass of shoot and roots at high concentrations, whereas there was no considerable effect on dry biomass in both the organs. The accumulation of Ni content in shoots and roots was increased in a dose-dependent manner, and the amount of Ni was higher in roots as compared to shoots at all treatments. The high Ni doses increased the levels of soluble protein in shoots and roots. Chlorosis was not observed in shoots of plants grown in Ni stress. TBARS and carotenoid contents were enhanced in the presence of stress. MDA content significantly increased in leaves and roots at high Ni treatments which were indicative of Ni-induced oxidative stress. Antioxidant enzyme might play a central role in cellular protection against the Ni-induced oxidative damage. Proline content proportionally increased with increased Ni concentration. *Z. xanthoxylon* revealed moderate Ni tolerance under tested culture conditions (Lu et al. 2010).

9.4 Medicinal Plants and Their Antioxidant Properties

In addition to providing defense, plants have been a source of natural antioxidants which become the component of our diet since ancient time. Two-thirds of the different plant species inhabiting earth are thought to have medicinal significance, and almost all of these possess excellent antioxidant prospective (Krishnaiah et al. 2011). In the evaluation of in vitro antioxidant activity of medicinal plants or their phytochemicals, several biochemical tests have been used. In ethanopharmacological and nutraceutical investigations, these assays are done to apprehend the mode of action of these antioxidant molecules and the probable mechanism underneath them (Antolovich et al. 2002) with the intention of minimizing the pathophysiology of diseases caused due to oxidative stress. To measure antioxidant activity of plants, there are several biochemical in vitro assays used to measure and confer to the activity to the plants. However, each of these has its own restrictions regarding applicability and use. These assays give the idea of the antioxidant potential of plants (Kasote 2012).

Since time immemorial, plants have been an important source of nutrition and medicines, either in the form of conventional preparations or as pure bioactive compounds. Most of the medicinal properties of plants have been attributed to their potent antioxidant properties. It has been proposed that free radicals are involved in the pathology of more than 50 human diseases, including aging (Halliwell 1991). Plants are rich storehouse of secondary metabolites, and the complex diversity of these metabolites makes them fascinating candidates for study. Plant antioxidant phytochemicals such as ascorbic acid, carotenoids, and flavonoids have been shown to be the best exogenous antioxidants. These compounds scavenge free radicals and confine ROS production and improve endogenous antioxidant defense system of the body (Halliwell 2006).

It is important to consider the fact that plants produce antioxidants for their own requirements to perform specific functions. Vitamin C is regarded as one of the most important dietary antioxidants that help to decrease free-radical-mediated damage. Humans cannot synthesize vitamin C due to the deficiency of gulonolactone oxidase (Putchala et al. 2013). Accordingly, vitamin C also shows prooxidant effect and works as an antioxidant by donating its two electrons, which prevents the oxidation of other compounds (Putchala et al. 2013; Heo et al. 2013). Vitamin E is another important and potent plant antioxidant that is normally present in lipid component of cells. This compound reacts with peroxyl free radicals to inhibit the progression of lipid peroxidation (Wagner et al. 1996). At high concentration, vitamin E also shows prooxidant effects like vitamin C. Vitamin E has also been shown to react with free radicals to become a reactive radical and also acts as prooxidant (Carocho and Ferreira 2013).

The antioxidant property of polyphenols is due to its chemical structure as they determine the conjugation reactions with methyl, sulfate, or glucuronide groups (Scalbert and Williamson 2000). Flavonoids are the most significant and abundant dietary polyphenols, with over 5000 reported to date (Ross and Kasum 2002; Dai and Mumper 2010). In addition to their remarkable antioxidant property, polyphenols have prooxidant properties also. Carotenoids, which are abundant in fruits and vegetables, neutralize free radicals and function as antioxidants (Jomova and Valko 2013). The antioxidant efficacy of dietary carotenoids such as beta-carotene and lycopene in living systems is dependent on the presence of other co-antioxidants and may lose their effectiveness as antioxidants at high concentrations or at high partial pressures of oxygen (Young and Lowe 2001).

9.5 Conclusion

Plants are exposed to harsh climatic and environmental conditions which lead to stress. Metal stresses overproduce ROS in plants which are extremely unstable and toxic to the cells and leads to oxidative damage. Manageable amount of ROS are produced during normal metabolic processes, but excessive amounts damage nucleic acids, lipids, and proteins, and they may lose their activity. Since plants are sessile, they need to be equipped with excellent antioxidant defense mechanisms to combat the harmful effects of ROS. The antioxidant defenses are categorized as enzymatic (e.g., SOD, CAT, POD, and GR) or nonenzymatic (e.g., GSH, GB, proline, α-tocopherols, phenols, carotenoids, and flavonoids).

The organelles like chloroplast, mitochondria, and peroxisomes play vital role in protecting the cells from ROS-induced oxidative stress, and antioxidant defense systems are well established in these organelles. Metal stresses induce the antioxidant machinery at the cellular level which in turn comes into action to provide defense. Many researchers have shown the overexpression of ROS-scavenging enzymes like isoforms of SOD, CAT, APX, GR, DHAR, GST, and GPX which resulted in providing tolerance to metal stress due to enhanced ROS-scavenging capacity. Therefore, the scope for the production of transgenic plants with improved components of antioxidant defense at gene level widens. In the future, the use of antisense and RNA interference technologies combined with selection of specific cDNA clones for isoenzymes may facilitate exploration of such enzyme-based studies.

Despite all the technological advancements, the knowledge of the antioxidative defense system of plants and stress biology is quite limited. Rigorous research must be carried out in order to elaborate the role of each component of antioxidant defense system by taking more mechanistic approach. The activities of antioxidant enzymes are seen to double in response to many stress situations which further needs optimization of their kinetics. However, several research groups have significantly supplemented earlier pure physiological or biochemical approaches by taking into consideration the omics and system biology approaches.

References

Acharya S, Sharma DK (2014) Study on the effects of heavy metals on seed germination and plant growth on *Jatropha curcas*. IJASR 3:031–034

Agbor AG, Ngogang YJ (2005) Toxicity of herbal preparations. Cam J Ethnobot 1:23–28

Ahire ML, Laxmi S, Walunj PR et al (2014) Effect of potassium chloride and calcium chloride induced stress on in vitro cultures of Bacopa monnieri L. Pennell and accumulation of medicinally important bacoside A. J Plant Biochem Biotechnol 23:366–378

Antolovich M, Prenzler PD, Patsalides E et al (2002) Methods for testing antioxidant activity. Analyst 127:183–198

Asada K (1994) Production and action of active oxygen species in photosynthetic tissues. In: Foyer CH, Mullineaux PM (eds) Causes of photooxidative stress and amelioration of defence systems in plants. CRC Press, Boca Raton, pp 77–104

Baxter-Roshek JL, Petrov AN, Dinman JD (2007) Optimization of ribosome structure and function by rRNA base modification. PLoS One 2:e174

Becana M, Moran JF, Iturbe OI (1998) Iron dependent oxygen free radical generation in plants subjected to environmental stress toxicity and antioxidant protection. Plant Soil 201:137–147

Bennoun P (1994) Chlororespiration revisited: mitochondrial plastid interactions in chlamydomonas. Biochim Biophys Acta 1186:59–66

Blokhina O, Fagerstedt KV (2010) Reactive oxygen species and nitric oxide in plant mitochondria: origin and redundant regulatory systems. Physiol Plant 138:447–462

Calgaroto NS, Castro GY, Cargnelutti D et al (2010) Antioxidant system activation by mercury in *Pfaffia glomerata* plantlets. Biometals 23:295–305

Calgaroto NS, Cargnelutti D, Rossato LV et al (2011) Zinc alleviates mercury-induced oxidative stress in *Pfaffia glomerata* (Spreng). Biometals 24:959–971

Carocho M, Ferreira IC (2013) A review on antioxidants, prooxidants and related controversy: natural and synthetic compounds, screening and analysis methodologies and future perspectives. Food ChemToxicol 51:15–25

Castillo FJ, Greppin H (1988) Extracellular ascorbic acid and enzyme activities related to ascorbic acid metabolism in *Sedum alba* L. leaves after ozone exposure. Environ Exp Bot 28:231–238

Dai J, Mumper RJ (2010) Plant phenolics: extraction, analysis and their antioxidant and anticancer properties. Molecules 15:7313–7352

Dehabadi SZ, Shoushtari A, Asrar Z (2013) Modulation of arsenic toxicity-induced oxidative damage by coronatine pretreatment in sweet basil (*Ocimum basilicum*) seedlings. Botany 91:442–448

del Rio LA, Sandalio LM, Corpas FJ et al (2006) Reactive oxygen species and reactive nitrogen species in peroxisomes: production, scavenging, and role in cell signaling. Plant Physiol 141:330–335

Desikan R, Mackerness SAH, Hancock JT et al (2001) Regulation of the *Arabidopsis transcriptome* by oxidative stress. Plant Physiol 127:159–172

Devi CM, Anu MS, Mahesh B et al (2014) A comparative study of heavy metal accumulation and antioxidant responses in *Jatropha curcas* (L.) IOSR-JESTFT 8:58–67

Dey S, Mazumder PB, Paul SB (2015) Copper-induced changes in growth and antioxidative mechanisms of tea plant (*Camellia sinensis* (L.) O. Kuntze). Afr J Biotechnol 14:582–592

Elstner EF, Wagner GA, Schutz W (1988) Activated oxygen in green plants in relation to stress situations. Curr Topics Plant Biochem Physiol 7:159–187

Fidalgo F, Azenha M, Silva AF (2013) Copper-induced stress in *Solanum nigrum* L. and antioxidant defense system responses. Food Ene Sec 2:70–80

Foyer CH (1997) Oxygen metabolism and electron transport in photosynthesis. In: Scandalios J (ed) Molecular biology of free radical scavenging systems. CSHL Press, New York, pp 587–621

Foyer CH, Harbinson J (1994) Oxygen metabolism and the regulation of photosynthetic electron transport. In: Foyer CH, Mullineaux PM (eds) Causes of photooxidative stresses and amelioration of defense systems in plants. CRC Press, Boca Raton, pp 1–42

Gao S, Yan R, Wu J et al (2009) Growth and antioxidant responses in *Jatropha curcas* cotyledons under lead stress. Z Naturforsch C 64:859–863

Gao Y, Miao C, Mao L et al (2010) Improvement of phytoextraction and antioxidative defense in *Solanum nigrum* L. under cadmium stress by application of cadmium-resistant strain and citric acid. J Haz Mat 181:771–777

Ghanati F, Morita A, Yokota H (2005) Effects of aluminium on the growth of tea plant and activation of antioxidant system. Plant Soil 276:133–141

Ghasemi H, Yousefirad M, Sepehr MF (2014) Effect of cadmium on oxidative enzymes activity in persian clover (*Trifolium resupinatum* L.) IJPP 5:1203–1208

Gill SS, Khan NA, Anjum NK et al (2011) Amelioration of cadmium stress in crop plants by nutrient management: morphological, physiological and biochemical aspects. Plant Stress 5:1–23

Gjorgieva D, Panovska TK, Ruskovska T (2013) Influence of heavy metal stress on antioxidant status and DNA damage in *Urtica dioica*. Biomed Res Int 2013:1–6

Gomes MP, Marques TCLLSM, Martins GA et al (2012) Cd-tolerance markers of *Pfaffia glomerata* (Spreng.) pedersen plants: anatomical and physiological features. Braz J Plant Physiol 24:293–304

Halliwell B (1991) Drug antioxidant effects. Drugs 42:569–505

Halliwell B (2006) Reactive species and antioxidants redox biology is a fundamental theme of aerobic life. Plant Physiol 141:312–322

Han C, Liu Q, Yang Y (2009) Short-term effects of experimental warming and enhanced ultraviolet-B radiation on photosynthesis and antioxidant defense of *Picea asperata* seedlings. Plant Growth Regul 58:153–162

Haribabu TE, Sudha PN (2011) Effect of heavy metals copper and cadmium exposure on the antioxidant properties of the plant *Cleome gynandra*. IJPAES 1:80–87

Heo JH (2013) The possible role of antioxidant vitamin C in Alzheimer's disease treatment and prevention. Am J Alzheimers Dis Other Demen 28:120–125

Heyno E, Mary V, Schopfer P et al (2011) Oxygen activation at the plasma membrane: relation between superoxide and hydroxyl radical production by isolated membranes. Planta 234:35–45

Hu WH, Song XS, Shi K et al (2008) Changes in electron transport, superoxide dismutase and ascorbate peroxidase isoenzymes in chloroplasts and mitochondria of cucumber leaves as influenced by chilling. Photosynthetica 46:581–588

Jomova K, Valko M (2013) Health protective effects of carotenoids and their interactions with other biological antioxidants. Eur J Med Chem 70:102–110

Kasote DM (2012) Flax seed phenolics as natural antioxidants. Int Food Res J 20:1797–1804

Khan MIR, Khan NA (2014) Ethylene reverses photosynthetic inhibition by nickel and zinc in mustard through changes in PS II activity, photosynthetic-nitrogen use efficiency and antioxidant metabolism. Protoplasma 251:1007–1019

Khan MIR, Asgher M, Khan NA (2014) Alleviation of salt-induced photosynthesis and growth inhibition by salicylic acid involves glycinebetaine and ethylene in mungbean (*Vigna radiata* L.) Plant Physiol Biochem 80:67–74

Khan MIR, Nazir F, Asgher M, Per TS, Khan NA (2015) Selenium and sulfur influence ethylene formation and alleviate cadmium-induced oxidative stress by improving proline and glutathione production in wheat. J Plant Physiol 178:9–18

Khan MIR, Iqbal N, Masood A, Mobin M, Anjum NA, Khan NA (2016a) Modulation and significance of nitrogen and sulfur metabolism in cadmium challenged plants. Plant Growth Regul 78:1–11

Khan MIR, Khan NA, Masood A, Per TS, Asgher M (2016b) Hydrogen peroxide alleviates nickel-inhibited photosynthetic responses through increase in use-efficiency of nitrogen and sulfur, and glutathione production in mustard. Front Plant Sci 7:44

Khatamipour M, Piri E, Esmaeilian Y et al (2011) Toxic effect of cadmium on germination, seedling growth and proline content of milk thistle (*Silybum marianum*). Ann Biol Res 2:527–532

Khatun S, Ali MB, Hahn EJ et al (2008) Copper toxicity in *Withania somnifera*: growth and antioxidant enzymes responses of in vitro grown plants. Environ Exp Bot 64:279–285

Khavari-Nejad RA, Najafi F, Angaji SA et al (2013) Molecular and physiological studies on basil (*Ocimum basilicum* L.) under cadmium stress. Am-Eurasian J Agri Environ Sci 13:754–762

Krishnaiah D, Sarbatly R, Nithyanandam R (2011) A review of the antioxidant potential of medicinal plant species. Food Bioprod Process 89:217–233

Li C, Xu H, Xu J et al (2011) Effects of aluminium on ultrastructure and antioxidant activity in leaves of tea plant. Acta Physiolo Plant 33:973–978

Liang J, Yang Z, Tang L et al (2012) Growth performance and tolerance responses of jatropha (*Jatropha curcas*) seedling subjected to isolated or combined cadmium and lead stresses. Int J Agric Biol 14:861–869

Lu Y, Li XR, He MZ et al (2010) Nickel effects on growth and antioxidative enzymes activities in desert plant *Zygophyllum xanthoxylon* (Bunge) Maxim. Sci Cold Arid Reg 2:436–444

Luo ZB, Xiao-Jia HE, Chen L et al (2010) Effects of zinc on growth and antioxidant responses in *Jatropha curcas* seedlings. Int J Agric Biol 12:119–124

Luwe MWF, Takkahama U, Heber U (1993) Role of ascorbate in detoxifying ozone in the apoplast of spinach (*Spinacia oleracea* L.) leaves. Plant Physiol 101:969–976

Maheshwari R, Dubey RS (2009) Nickel-induced oxidative stress and the role of antioxidant defence in rice seedlings. Plant Growth Regul 59:37–49

Maiga A, Diallo D, Bye R et al (2005) Determination of some toxic and essential metal ions in medicinal and edible plants from Mali. JAFC 53:2316–2321

Maleva MG, Nekrasova GF, Borisova GG et al (2012) Effect of heavy metals on photosynthetic apparatus and antioxidant status of *Elodea densa* Planch. Russ J Plant Physiol 59:190–197

Marques TCLLSM, Soares AM (2011) Antioxidant system of ginseng under stress by cadmium. Sci Agric (Piracicaba, Braz) 68:482–488

Meriga B, Reddy BK, Rao KR et al (2004) Aluminium-induced production of oxygen radicals, lipid peroxidation and DNA damage in seedlings of rice (*Oryza sativa*). J Plant Physiol 161:63–68

Mishra S, Jha AB, Dubey RS (2011) Arsenite treatment induces oxidative stress, upregulates antioxidant system, and causes phytochelatin synthesis in rice seedlings. Protoplasma 248:565–577

Mittler R (2002) Oxidative stress antioxidants and stress tolerance. Trends Plant Sci 7:405–410

Mittler R, Vanderauwera S, Gollery M et al (2004) Reactive oxygen gene network of plants. Trends Plant Sci 9:490–498

Morsy AA, Salama KHA, Kamel HA et al (2012) Effect of heavy metals on plasma membrane lipids and antioxidant enzymes of *Zygophyllum* species. Eurasia J Biosci 6:1–10

Mukhopadyay M, Bantawa P, Das A et al (2012) Changes of growth, photosynthesis and alteration of leaf antioxidative defence system of tea [*Camellia sinensis* (L.) O. Kuntze] seedlings under aluminum stress. Biometals 25:1141–1154

Nadgorska-Socha A, Ptasinski B, Kita A (2013) Heavy metal bioaccumulation and antioxidative responses in *Cardaminopsis arenosa* and *Plantago lanceolata* leaves from metalliferous and non-metalliferous sites: a field study. Ecotoxicol 22:1422–1434

Neill S, Desikan R, Hancock J (2002) Hydrogen peroxide signaling. Curr Opin Plant Biol 5:388–395

Ou-yang C, Gao S, Mei L et al (2014) Effect of aluminum toxicity on the growth and antioxidant status in *Jatropha curcas* seedlings. J Med Plants Res 8:178–185

Ozcan MM, Akbulut M (2008) Estimation of minerals, nitrate and nitrite contents of medicinal and aromatic plants used as spices, condiments and herbal tea. Food Chem 106:852–858

Patel A, Pandey V, Patra DD (2016) Metal absorption properties of *Mentha spicata* grown under tannery sludge amended soil- its effect on antioxidant system and quality. Chemosphere 147:67–73
Pauln S, Shakya K (2013) Arsenic, chromium and NaCl induced artemisinin biosynthesis in *Artemisia annua* L.: a valuable antimalarial plant. Ecotoxicol Environ Saf 98:59–6560
Putchala MC, Ramani P, Sherlin HJ et al (2013) Ascorbic acid and its pro-oxidant activity as a therapy for tumours of oral cavity: a systematic review. Arch Oral Biol 58:563–574
Raj S, Mohan S (2016) Impact on proline content of *Jatropha curcas* in fly ash amended soil with respect to heavy metals. Int J Pharm Pharm Sci 8:244–247
Rimbach G, Fuchs J, Packer L (2005) Application of nutrigenomics tools to analyze the role of oxidants and antioxidants in gene expression. In: Rimbach G, Fuchs J, Packer L (eds) Nutrigenomics. Taylor and Francis Boca Raton Publishers, Boca Raton, pp 1–12
Ross JA, Kasum CM (2002) Dietary flavonoids bioavailability, metabolic effects, and safety. Annu Rev Nut 22:19–34
Rout JR, Behera S, Keshari N et al (2015) Effect of iron stress on *Withania somnifera* L.: antioxidant enzyme response and nutrient elemental uptake of in vitro grown plants. Ecotoxicology 24:401–413
Saha D, Mandal S, Saha A (2012) Copper induced oxidative stress in tea (*Camellia sinensis*) leave. J Environ Biol 33:861–866
Sandmann G, Gonzales HG (1989) Peroxidative processes induced in bean leaves by fumigation with sulphur dioxide. Environ Pollut 56:145–154
Scalbert A, Williamson G (2000) Dietary intake and bioavailability of polyphenols. J Nut 130:2073S–2085S
Shah K, Kumar RG, Verma S et al (2001) Effect of cadmium on lipid peroxidation, superoxide anion generation and activities of antioxidant enzymes in growing rice seedlings. Plant Sci 161:1135–1144
Sharma P, Dubey RS (2005) Drought induces oxidative stress and enhances the activities of antioxidant enzymes in growing rice seedlings. Plant Growth Regul 46:209–221
Sharma P, Dubey RS (2007) Involvement of oxidative stress and role of antioxidative defense system in growing rice seedlings exposed to toxic concentrations of aluminum. Plant Cell Rep 26:2027–2038
Sharma A, Gontia I, Agarwal PK et al (2010) Accumulation of heavy metals and its biochemical responses in *Salicornia brachiata*, an extreme halophyte. Mar Biol Res 6:511–518
Shengjun MA, Wenquan W, Runqing DU et al (2015) Effects of manganese stress on physiological and growth characteristics of *Glycyrrhiza uralensis* Fisch. Northwest Pharma J 30:1–7
Shu X, Zhang QF, Wang WB (2014) Lead induced changes in growth and micronutrient uptake of *Jatropha curcas* (L.) Bull Environ Contam Toxicol 93:611–617
Singh S, Eapen S, D'Souza SF (2006) Cadmium accumulation and its influence on lipid peroxidation and antioxidative system in an aquatic plant, *Bacopa monnieri* L. Chemosphere 62:233–246
Singh S, Anjum NA, Khan NA et al (2008) Metal-binding peptides and antioxidant defence system in plants: significance in cadmium tolerance. In: Khan NA, Singh S (eds) Abiotic stress and plant responses. IK International, New Delhi, pp 159–189
Singh A, Lawrencea K, Pandita S et al (2014) Response of leaves, stems and roots of *Withania somnifera* to copper stress. IJPAES 4:60–67
Sinha S, Saxena R (2006) Effect of iron on lipid peroxidation, and enzymatic and non-enzymatic antioxidants and bacoside-A content in medicinal plant *Bacopa monnieri* (L.) Chemosphere 62:1340–1350
Somashekaraiah BV, Padmaja K, Prasad ARK et al (1992) Phytotoxicity of cadmium ions on germinating seedlings of mung bean (*Phaseolus vulgaris*) involvement of lipid peroxides in chlorophyll degradation. Plant Physiol 85:85–89

Srivastava S, Dubey RS (2011) Manganese-excess induces oxidative stress, lowers the pool of antioxidants and elevates activities of key antioxidative enzymes in rice seedlings. Plant Growth Regul 64:1–16

Srivastava S, Tripathi D, Raghavan G et al (2006) Phytochelatin synthesis and responses of antioxidants during cadmium stress in *Bacopa monnieri* (L.) Plant Physiol Biochem 44:25–37

Stancheva I, Geneva M, Markovska Y et al (2014) A comparative study on plant morphology, gas exchange parameters, and antioxidant response of *Ocimum basilicum* L. and *Origanum vulgare* L. grown on industrially polluted soil. Turk J Biol 38:89–102

Streb P, Michael KA, Feierabend J (1993) Referential inactivation of catalase and photo inhibition of PS II are common early symptoms under various osmotic and chemical stress conditions. Plant Physiol 88:590–598

Takahashi S, Murata N (2008) How do environmental stresses accelerate photoinhibition? Trends Plant Sci 13:178–182

Tanou G, Molassiotis A, Diamantidis G (2009) Induction of reactive oxygen species and necrotic death-like destruction in strawberry leaves by salinity. Environ Exp Bot 65:270–281

Tarhan L, Kavakcioglu B (2016) Glutathione metabolism in *Urtica dioica* in response to cadmium based oxidative stress. Biol Planta 60:163–172

Tuteja N (2007) Mechanisms of high salinity tolerance in plants. Meth Enzymol: Osmosens Osmosignal 428:419–438

Uddin I, Bano A, Masood S (2015) Chromium toxicity tolerance of *Solanum nigrum* L. and *Parthenium hysterophorus* L. plants with reference to ion pattern, antioxidation activity and root exudation. Ecotoxicol Environ Saf 113:271–278

Upadhyaya H, Panda SK (2013) Abiotic stress responses in tea (*Camellia sinensis* L): an overview. RAS 1:1–10

Van AF, Clijsters H (1990) Effects of metal on enzyme activity in plants. Plant Cell Environ 13:195–206

Verma S, Dubey RS (2003) Lead toxicity induces lipid peroxidation and alters the activities of antioxidant enzymes in growing rice plants. Plant Sci 164:645–655

Vijayalakshmi VK, Revathi K, Sudha PN (2010) Comparative studies on the effect of antioxidant properties of the plants *Helianthus annuus* and *Solanum nigrum* exposed to the heavy metal chromium. J Pharm Sci Res 2:889–895

Wagner BA, Buettner GR, Burns CP (1996) Vitamin E slows the rate of free radical-mediated lipid peroxidation in cells. Arch Biochem Biophys 334:261–267

Woolhouse HW (1983) Toxicity and tolerance in the responses of plants to metals. In: Lange OL, Nobel PS, Osmond CB et al (eds) Encyclopedia of plant physiology, New series, 12C. Physiological plant ecology III. Springer-Verlag, Berlin, pp 245–300

Xu J, Yin HX, Li X (2009) Protective effects of proline against cadmium toxicity in micropropagated hyperaccumulator, *Solanum nigrum* (L.) Plant Cell Rep 28:325–333

Yan J, Tsuichihara N, Etoh T et al (2007) Reactive oxygen species and nitric oxide are involved in ABA inhibition of stomatal opening. Plant Cell Environ 30:1320–1325

Yan R, Gao S, Yang W et al (2008) Nickel toxicity induced antioxidant enzyme and phenylalanine ammonia-lyase activities in *Jatropha curcas* L. cotyledons. Plant Soil Environ 54:294–300

Yildiztugay E, Ozfidan-Konakci C (2015) Profiling of rutin-mediated alleviation of cadmium-induced oxidative stress in *Zygophyllum fabago*. Environ Toxicol 30:816–835

Young AJ, Lowe GM (2001) Antioxidant and prooxidant properties of carotenoids. Arch Biochem Biophys 385:20–27

Zhang FQ, Wang YS, Lou ZP et al (2007) Effect of heavy metal stress on antioxidative enzymes and lipid peroxidation in leaves and roots of two mangrove plant seedlings (*Kandelia candel* and *Bruguiera gymnorrhiza*). Chemosphere 67:44–50

Zheng G, Lv HP, Gao S et al (2010) Effect of cadmium on growth and antioxidant responses in *Glycyrrhiza uralensis* seedlings. Plant Soil Environ 56:508–515

Zouari M, Ahmed CB, Elloumi N et al (2016) Impact of proline application on cadmium accumulation, mineral nutrition and enzymatic antioxidant defense system of *Olea europaea* L. cv chemlali exposed to cadmium stress. Ecotoxicol Environ Saf 128:195–205

Responses, Adaptation, and ROS Metabolism in Plants Exposed to Waterlogging Stress

10

Mirza Hasanuzzaman, Jubayer Al Mahmud, Kamrun Nahar, Taufika Islam Anee, Masashi Inafuku, Hirosuke Oku, and Masayuki Fujita

Abstract

Waterlogging condition imposes serious threat to plant survival by disturbing normal growth and development by hampering different physiological and metabolic activities, which includes reduction in stomatal conductance, CO_2 assimilation rate, photosynthesis rate, and nutritional imbalance resulting in crop yield loss. However, plant develops different morphological, anatomical, and physiological adaptive features, for instance, formation of adventitious roots and aerenchyma, enhances the production of ethylene, and increases ADH activities and proline content under flooding condition. Under flooding, plants show oxidative damage condition due to the excess ROS production. ROS interferes with normal metabolism through oxidation of protein, nucleic acid, and DNA

M. Hasanuzzaman (✉)
Department of Agronomy, Faculty of Agriculture, Sher-e-Bangla Agricultural University, Dhaka, Bangladesh

Molecular Biotechnology Group, Center of Molecular Biosciences (COMB), Tropical Biosphere Research Center, University of the Ryukyus, 1 Senbaru, Nishihara, Okinawa 903-0213, Japan
e-mail: mhzsauag@yahoo.com

J. Al Mahmud
Laboratory of Plant Stress Responses, Department of Applied Biological Science, Faculty of Agriculture, Kagawa University, Miki-cho, Kita-gun, Kagawa, Japan

Department of Agroforestry and Environmental Science, Faculty of Agriculture, Sher-e-Bangla Agricultural University, Dhaka, Bangladesh

K. Nahar
Laboratory of Plant Stress Responses, Department of Applied Biological Science, Faculty of Agriculture, Kagawa University, Miki-cho, Kita-gun, Kagawa, Japan

Department of Agricultural Botany, Faculty of Agriculture, Sher-e-Bangla Agricultural University, Dhaka, Bangladesh

M.I.R. Khan, N.A. Khan (eds.), *Reactive Oxygen Species and Antioxidant Systems in Plants: Role and Regulation under Abiotic Stress*,
DOI 10.1007/978-981-10-5254-5_10

and reduces membrane integrity. Under this condition, plants itself develop antioxidant defense system including both enzymatic and non-enzymatic antioxidants which scavenges excess ROS. In this chapter, we tried to focus on different types of strategies such as hormonal regulation, osmoprotectants, and mineral nutrients to cope up with excess ROS production and suggested the future need of research in hormonal interaction and significance of osmolytes in protecting crop plants under waterlogging environment.

Keywords
Anaerobic soils • Antioxidant defense • Flooding • Nutrient imbalance • Oxidative Stress • Plant Metabolism

Abbreviations

ABA	Abscisic acid
ADH	Alcohol dehydrogenase
ALA	5-Aminolevulinic acid
APX	Ascorbate peroxidase
AsA	Ascorbic acid
ATP	Adenosine triphosphate
CAT	Catalase
DHAR	Dehydroascorbate reductase
Eh	Redox potential
ETC	Electron transport chain
GA	Gibberellic acid
GR	Glutathione reductase
g_s	Stomatal conductance
GSH	Glutathione
GST	Glutathione S-transferase

T.I. Anee
Department of Agronomy, Faculty of Agriculture, Sher-e-Bangla Agricultural University, Dhaka, Bangladesh

Laboratory of Plant Stress Responses, Department of Applied Biological Science, Faculty of Agriculture, Kagawa University, Miki-cho, Kita-gun, Kagawa, Japan

M. Inafuku • H. Oku
Molecular Biotechnology Group, Center of Molecular Biosciences (COMB), Tropical Biosphere Research Center, University of the Ryukyus, 1 Senbaru, Nishihara, Okinawa 903-0213, Japan

M. Fujita
Laboratory of Plant Stress Responses, Department of Applied Biological Science, Faculty of Agriculture, Kagawa University, Miki-cho, Kita-gun, Kagawa, Japan

MDA	Malondialdehyde
MDHA	Monodehydroascorbate reductase
NADPH	Nicotinamide adenine dinucleotide phosphate
POD	Peroxidase
Pro	Proline
PSII	Photosystem II
ROS	Reactive oxygen species
RuBisCO	Ribulose-1,5-bisphosphate carboxylase/oxygenase
SOD	Superoxide dismutase

10.1 Introduction

Global climate change results in frequent flooding events in different unexpected regimes of the world which brings modification in natural ecosystems. Plant distribution is supposed to alter due to waterlogging condition and at extreme and gradual/frequent flooding/waterlogging cause extinction of plant species (Bailey-Serres and Voesenek 2008). Like other aerobic organisms, oxygen (O_2) is the requisite for maintaining the life of terrestrial plants. Waterlogging condition decreases the available O_2 for plants (Capon et al. 2009). Waterlogging stress starts with the hypoxic (deficiency of O_2) stress indicating the availability of low oxygen levels limiting mitochondrial respiration. With the succession of time, this stress creates anoxic (absence of O_2) stress which results in the inhibition of respiration (Wegner 2010). Waterlogging reduces soil redox potential and enhances generation of toxic compounds (Fiedler et al. 2007). Inhibitions of respiration together with generation of toxic compounds potentially hamper the metabolic processes of plants growing under waterlogging condition.

Stunted growth and reduced biomass have been reported in plants growing under waterlogged environment. Alteration in root hydraulic conductivity, light interception, stomatal conductance, CO_2 assimilation, drastic reduction of photosynthesis, alteration in respiration, and generation of a range of secondary metabolites are responsible for reducing plant's potential to grow and showing their productivity under waterlogging stress (Ashraf 2012). Waterlogging condition causes oxygen deficiency, light deficiency, altered stomatal conductance, and CO_2 assimilation which results in overproduction of reactive oxygen species (ROS) including superoxide ($O_2^{\bullet-}$), hydrogen peroxide (H_2O_2), and hydroxyl radical ($OH^{\bullet}$) which are responsible for photooxidative damage (Yordanova et al. 2004). Together with altered physiological and growth processes, waterlogging severely affects the reproductive development which causes yield reduction as reported in a number of plant species. However, plants subjected to waterlogged condition tend to survive through a range of biochemical, physiological, anatomical, and morphological adaptations. Production of ethylene and formation of adventitious root are some of the major adaptive responses under waterlogging stress (Pezeshki 2001). Formation of aerenchyma close to the water surface is a common response to diminish O_2 deficiency (Suralta and Yamauchi 2008). Aerobic respiration of plants is converted to anaerobic respiration to ensure continuous supply of ATP (Ashraf 2012). Increased synthesis of soluble sugars and regulation of associated pathways including phytohormone synthesis and antioxidant defense system make the plants fit for such situations like hypoxia/anoxia (Sairam et al. 2008). Waterlogging is a potential threat among the abiotic stresses which is tended to be very frequent and unpredictable due to the adverse effect of climate change. The development of plant species tolerant against waterlogging is very much important to sustain plant productivity to fulfill the demand of food, other products, and services derived from plant. The mechanisms of damage and adaptation under waterlogging stress have been studied in several experiments. But the strategies for developing waterlogging-tolerant plant species have not yet been demonstrated extensively. This chapter highlights the major metabolic, morphological, physiological, and hormonal changes/adaptations of plants under waterlogging stress and approaches for improving waterlogging tolerance from accessible literatures. Special emphasis has been provided on the ROS metabolism under waterlogging.

10.2 Plant Responses to Waterlogging

Excess rainfall, tides, floods, and lack of proper drainage facilities are the reasons causing waterlogging stress in plants. The primary effect of waterlogging in crop plants is anoxia or oxygen deprivation. Plants need oxygen for several unavoidable processes including cell division and growth, respiration, and uptake and transportation of nutrients. Chlorosis, necrosis, defoliation, growth reduction, reduced N fixation, yield loss, and plant death are effects of waterlogging stress in plants which are often occurring at various vegetative and reproductive stages (Hasanuzzaman et al. 2016).

10.2.1 Growth

Reduction of growth rate is one of the primary effects of waterlogging. Waterlogging condition dramatically affects plant growth, development, and survival in various ways (Parent et al. 2008). Height and diameter of plant, number of leaf, and total growing period of maize plant were harshly affected by waterlogging (2 cm standing water) (Ali et al. 1999). In their study, Mauchamp et al. (2001) showed that partial submergence (50 and 80%) considerably improved accumulation of biomass and growth, while full submergence reduced the values of same parameters in *Phragmites australis* plant. Later on Mensah et al. (2006) showed flooding reduced dry matter accumulation and number of leaves per plant in sesame. Long-term flooding induced chlorosis, abortion of flower, and reduced time of maturity. The overall growth of sesame plant was negatively affected by continuous flooding. Gonzalez et al. (2009) carried an experiment with Quinoa (*Chenopodium quinoa* Wild.) plants and found specific leaf area, and leaf areas were reduced under waterlogging stress by 26% and 36%, respectively, in contrast to control. Lone and Warsi (2009) demonstrated that plant height and ear height of maize were severely affected under flooding. Almost all tested genotypes showed reduction in plant height due to flooding. Ezin et al. (2010) observed differential growth responses in two contrasting varieties of cultivated and wild origin. Plant height, leaf length, and leaf number were decreased, but formation of adventitious root was increased with increased duration of waterlogging in all cultivars. Among the tested tomato genotypes, LA1579 showed sensitiveness to flood. The CLN2498E and CA4 genotypes exhibited high tolerance to flooding stress which is followed by LA1421 genotype. Paltaa et al. (2010) also observed that the leaf area of *Cicer arietinum* reduced by 56–70% compared to well drained and depending on the cultivars. Compared to desi cultivar (Rupali), the growth retardation was less in kabuli cultivar (Almaz). Dry weight of shoot under waterlogging condition reduced by 70 and 56% in the cultivars Rupali and Almaz, respectively. The number of branches decreased by 50% in both chickpea cultivars under waterlogging stress. Similar alteration of root growth and activity was also observed under extended waterlogging. Prasanna and Rao (2014) conducted an experiment with green gram and reported significant difference on growth parameters due to waterlogging. Plant height, leaf area, number of leaves, and total dry matter were significantly affected by waterlogging throughout the life cycle. The effect of 4-d waterlogging was more severe in comparison with 2-d waterlogging treatment over the control. In waterlogged plants, plant height, number of branches, number of leaves, leaf area, and total dry matter were reduced by 30–34% compared to well-watered condition. Ren et al. (2014) performed a study in a field to elucidate the responses of waterlogging for different growth and reproductive stage of summer maize and observed that waterlogging negatively affected the overall growth and development of crop.

10.2.2 Physiology and Metabolism

Physiology and metabolism of plants are disrupted under waterlogging condition. Impaired stomatal conductance, gas exchange, CO_2 assimilation, and hydraulic conductivity of roots are some of the primary consequence in waterlogged plants (Ashraf 2012). Lone and Warsi (2009) observed that waterlogging stress had a great impact on few physiological traits in some crops. But leaf temperature remained mostly constant under both normal and waterlogged conditions in both winter and summer crop. Reduction of the net photosynthetic rate of wheat flag leaf was notable when the plants were remained under waterlogging condition (Zheng et al. 2009). Flooding frequently induces stomatal closing mostly in C_3 plants (Akhtar and Nazir 2013) which hampers photosynthesis rate. Liao and Lin (1996) observed that the reduction of leaf photosynthesis in bitter melon under waterlogging stress was associated with the decrease in transpiration, stomatal conductance, and the activity of ribulose-1,5-bisphosphate carboxylase/oxygenase (RuBisCO). Kumar et al. (2013) reported that waterlogging results in the reduction of relative water content (RWC) and membrane stability index. However, these effects were more prominent in sensitive genotypes (MH-1K-24 and Pusa Baisakhi) than tolerant genotypes (T-44 and MH-96-1). Photosynthesis rate decreased in all tested genotypes under waterlogging stress, and these inhibitions increased in duration-dependent manner. Like photosynthesis rate, stomatal conductance also exhibited the same trend. In comparison with control plants, the respiration of leaf increased after 48 h of waterlogging. Sensitive genotypes also showed increased foliar respiration, while the minor reduction of respiration rate was recorded at sixth and ninth day of waterlogging. Moreover, foliar respiration in tolerant genotype was unchanged at any growth duration. Therefore, they concluded that respiration of leaf did not reduce due to any tested duration of waterlogging.

Flood stress involves different stressful situations for vegetation. Depth and duration of water during flooding determine the intensity of flood stress. Tolerant species of plants can survive under flooding/waterlogging condition through oxygenation of submerged tissues (roots and shoots) and above water leaves continuing carbon fixation. The uptake of CO_2 for carbon fixation maintained on mild days by active stomatal conductance even though stomatal closure for regulating plant water homeostasis at high atmospheric evaporative condition which helps to keep the balance between water uptake and transpiration loss. Plants can be adopted to complete submerged condition by two main approaches, viz., low-oxygen quiescence syndrome (LOQS) and low-oxygen escape syndrome (LOES). The former is an effective strategy for plant survival, where the plant uses its reserve carbohydrates conservatively during the submergence period. When water falls down, plants showing LOQS restart their growth and development. The LOES involves upward elongation of shoots, which assists leaves exposed to the atmosphere. The plant species (or ecotypes) evoluted to shallow prolonged flooded environments showed these phenomena (Striker 2012).

The detrimental effect of waterlogging on water-loving plants has been reported in some studies. Anandana et al. (2015) recorded some damages in rice plant under long-time flooding stress. *Oryza sativa* cv. Puzhuthiikar showed notable enhancement of leaf blade length, sheath length, and area but reduction of leaf blade area. Accordingly, transpiration, rate of photosynthesis, and intercellular CO_2 were augmented in rice plant. In addition Pezeshki (2001) reported wetland vegetations acquire diverse characteristics that facilitate them to stay alive under saturated soil by adjusting soil chemistry such as redox potential (Eh). Decrease in Eh creates a situation where plant roots need to prepare themselves to uptake higher amount of oxygen. However, these phenomena are very common in wetland species, while the actual relationship between soil chemistry and plant physiological processes and the ultimate conclusion is still under study. Undoubtedly, extreme reduction in soil elevated oxygen demand exerts insightful influence on transportation and release of oxygen to the rhizosphere.

Vandoorne et al. (2014) worked with *Cichorium intybus* plant and found that flooding stress enhanced the number of leaves but decreased the photosynthetic activity by closing stomata and reducing the efficiency of photosystem II (PSII). Under flooding condition, the roots of plant became shorter with no changes in fresh weight and dry weight. Roots of flooded plant accumulated glucose, fructose, sucrose, and 1-kestotriose but leaves accumulated organic acids and reducing sugars. Activities of invertase and sucrose synthase increased in both leaf and root, while the activity of sucrose-phosphate synthase remained unaltered due to flood stress. Synthesis of inulin was delayed in roots of flooded plants, and its mean degree of polymerization decreased as a consequence of the inhibition of fructan: fructan 1-fructosyltransferase.

10.2.3 Anatomical Changes

Plants exhibit several anatomical changes under flooding/waterlogging condition. Under flooding/waterlogging condition, plants initiate adventitious root, hypertrophied lenticel, and/or aerenchyma (Ashraf 2012) for adjusting with the unfavorable condition. Although the exact physiological functions of lenticels is not yet clear, their connection with waterlogging tolerance in plants is evident (Parelle et al. 2006). It is thought that lenticels play a role in the diffusion of O_2 and different gaseous substances involved in anaerobic metabolism which includes C_2H_2, CO_2, and CH_4. Plants exposed to waterlogging resulted in distinct reduction in the seminal root diameter, stele diameter, and cortex thickness (Grzesiak et al. 1999). Vasellati et al. (2001) conducted an experiment with *Paspalum dilatatum* plants to elucidate the anatomical changes of plants under flooding condition. They observed few plastic responses under waterlogging: overgrowth of aerenchymatous tissue in the root cortex and the leaf sheaths and reduction in the number of root hairs and root length. In contrary, Corrêa de Souzaa et al. (2013) reported that root cortex of flooded seedlings did not show significant difference, but the width of exodermis was increased by 24% compared to well-aerated plants. Flooded condition also increased the width of phloem tissue and number of xylem vessel in the roots, while the area of xylem fibers decreased considerably.

Among the different mechanism of maintaining root function of plants under flood stress, founding of tangential diffusion barriers is very important due to the reduction of radial O_2 from the roots of waterlogged soil (Sauter 2013). However, maintaining O_2 in the cells is necessary for the survival where aerenchyma takes part of a great role in allocating O_2 from the shoot to the root. Waterlogging enhances the construction of aerenchymatous tissue in both wetland and dryland species (Colmer and Voesenek 2009; Shiono et al. 2011) which make them adaptive under waterlogging condition (Striker 2012).

10.2.4 Nutrient Availability

Flooding/waterlogging can break the balance of availability of different essential nutrients in soil that negatively affect numerous plant processes due to the unavailability or inefficiency of essential nutrients like N, K, Ca, Mg, etc. (Ashraf 2012). Reddy and Mittra (1985) reported that submergence led to the enhancement of N content of the plant tissues in duration-dependent manner. However, the submerged plants showed small amounts of P and K contents at any growth stages of the plant. Submergence also decreased the production of carbohydrates associated with the decomposition of proteins due to enhanced proteolytic activity. Since P and K deficiencies hold back, the synthesis of protein in plants accumulates some non-protein nitrogenous compounds which lead to increased N content in the tissue. Water stress-induced decrease in endogenous N, P, and K levels was reported in maize (Atwell and Steer 1990) and canola (Boem et al. 1996). Tarekegne et al. (2000) found great reduction of the uptake of Zn, Cu, K, and P in susceptible wheat genotype grown under waterlogging. Due to the above-mentioned reasons, waterlogging results in nutrient deficiencies (Smethurst et al. 2005). *Medicago sativa* showed a noticeable reduction in the nutrient composition (K, P, Cu, Ca, Mg, Zn, and B) of leaf and root under waterlogging condition (Smethurst et al. 2005). Akhtar and Nazir (2013) also reported flooding potentially affected growth of plant and uptake of both macro- and micronutrients.

10.2.5 Yield

Except some water-loving plants, significant yield reduction occurs due to flooding or waterlogging. Flooding/waterlogging at any growth stage of plant ultimately hampers the yield of plant. Reddy and Mittra (1985) conducted an experiment with rice plant and showed that grain yield was adversely affected by full submergence. Maximum yield reduction (58%) was recorded from complete submergence at flowering stage. This reduction of grain yield was observed due to impaired anthesis as well as high sterility of flower. Around 69% unfilled spikelets were recorded in the experiment. Full submergence at seedling establishment stage reduced the grain yield by 29%, and submergence at maximum tillering stage reduced the grain yield by 18% in rice plant. Later on, Mensah et al. (2006) observed similar result in sesame plant. They observed that continuous flooding reduced the growth and seed yield of sesame plant. Flooding decreased dry matter accumulation, leaf number per plant, and ultimately the yield of sesame. Long-time flooding decreased the maturity time of sesame plant. It also stimulated chlorosis and abortion of flower. Lone and Warsi (2009) noted that grain yield of maize showed drastic reduction under excess soil moisture conditions in both winter and summer seasons. In winter trial, overall yield was higher in all the genotypes in both sets of experiments, but still yield loss (%) was higher. The yield reduction varies from 19% in YHPP45 (tolerant genotype) to 53% in Pop 3121 × YHPP45. While in case of summer trial, overall yields were lower, but highest reduction in yield (66%) was detected in Tarun 83 (susceptible genotype), and lowest reduction (2%) was detected in YHPP45 genotype. Ezin et al. (2010) conducted an experiment and showed that the reaction of four tomato genotypes is quite different from each other under waterlogging condition. They found significant differences in the phenology of four genotypes under waterlogging condition. Flowering time and fruit set time were earlier in CLN2498E genotype than the genotypes CA4, LA1241, and LA1579. Compared to control, the average fruit weight in LA1421 and CA4 genotypes varied noticeably when exposed to 8 days of waterlogging. Fruit-bearing habit and fruit yield were also negatively affected by waterlogging depending on the duration. Paltaa et al. (2010) detected that the temporary waterlogging decreased the seed yield by 54 and 44% in kabuli (Almaz) and desi (Rupali) chickpea cultivars, respectively. The lower seed yield in kabuli cultivars was mainly due to lesser pod number and number of seeds pod^{-1} under waterlogging condition. On the other hand, Yaduvanshi et al. (2010) reported that waterlogging imposition decreased grain yield in contrast to drained soils. However, the responses were greatly varied with genotypes and soil types. This differential response of wheat genotypes might be due to the function of different tolerance mechanisms under waterlogging. Rasaei et al. (2012) showed a significant difference among all the durations (10, 20, and 30 days) of the waterlogging stresses in wheat. The yield reduction up to 20 days was reasonable, but 30 days of waterlogging decreased the yield to 45% compared to well-drained condition. The yield of wheat under non-waterlogging condition and 10, 20, and 30 days of waterlogging were 7518.4 kg ha^{-1}, 6,815.5 kg ha^{-1}, 5,587 kg ha^{-1}, and 4,138.6 kg ha^{-1}, respectively. Kumar et al. (2013) carried an experiment with different genotypes of mung bean under waterlogging stress and found that yield was affected by waterlogging in all the genotypes. At vegetative stage, yield loss enhanced with duration-dependent manner. In average, grain yield losses in all mung bean cultivars were 20.01%, 33.79%, and 52% due to 3, 6, and 9 days of waterlogging, respectively. The tolerant genotypes were able to recover the grain yield losses caused by 3 days of waterlogging. On the other hand, for sensitive genotypes, even 3 days of waterlogging decreased the yield (up to 20%). In sensitive genotypes, grain yield losses were estimated at 70% (Pusa Baisakhi) to 85% (MH–1K–24) after 9 days of waterlogging in comparison with control plants. Tolerant genotypes confirmed comparatively lower yield reduction even after 9 days of waterlogging. Amri et al. (2014) applied flooding stress to six genotypes of wheat for 28 days and noted that the grain yield affected significantly in all evaluated genotypes. Compared to rainfed conditions (control), waterlogging induced an average decrease of grain yield by 56% with a maximum of 74% recorded for cv. Ariana and cv. Vaga against a lowest decrease of 39% recorded for the cultivars Salammbô and Utique. The two cultivars FxA and Haïdra showed an intermediate behavior with respective decreases of 60% and 48%. Prasanna and Rao (2014) conducted an experiment with waterlogging in green gram. They observed that the waterlogging condition for 2 days and 4 days considerably decreased the yield components and finally yield of green gram. The number of pods plant^{-1}, number of seeds pod^{-1}, 100 seeds weight, harvest index, and yield of green gram decreased by 51%, 27%, 3%, 33%, and 71%, respectively, due to 4 days of waterlogging and by 29%, 14%, 1%, 13%, and 25%, respectively, due to 2 days of waterlogging in contrast to control plants. Very recently Ren et al. (2014) performed an experiment in the field for studying the effects of waterlogging for different durations (3 and 6 days) on the yield and growth of summer maize at the three-leaf stage (V3), six-leaf stage (V6), and the tenth day after the tasseling stage (10VT). The results after 2 years indicated that maize development and grain yield responses to waterlogging depended on both stress severity (intensity and duration) and different growth stages. Yield decreased significantly with an increased waterlogging duration during V3 and V6. The yields of maize hybrid Denghai 605 (DH605) in treatments V3-3, V3-6, V6-3, V6-6, 10VT-3, and 10VT-6 were 23%, 32%, 20%, 24%, 8%, and 18% lower than those of the control, respectively; yields of Zhengdan 958 (ZD958) were decreased by 21%, 35%, 15%, 33%, 7%, and 12%, respectively, compared to control, under same treatments.

10.3 Plant's Adaptation to Waterlogging

10.3.1 Morphological Adaptations

Plants show a range of morphological adaptations at their different growth stages under flooding stress condition. Rice seeds are able to germinate anaerobically. After germination, the coleoptile of rice seedlings rapidly elongate to avoid the anaerobic condition derived from waterlogged condition (Magneschi and Perata 2009). Adventitious root formation is one of the major adaptation responses under flooding stress. Tolerant plant varieties are characterized with higher generation of adventitious root, compared to susceptible plant varieties. Susceptible chrysanthemum cultivar (13-13) showed wilting and leaf chlorosis symptoms (at 4 days of waterlogging) with small number of adventitious root (at 20 days of waterlogging stress). In contrast, tolerant chrysanthemum cultivar (53-4) produced vigorous adventitious roots and showed less damage effects on leaves (Yin et al. 2009). *Sesbania javanica* has special features in stem (hypertrophy) and adventitious root formation which contributed to *S. javanica*'s tolerance to flooding (Jackson 2006). The formation of ethylene and adventitious root formation were documented in waterlogged tomato, *Rumex* spp., and *Lepidium latifolium* (Wang and Arteca 1992; Banga et al. 1997; Chen et al. 2002). *Zea mays* plants exposed to waterlogging showed reduction of seminal root diameter (Grzesiak et al. 1999). Root hairs in *Paspalum dilatatum* was found to be decreased upon waterlogging (Vasellati et al. 2001). Rapid elongations of submerged shoots and leaf are observed in flood-affected plants which bring plants to contact with air to mitigate oxygen deficiency (Jackson 2008). Stem elongation is a commonly occurred adaptation mechanism in submerged rice cultivars (Kende et al. 1998), whereas rapid growth of petioles and leaf blade has been demonstrated in *Rumex palustris* (Voesenek et al. 2003). Aquatic leaves are often filamentous, dissected leaves with reduced number of stomata (Sculthorpe 1967). In general, amphibious plants possess elongated and thinner leaves with higher specific leaf areas when grown under aquatic condition to provide sufficient gas exchange area (Mommer et al. 2004).

10.3.2 Anatomical Adaptations

The formation of internal gas spaces or aerenchyma is one of the most common adaptive features of plants under both waterlogging and flooding that facilitates flow and distribution of oxygen from root to shoot (Colmer and Voesenek 2009; Shiono et al. 2011). Lysigenous aerenchyma in roots is generated by ethylene accumulation with subsequent programmed cell death (Shiono et al. 2008). The gas-filled spaces in cell or aerenchyma have been demonstrated in swollen stem and roots of *S. javanica* (Jackson 2006). Establishment of a lateral diffusion barrier is another important adaptive feature of flood-affected plants because it decreases radial oxygen loss from flooded roots (Sauter 2013). 1-Aminocyclopropane-1-carboxylic acid is converted to ethylene that is responsible for adventitious root formation (Shiono et al. 2008). Adventitious roots generate near water surface where the stem produces aerenchyma to get oxygen (Suralta and Yamauchi 2008). Porosity of roots of some plant species under flooding condition can be increased up to 53% to assist entry of oxygen (Justin and Armstrong 1987).

Plants growing under flooded condition are characterized with an initiation of hypertrophied lenticels (Ashraf 2012). The development of lenticels in plants improved their waterlogging tolerance. Lenticels are believed to be involved in the downward diffusion of O_2 and transportation of by-products of anaerobic metabolism such as ethanol, CO_2, and CH_4 (Parelle et al. 2006). Reductions in stele diameter and cortex thickness have been reported in *Z. mays* plants exposed to waterlogging (Grzesiak et al. 1999). *P. dilatatum* plants showed increased aerenchymatous tissue in root cortex and leaf sheaths under flooding condition. High proportion of aerenchyma with the development of sclerenchyma was also noticed in *P. dilatatum*. The development of sclerenchyma is advantageous to prevent endodermis and stelar tissues, root, and leaf sheath from desiccation and also from collapse by soil compaction due to water pressure under flooded condition (Vasellati et al. 2001). In *Garcinia brasiliensis* (Clusiaceae), the thickness of exodermis increased by 23.70% under flooding condition compared to control. Flooding condition also increased phloem width and number of xylem vessels of roots (Corrêa de Souzaa et al. 2013).

10.3.3 Biochemical Adaptation and Hormonal Regulations

Unlike other stresses, flooding or waterlogging leads to a deprivation of oxygen in plants and creates hypoxia or anoxia. Both morphological and anatomical adaptations are associated with a series of biochemical changes in plant cells which make them tolerant to excess water stress. In oxygen-deprived condition, the metabolic activities are adversely affected; in most of the cases, plant shifts its metabolism to anaerobic mode. In such case, ADH is the key enzyme involved in anaerobic fermentation that converts acetaldehyde into ethanol with the oxidation of NADH into NAD^+. Although this process cannot produce much ATP, it is one of the adaptive mechanisms in plants during waterlogging. Increase in ADH activity under waterlogging condition has been reported in many plants (Christine and Musgrave 1994; Zaidi et al. 2004; Srivastava et al. 2007; Sairam et al. 2008).

Bansal and Srivastava (2015) found higher activity of ADH in *Cajanus cajan* genotypes under waterlogging condition. However, the tolerant genotypes showed higher activity of ADH than sensitive ones. After 6 days of waterlogging, ADH activity increased more than fourfolds in ICPL84023 and more than twofolds in MAL18 as compared to control. Bajpai and Chandra (2015) indicated that proline (Pro) might be one of the biochemical adaptive mechanisms of plants under waterlogging. In their experiment, they observed a marked increase in Pro content after 96 h of waterlogging, compared to control. Moreover, the tolerant genotype showed higher accumulation of Pro, compared to sensitive ones. Under waterlogging condition, increased Pro content was also reported in *Casuarina* (Olgun et al. 2008) and in wheat (Carter et al. 2006). Waterlogging-induced increase in soluble protein content is observed in maize (Rai et al. 2004). While investigating *Corchorus capsularis* – a moderate waterlogging-tolerant fiber crop, Parvin and Karmoker (2013) observed that total sugar content in the root increased by 24% after 21 days of waterlogging, while in the stem and leaves, it was increased up to 15% and 17%, respectively. After 3 days of waterlogging, Pro content was increased by 50% (Parvin and Karmoker 2013).

Sairam et al. (2009) found different adaptive responses in terms of sucrose synthase and ADH activity in two cultivated *V. radiata* genotypes, viz., T44 (tolerant) and Pusa Baisakhi (susceptible) and *V. luteola* (wild). Waterlogging declined total and non-reducing sugars in all the genotypes and reducing sugars in Pusa Baisakhi, while the content of reducing sugar increased in *V. luteola* and T44. Waterlogging condition for 8 days resulted in 20 and 15 times more ADH activity than initial stage in *V. luteola* and T44, while this increase was sustained only up to the 4 days in Pusa Baisakhi (Sairam et al. 2009). This indicated a carbohydrate-based tolerance mechanism in *V. luteola* and T44 genotype. This increased activity of soluble sugar in tolerant genotypes enhanced the availability of reducing sugars, which sustained their energy requirement because it is a key enzyme responsible for the hydrolysis of sucrose to fructose and glucose under oxygen deprivation (Ruan et al. 2003). Sairam et al. (2009) also concluded that greater ADH activity is a factor contributing to their better survival under waterlogged condition.

Hormonal regulation is one of the adaptive responses in plants under waterlogging. Some morphological and anatomical changes are also directly related to hormonal changes. For instance, being a flood-sensitive species, *S. lycopersicum* showed its ability to produce adventitious roots to survive under anoxia which was associated with ethylene synthesis (Vidoz et al. 2010). Apart from the direct effect, ethylene also coordinates the balance of other phytohormones such as gibberellic acid (GA) and abscisic acid (ABA) and those that have specific adaptive function in plants including shoot elongation and adventitious root formation (Fukao and Bailey-Serres 2008).

10.4 ROS Metabolism Under Waterlogging

10.4.1 Oxidative Stress

Flooding/waterlogging stress causes oxidative damages due to generation of ROS (Bowler et al. 1992; Sairam and Srivastva 2002; Blokhina et al. 2003; Table 10.1). Under flooding/waterlogging condition, the photosynthetic electron transport chain (ETC) becomes over-reduced, causing the generation of several ROS and as a result the oxidative stress (Ahmed et al. 2002). Productions of ROS take places through different mechanisms, for example, when molecules of aerobic system come in contact with the ionizing radiations; this interaction results in the production of ROS. Flowing of electrons through ETC may leak from their appropriate route and in the absence of any electron acceptor; these electrons react with oxygen for producing ROS (Ashraf 2009; Fig. 10.1). Different cellular organelles such as mitochondria, chloroplasts, and peroxisomes are known as the sites for generation of ROS (Sairam and Srivastva 2002). It is evident that oxidative damage is more severe after desubmergence than submergence condition. The re-entry of air after desubmergence introduces higher oxygen concentration relative to the very low concentration under water. Injury of the submerged plant generally develops after desubmergence and is possibly caused by ROS (Reviewed in Ito et al. 1999; Fig. 10.2). Both waterlogging and flooding condition create anaerobic environment which in turn lead to hypoxic and anoxic conditions for plants and as a result produce ROS in transition which is many fold higher than normal growing condition causing severe damage to plant cell (Ashraf 2009; Irfan et al. 2010; Halliwell and Gutteridge 1999). The $OH^{\bullet}$ and their dismutation product, H_2O_2, can directly attack membrane lipids resulting in higher malondialdehyde (MDA) content and also inactivate SH-containing enzymes (Ahmed et al. 2002; Hossain et al. 2009). Elevated level of H_2O_2 also can inhibit the activities of Calvin cycle (Ashraf 2012).

Flooding/waterlogging condition affects nutrient and water uptake of plants. As a result, plants show wilting even in excess of water environment. Because under waterlogging condition, oxygen becomes unavailable for plant (Sairam et al. 2008). Zheng et al. (2009) reported that MDA content in chloroplasts of plant generally increased under water stresses indicating a more severe lipid peroxidation in chloroplasts due to the enhanced H_2O_2 production and of $O_2^{\bullet -}$ release and depressed activities of antioxidant enzymes in comparison with the control. Komatsu et al. (2010) showed differential ROS production in different *G. max* genotypes due to the variations of lining of germin-like protein with oxalate oxidase or superoxide dismutase (SOD) activity within the cell wall under waterlogging stress. According to Sairam et al. (2011), *V. radiata* plants showed enhanced production of H_2O_2 in time-dependent manners where tolerant genotypes produced significantly lower amount of H_2O_2 than the susceptible one. Similar result was observed in *S. indicum* by Xu et al. (2012) where MDA contents of leaves increased by 8%, 2%, and 29% in cv. WTG-2541, WTG-2413, and WSG-EZhi2, respectively.

Table 10.1 Oxidative stress in plants under flooding/waterlogging

Crop species	Stress duration	Oxidative stress	References
H. vulgare cv. Xiumai 3	18 days	Increased membrane lipid peroxidation	Zhang et al. (2007)
T. aestivum cv. Yumai 34	15 days	Content of MDA increased by 22%	Tan et al. (2008)
C. cajan cv. ICPL 84023 and ICP 7035	6 days	Increased H_2O_2 and $OH^•$ contents	Sairam et al. (2009)
A. fistulosum	10 days	Rate of O_2 generation increased by 240% and H_2O_2 content by 290%	Yiu et al. (2009)
T. aestivum cv. Huaimai 17 and Yangmai 12	5 days	Enhanced lipid peroxidation	Zheng et al. (2009)
		Depressed ATP synthesis in the chloroplasts	
Z. mays cv. HZ 32 and K 12	10 days	Lipid peroxidation enhanced significantly	Tang et al. (2010)
T. aestivum cv. Yangmai 9	7 days	Increased MDA content	Li et al. (2011)
S. indicum cv. WSG-EZhi2	2 days	MDA content increased by 29%	Xu et al. (2012)
G. brasiliensis	60 days	Increased the level of H_2O_2 significantly	Corrêa de Souzaa et al. (2013)
V. unguiculata	30 days	Increased MDA content by 32% and H_2O_2 43% in root	El-Enany et al. (2013)
S. indicum cv. Ezhi-2	6 days	Content of MDA increased by 1.8 fold	Wei et al. (2013)

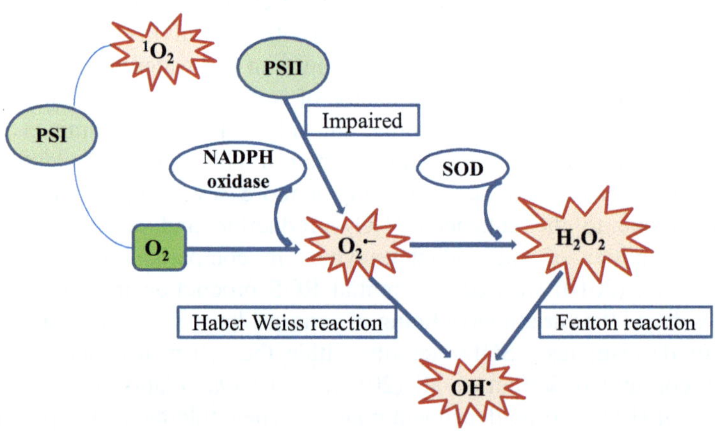

Fig. 10.1 Production of ROS ($^1O_2$, $O_2^{•-}$, $OH^•$, and H_2O_2) under flood/waterlogging stress

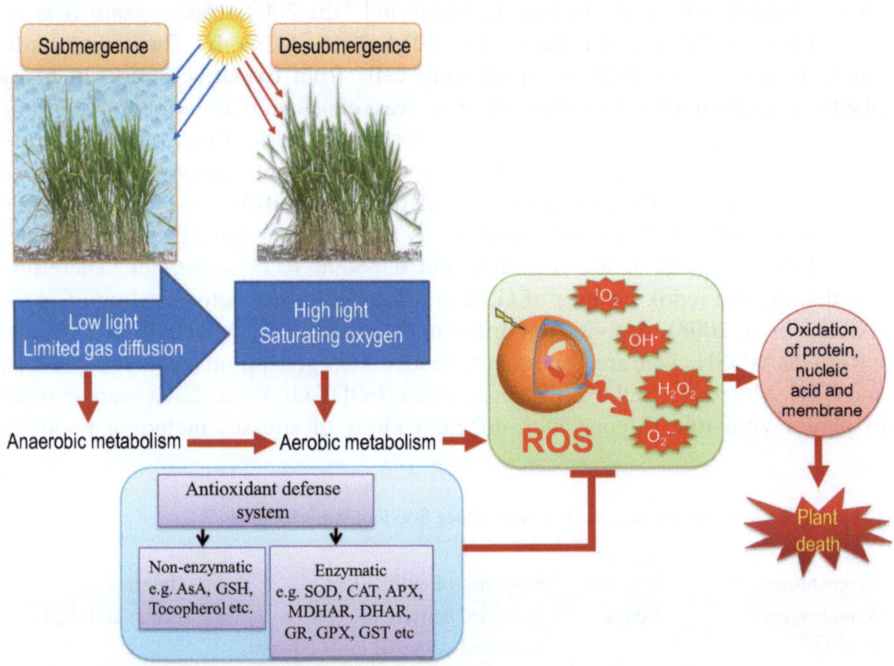

Fig. 10.2 ROS generation and detoxification under submergence and afterward (Modified from Ito et al. 1999, with Permission from Elsevier)

Ashraf (2012) reported that an enhanced ROS generation disrupted the integrity of membranes and reduced the efficiency of PSII under waterlogged condition. Increased MDA and H_2O_2 contents were monitored in root of *Vigna sinensis* plants under waterlogging stress (El-Enany et al. 2013). Most recently, Corrêa de Souzaa et al. (2013) observed that H_2O_2 content significantly increased after 40 and 55 days of flooding.

10.4.2 Antioxidant Defense

The excess amount of ROS inside the plant cells generated due to waterlogging/flooding stress should be kept in a balance state to continue normal metabolic processes of plant cell. An imbalance between the scavenging of ROS by plant's antioxidant defense system and production of ROS can result in excess accumulation of ROS in plant cell (Irfan et al. 2010). Plant's antioxidant defense system includes both the enzymatic antioxidants, i.e., CAT, SOD, ascorbate peroxidase (APX), monodehydroascorbate reductase (MDHAR), dehydroascorbate reductase (DHAR), glutathione *S*-transferase (GST), glutathione reductase (GR), and peroxidase (POX) and nonenzymatic antioxidants, i.e., glutathione (GSH), ascorbate

(AsA), tocopherols, and carotenoids (Apel and Hirt 2004; Hasanuzzaman et al. 2012; Khan and Khan 2014; Khan et al. 2014, 2015, 2016a, b). These act coordinately in scavenging ROS and protecting cells from oxidative stress. In many plants, activation of antioxidant defense was considered as one of the prime prerequisite adaptive mechanisms under waterlogging stress. However, this defense system is largely dependent on the stress duration and plant genotypes (Table 10.2).

Ascorbate and GSH are components of the Halliwell-Asada cycle. Ascorbate reacts with different ROS and functions as the substrate for APX. On the other hand, GSH is a cell redox regulator and a potent ROS scavenger. Therefore, maintaining the redox balance of GSH is effective for the detoxification of H_2O_2 (Arbona et al. 2008). Activities of different enzymes (APX, MDHAR, DHAR, and GR) of AsA-GSH cycle are important to reduce ROS generation and to recycle AsA and GSH. Moreover, SOD, CAT, peroxidases (POD), GPX, etc. have been reported to play a vital role to cope with different kinds of stresses including flooding/

Table 10.2 Antioxidant defense in plants under flooding/waterlogging

Crop species	Stress duration	Response of antioxidant	References
S. melongena cv. EG203	3 days	Increased activity of APX	Lin et al. (2004)
		Increased AsA and GSH contents	
L. pimpinellifolium cv. L4422	3 days	Increased activity of SOD	Lin et al. (2004)
		Increased total AsA and GSH contents	
H. vulgare cv. Xiumai 3	18 days	Increased activities of SOD, POX, CAT, and GR	Zhang et al. (2007)
Citrus spp.	45 days	Increased activities of SOD, CAT, APX, and GR	Arbona et al. (2008)
		Increased AsA and GSH contents	
T. aestivum cv. Yumai 34	15 days	Increased activity of POX	Tan et al. (2008)
C. cajan cv. ICPL 84023	6 days	Increased activities of SOD, APX, GR, and CAT	Sairam et al. (2009)
C. indicum	8 days	Increased activities of APX, CAT, and SOD	Yin et al. (2009)
A. fistulosum	10 days	Increased activities of SOD, POX, CAT, and GR	Yiu et al. (2009)
O. sativa cv. FR13A	8 days	Increased activities of APX and SOD	Damanik et al. (2010)
Z. mays cv. HZ32	10 days	Increased activities of SOD, POX, CAT, APX, and GR	Bin et al. (2010)
V. radiata cv. T44	8 days	Increased activities of SOD, APX, and GR	Sairam et al. (2011)
S. indicum cv. Ezhi-2	8 days	Increased activities of SOD and APX	Wei et al. (2013)
G. brasiliensis	80 days	Increased activities of SOD and APX	Corrêa de Souzaa et al. (2013)

waterlogging. Marked increase in SOD, CAT, and POD activities were observed in *C. cajan* subjected to 2–8 days of waterlogging, while the activities were variable according to the duration of stress (Sairam et al. 2009). Later on Sairam et al. (2011) observed the enhancement of the activities of GR, SOD, and APX in *V. radiata*. Similarly, increases in the activities of different enzymes were recorded in *Z. mays* seedlings when subjected to varying degrees of waterlogging stress (Bin et al. 2010). In *O. sativa*, the tolerance to submergence stress (8 days) was positively correlated with the capacity of the antioxidant defense system (Damanik et al. 2010). Higher activities of SOD, CAT, APX, and GR were observed in tolerant varieties, compared to susceptible varieties. Similar results were also observed by Bin et al. (2010) in *Z. mays* seedlings. Zhang et al. (2007) reported that enhanced POX and CAT activities were considered as a trait of tolerant plant species under waterlogging stress condition which enable them to protect themselves against oxidative stress. Tan et al. (2008) detected increased POX activity in *T. aestivum*. Citrus leaves showed an increase in AsA redox system when exposed to waterlogging (Arbona et al. 2008). In contrary, Ahmed et al. (2002) reported that waterlogging decreased the activities of SOD, GR, CAT, and APX in *V. radiata*.

10.5 Use of Exogenous Protectants in Mitigating Waterlogging/ Flooding Stress

Exogenous protectants of different groups are widely being used in modern approaches of alleviating damage effects caused by abiotic stresses in plants. Unlike other abiotic stresses, the mechanism of waterlogging or flooding stress tolerance has been demonstrated as a more complex trait by several researchers (Setter and Waters 2003; Setter et al. 2009; Pang and Shabala 2010; Zhou 2010; Table 10.3). The reason underlying this fact is that waterlogging stress is correlated with a number of factors like climate, soil properties, plant species, plant development stage, duration and severity of waterlogging, and so on (Leul and Zhou 1999; Setter and Waters 2003; Setter et al. 2009; Arru and Fornaciari 2010). As a consequence of this complexity in understanding, there are hardly any studies that evinced the role of exogenous protectants in diminishing the damages caused by waterlogging stress. However, some derivatives of triazole have been used to serve this purpose. There are reports indicating their ability to protect plants from several stresses, and hence this triazole compounds were termed as "multi-protectants" (Fletcher and Hofstra 1990; Pinhero and Fletcher 1994; Lin et al. 2006). In 1996, Webb and Fletcher used paclobutrazol ($C_{15}H_{20}ClN_3O$) as a seed priming agent in *T. aestivum* and exposed the seedlings to waterlogged condition for 4 weeks. They measured some growth parameters and photosynthetic pigments at three different height levels of the stressed plants. Paclobutrazol (PBZ)-treated seedlings showed higher chlorophyll (chl) contents and lower damage symptoms compared to non-treated stressed seedlings. Their study concluded that PBZ protects *T. aestivum* seedlings from waterlogging-induced injuries. Leul and Zhou (1998, 1999) used another triazole named uniconazole (UCZ, $C_{15}H_{18}ClN_3O$) to

Table 10.3 Role of exogenous protectants in alleviating damage effects of waterlogging/flooding stress

Plant species	Duration of waterlogging	Dose of protectant	Protective effects	References
T. aestivum	28 days	100 mg L^{-1} PBZ	Increased chl and carotenoid contents	Webb and Fletcher (1996)
			Increased root to shoot ratio	
			Decreased leakiness of shoot tissue	
B. napus	21 days	50 mg L^{-1} UCZ	Increased growth and yield	Leul and Zhou (1998)
			Increased leaf photosynthetic rate	
			Decreased erucic acid content	
B. napus	21 days	50 mg L^{-1} UCZ	Increased activities of SOD, CAT, and POD	Leul and Zhou (1999)
			Increased endogenous Pro content	
			Decreased lipid peroxidation and electrolyte leakage rate	
I. batatas	5 days	0.5 mg PBZ $plant^{-1}$	Increased SOD, CAT, GR, and APX activities	Lin et al. (2006)
			Increased GSH, oxidized AsA and total AsA contents	
			Decreased MDA content	
B. napus	14 days	50 mg L^{-1} tricyclazole	Increased chl and carotenoid contents	Habibzadeh et al. (2012)
			Increased yield	
F. carica	2–6 days	5–20 mg L^{-1} ALA	Increased activities of SOD and POD	An et al. (2016)
			Decreased $O_2^{\bullet -}$ and MDA contents	

alleviate the damage of waterlogging in *B. napus* seedlings. They used it as foliar spray on waterlogged seedlings which resulted in increased growth, yield, enzyme (SOD, CAT, POD) activities, and Pro content and decreased lipid peroxidation, electrolyte leakage rate, and erucic acid content compared to the waterlogged seedlings alone (Leul and Zhou 1998, 1999). In *Ipomoea batatas* L. seedlings, PBZ application resulted in higher activities of antioxidative enzymes (SOD, CAT, APX, GR) and GSH and AsA content under waterlogging (5 days) condition (Lin et al. 2006). Later on, Habibzadeh et al. (2012) reported the role of tricyclazole ($C_9H_7N_3S$) on the morphological and biochemical characteristics of waterlogged *B. napus* seedlings. The foliar spray of tricyclazole increased the amount of photosynthetic pigments and yield attributes under waterlogging. However, above-mentioned studies reveal that triazole group compounds have regulatory effects on the mitigation of stress damage and induction of tolerance against waterlogging. They can enhance the activities of antioxidant enzymes, increase chl content, and improve growth and yield (Webb and Fletcher 1996; Leul and Zhou 1999; Lin et al. 2006). Triazoles also increased the photosynthetic rate (Leul and Zhou 1998), protein, and soluble sugar contents (Yang et al. 1994) and decreased lipid peroxidation, electrolyte leakage, and harmful metabolite production (Leul and Zhou 1999; Lin et al. 2006). There are numerous studies that exhibited the improvement of genetic and molecular strategies to develop varieties tolerant to waterlogging (Zhou 2010). But there is hardly any literature available about protectants that possess the evidence of promoting waterlogging stress tolerance in plants other than these few triazole compounds. However, An et al. (2016) investigated the effect of 5-aminolevulinic acid (ALA) in improving waterlogging-induced oxidative damage in *Ficus carica* plants. The plants grown under waterlogging condition for 2–6 days resulted in higher production of $O_2^{\bullet-}$ and lipid peroxidation. On the other hand, the plants which were pretreated with 5–20 mg L^{-1} of ALA reduced leaf $O_2^{\bullet-}$ production rate and MDA content which was associated with the enhancement of the activities SOD and POD in ALA-pretreated plants (An et al. 2016). So, comprehensive studies on the mechanisms of plant responses to waterlogging are needed to discover the possible ways to alleviate the damages caused by waterlogging. It will also help in understanding the required characteristics and selection criteria of the protectants to be used against waterlogging stress.

10.6 Conclusions and Future Perspectives

In many Asian countries, flooding in rainfed lowland areas during the monsoon season is a very common phenomenon. In these areas, crop production is largely affected by waterlogging. There is hardly any crop species except rice that can withstand long duration of flooding or even waterlogging. Thus finding the ways to develop waterlogging tolerance is one of the urgent tasks for researchers in those areas. Although the morphological and anatomical changes under waterlogging/flooding have been studied in many plants, the actual biochemical or molecular basis of those changes is still unclear. In addition, these responses are largely dependent on stress duration, level of water, and plant genotypes. Many plants also showed complex phenomenon after removal of waterlogging. Thus, the actual signal mechanisms under waterlogging/flooding would be one of the interesting fields of study. The hormonal regulation and interactions among phytohormones in conferring waterlogging stress are still unclear. Plant physiologists and agronomist may find the actual stages in specific crop plants which are sensitive to waterlogging. Tailoring the genes associated with antioxidant defense systems would be the best strategies in developing oxidative stress tolerance in plants under waterlogging.

Acknowledgments First of all, the authors wish to thank Khursheda Parvin, Dr. Anisur Rahman, Tasnim Farha Bhuiyan, and Mazhar Ul Alam, Faculty of Agriculture, Sher-e-Bangla Agricultural University, Dhaka, Bangladesh, for their critical reading and suggestions in improving the manuscript. The first author acknowledges the Japan Society for the Promotion of Science (JSPS) for funding in his research. We are also highly thankful to Mr. Md. Rumainul Islam, Environment Engineering Program, Chongqing University, and Md. Mosfeq-Ul-Hasan, Zhejiang University, Hangzhou, China, for providing us several supporting articles.

References

Ahmed S, Nawata E, Hosokawa M, Domae Y, Sakuratani T (2002) Alterations in photosynthesis and some antioxidant enzymatic activities of mung bean subjected to waterlogging. Plant Sci 163:117–123

Akhtar I, Nazir N (2013) Effect of waterlogging and drought stress in plants. Intl J Water Resour Environ Sci 2:34–40

Ali MA, Sarwern AKMG, Prodhan AKMA (1999) Effect of water stress on the growth features of different maize (*Zea mays* L.) cultivars. Pak J Bot 31:455–460

Amri M, El Ouni MHM, Salem B (2014) Waterlogging affect the development, yield and components, chlorophyll content and chlorophyll fluorescence of six bread wheat genotypes (*Triticum aestivum* L.) Bulgarian J Agr Sci 20:647–657

An Y, Qi L, Wang L (2016) ALA pretreatment improves waterlogging tolerance of fig plants. PLoS ONE 11(1):e0147202. doi:10.1371/ journal.pone.0147202

Anandana A, Pradhan SK, Das SK, Behera L, Sangeetha G (2015) Differential responses of rice genotypes and physiological mechanism under prolonged deepwater flooding. Field Crop Res 172:153–163

Apel K, Hirt H (2004) Reactive oxygen species: metabolism oxidative stress and signal transduction. Ann rev Plant Mol Biol 55:373–399

Arbona V, Hossain Z, López-Climent MF, Pérez-Clemente RM, Gómez-Cadenas A (2008) Antioxidant enzymatic activity is linked to waterlogging stress tolerance in citrus. Physiol Plant 132:452–466

Arru L, Fornaciari S (2010) Root oxygen deprivation and leaf biochemistry in trees. In: Mancuso S, Shabala S (eds) Waterlogging signalling and tolerance in plants. Springer, Berlin, pp 197–213

Ashraf M (2009) Biotechnological approach of improving plant salt tolerance using antioxidants as markers. Biotechnol Adv 27:84–93

Ashraf MA (2012) Waterlogging stress in plants: a review. Afr J Agric Res 7:1976–1981

Atwell BJ, Steer BT (1990) The effect of oxygen deficiency on uptake and distribution of nutrients in maize plants. Plant Soil 122:1–8

Bailey-Serres J, Voesenek LACJ (2008) Flooding stress: acclimations and genetic diversity. Annu Rev Plant Biol 59:313–339

Bajpai S, Chandra R (2015) Effect of waterlogging stress on growth characteristics and SOD gene expression in sugarcane. Int J Sci Res Pub 5:1–8

Banga M, Bogemann GM, Blom CWPM, Voesenek LACJ (1997) Flooding resistance of *Rumex* species strongly depends on their response to ethylene: rapid shoot elongation or foliar senescence. Physiol Plant 99:415–422

Bansal R, Srivastava JP (2015) Effect of waterlogging on photosynthetic and biochemical parameters in pigeonpea. Russ J Plant Physiol 62:322–327

Bin T, Shang-zhong XU, Zou XL, Zheng YL, Qi FZ (2010) Changes of antioxidative enzymes and lipid peroxidation in leaves and roots of waterlogging-tolerant and waterlogging-sensitive maize genotypes at seedling stage. Agric Sci China 9:651–661

Blokhnia O, Virolainen E, Fagerstedt V (2003) Antioxidants, oxidative damage and oxygen deprivation stress: a review. Ann Bot 91:179–194

Boem FHG, Lavado RS, Porcelli CA (1996) Note on the effects of winter and spring waterlogging on growth, chemical composition and yield of rapeseed. Field Crop Res 47:175–179

Bowler C, Montagu MV, Inze D (1992) Superoxide dismutase and stress tolerance. Annu Rev Plant Physiol Plant Mol Biol 43:83–116

Capon SJ, Jamesb CS, Williams L, Quinnc GP (2009) Responses to flooding and drying in seedlings of a common Australian desert floodplain shrub: *Muehlenbeckia florulenta* Meisn. Environ Exp Bot 66:178–185

Carter JL, Colmer TD, Veneklaas EJ (2006) Variable tolerance of wetland tree species to combined salinity and waterlogging is related to regulation of ion uptake and production of organic solutes. New Phytol 169:123–134

Chen HJ, Qualls RG, Miller GC (2002) Adaptive responses of *Lepidium latifolium* to soil flooding: biomass allocation, adventitious rooting, aerenchyma formation and ethylene production. Environ Exp Bot 48:119–128

Christine JD, Musgrave ME (1994) Characterization of population of rapid cycling *Brassica rapa* L. selected for differential waterlogging tolerance. J Exp Bot 45:385–392

Colmer TD, Voesenek LAC (2009) Flooding tolerance: suites of plant traits in variable environments. Funct Plant Biol 36:665–681

Corrêa de Souzaa T, Souza ES, Dousseau S, Mauro de Castroa E, Magalhães PC (2013) Seedlings of *Garcinia brasiliensis* (Clusiaceae) subjected to root flooding: physiological, morphoanatomical, and antioxidant responses to the stress. Aquat Bot 111:43–49

Damanik RI, Maziah M, Ismail MR, Ahmad S, Zain AM (2010) Responses of the antioxidative enzymes in Malaysian rice (*Oryza sativa* L.) cultivars under submergence condition. Acta Physiol Plant 32:739–747

El-Enany AE, Al-Anazi AD, Dief N, Al-Taisan WA (2013) Role of antioxidant enzymes in amelioration of water deficit and waterlogging stresses on *Vigna sinensis* plants. J Biol Earth Sci 3:B144–B153

Ezin V, Pena RDL, Ahanchede A (2010) Flooding tolerance of tomato genotypes during vegetative and reproductive stages. Braz J Plant Physiol 22:131–142

Fiedler S, Vepraskas MJ, Richardson JL (2007) Soil redox potential: importance, field measurements, and observations. Adv Agron 94:2–56

Fletcher RA, Hofstra G (1990) Improvement of uniconazole-induced protection in wheat seedlings. J Plant Growth Regul 19:207–212

Fukao T, Bailey-Serres J (2008) Ethylene—a key regulator of submergence responses in rice. Plant Sci 175:43–51

González JA, Gallardo M, Hilal M, Rosa M, Prado FE (2009) Physiological responses of quinoa (*Chenopodium quinoa* Willd.) to drought and waterlogging stresses: dry matter partitioning. Bot Stud 50:35–42

Grzesiak S, Hura T, Grzesiak MT, Piefikowski S (1999) The impact of limited soil moisture and waterlogging stress conditions on morphological and anatomical root traits in maize (*Zea mays* L.) hybrids of different drought tolerance. Acta Physiol Plant 21:305–315

Habibzadeh F, Sorooshzadeh A, Pirdashti H, Sanavy SAMM (2012) Effect of nitrogen compounds and tricyclazole on some biochemical and morphological characteristics of waterlogged-canola. Intl Res J Appl Basic Sci 3:77–84

Halliwell B, Gutteridge JMC (1999) Free radicals in biology and medicine, 3rd edn. Oxford University Press, Oxford

Hasanuzzaman M, Hosssain MA, Silva JAT, Fujita M (2012) Plant response and tolerance to abiotic oxidative stress: antioxidant defense is a key factor. In: Venkateswarlu B, Shanker AK, Shanker C, Maheswari M (eds) Crop stress and it's management: perspectives and strategies. Springer, Berlin, pp 261–316

Hasanuzzaman M, Nahar K, Rahman A, Mahmud JA, Hossain MS, Fujita M (2016) Soybean production and environmental stresses. In: Miransari M (ed) Environmental stresses in soybean production. Academic, New York, pp 61–102

Hossain Z, López-Climent MF, Arbona V, Pérez-Clemente RM, Gómez-Cadenas A (2009) Modulation of the antioxidant system in citrus under waterlogging and subsequent drainage. J Plant Physiol 166:1391–1404

Irfan M, Hayat S, Hayat O, Afroz S, Ahmad A (2010) Physiological and biochemical changes in plants under waterlogging. Protoplasma 241:3–17

Ito O, Ella E, Kawano N (1999) Physiological basis of submergence tolerance in rainfed lowland rice ecosystem. Field Crop Res 64:75–90

Jackson MB (2006) Plant survival in wet environments: resilience and escape mediated by shoot systems. In: Bobbink R, Beltman B, Verhoeven JTA, Whigham DF (eds) Wetlands: functioning, biodiversity, conservation and restoration. Springer, Berlin, pp 16–36

Jackson MB (2008) Ethylene-promoted elongation: an adaptation to submergence stress. Ann Bot 101:229–248

Justin SHFW, Armstrong W (1987) The anatomical characteristics of roots and plant response to soil flooding. New Phytol 106:465–495

Kende H, Van der Knaap E, Cho HT (1998) Deepwater rice: a model plant to study stem elongation. Plant Physiol 118:1105–1110

Khan MIR, Khan NA (2014) Ethylene reverses photosynthetic inhibition by nickel and zinc in mustard through changes in PS II activity, photosynthetic-nitrogen use efficiency and antioxidant metabolism. Protoplasma 251:1007–1019

Khan MIR, Asgher M, Khan NA (2014) Alleviation of salt-induced photosynthesis and growth inhibition by salicylic acid involves glycinebetaine and ethylene in mungbean (*Vigna radiata* L.) Plant Physiol Biochem 80:67–74

Khan MIR, Nazir F, Asgher M, Per TS, Khan NA (2015) Selenium and sulfur influence ethylene formation and alleviate cadmium-induced oxidative stress by improving proline and glutathione production in wheat. J Plant Physiol 178:9–18

Khan MIR, Iqbal N, Masood A, Mobin M, Anjum NA, Khan NA (2016a) Modulation and significance of nitrogen and sulfur metabolism in cadmium challenged plants. Plant Growth Regul 78:1–11

Khan MIR, Khan NA, Masood A, Per TS, Asgher M (2016b) Hydrogen peroxide alleviates nickel-inhibited photosynthetic responses through increase in use-efficiency of nitrogen and sulfur, and glutathione production in mustard. Front Plant Sci 7:44

Komatsu S, Kobayashi Y, Nishizawa K, Nanjo Y, Furukawa K (2010) Comparative proteomics analysis of differentially expressed proteins in soybean cell wall during flooding stress. Amino Acids 39:1435–1449

Kumar P, Pal M, Joshi R, Sairam RK (2013) Yield, growth and physiological responses of mung bean [*Vigna radiata* (L.) Wilczek] genotypes to waterlogging at vegetative stage. Physiol Mol Biol Plants 19:209–220

Leul M, Zhou W (1998) Alleviation of waterlogging damage in winter rape by application of uniconazole: effects on morphological characteristics, hormones and photosynthesis. Field Crop Res 59:121–127

Leul M, Zhou W (1999) Alleviation of waterlogging damage in winter rape by uniconazole application: effects on enzyme activity, lipid peroxidation, and membrane integrity. J Plant Growth Regul 18:9–14

Li C, Jiang D, Wollenweber B, Li Y, Daia T, Caoa W (2011) Waterlogging pretreatment during vegetative growth improves tolerance to waterlogging after anthesis in wheat. Plant Sci 180:672–678

Liao CT, Lin CH (1996) Photosynthesis responses of grafted bitter melon seedlings to flood stress. Environ Exp Bot 36:167–172

Lin KR, Weng C, Lo H, Chen J (2004) Study of the root antioxidative system of tomatoes and eggplants under waterlogged conditions. Plant Sci 167:355–365

Lin KH, Tsou CG, Hwang SY, Lo HF (2006) Paclobutrazol pre-treatment enhanced flooding tolerance of sweet potato. J Plant Physiol 163:750–760

Lone AA, Warsi MZK (2009) Response of maize (*Zea mays* L.) to excess soil moisture (ESM) tolerance at different stages of life cycle. Bot Res Int 2:211–217

Magneschi L, Perata P (2009) Rice germination and seedling growth in the absence of oxygen. Ann Bot 103:189–196

Mauchamp A, Blanch S, Grillas P (2001) Effects of submergence on the growth of *Phragmites australis* seedlings. Aquat Bot 69:147–164

Mensah JK, Obadoni BO, Eruotor PG, Onome-Irieguna F (2006) Simulated flooding and drought effects on germination, growth, and yield parameters of sesame (*Sesamum indicum* L.) Afr J Biotechnol 5:1249–1253

Mommer L, Pedersen O, Visser EJW (2004) Acclimation of a terrestrial plant to submergence facilitates gas exchange under water. Plant Cell Environ 27:1281–1287

Olgun M, Kumlay AM, Adiguzel MC, Caglar A (2008) Effects of waterlogging in wheat (*T. aestivum* L.) Acta Agric Scand B Plant Soil Sci 58:193–198

Paltaa JA, Ganjealic A, Turnerb NC, Siddique KHM (2010) Effects of transient subsurface waterlogging on root growth, plant biomass and yield of chickpea. Agric Water Manag 97:1469–1476

Pang J, Shabala S (2010) Membrane transporters and waterlogging tolerance. In: Mancuso S, Shabala S (eds) Waterlogging signalling and tolerance in plants. Springer, Berlin, pp 197–213

Parelle J, Roudaut JP, Ducrey M (2006) Light acclimation and photosynthetic response of beech (*Fagus sylvatica* L.) saplings under artificial shading or natural Mediterranean conditions. Ann For Sci 63:257–266

Parent C, Capelli N, Berger A, Cravecoeur M, Dat JF (2008) An overview of plant responses to soil waterlogging. Plant Stress 2:20–27

Parvin D, Karmoker JL (2013) Effects of waterlogging onion accumulation and sugar, protein and proline contents in *corchorus capsularis* L. Bangladesh J bot 42:55–63

Pezeshki SR (2001) Wetland plant responses to soil flooding. Environ Exp Bot 46:299–312

Pinhero RG, Fletcher RA (1994) Paclobutrazol and ancymidol protect corn seedlings from high and low temperature stresses. Plant Growth Regul 15:47–53

Prasanna YL, Rao GR (2014) Effect of waterlogging on growth and seed yield in greengram genotypes. Int J Food Agric Vet Sci 4:124–128

Rai RK, Srivastava JP, Shahi JP (2004) Effects of waterlogging on some biochemical parameters during early growth stages of maize. Indian J Plant Physiol 9:65–68

Rasaei A, Ghobadi ME, Jalali-Honarmand S, Ghobadi M, Saeidi M (2012) Impacts of waterlogging on shoot apex development and recovery effects of nitrogen on grain yield of wheat. Eur J Exp Biol 2:1000–1007

Reddy MD, Mittra BN (1985) Effect of complete plant submergence at different growth stages on grain yield, yield components and nutrient content of rice. Plant Soil 86:379–386

Ren B, Zhang J, Li X, Fan X, Dong S, Liu P, Zhao B (2014) Effects of waterlogging on the yield and growth of summer maize under field conditions. Can J Plant Sci 94:23–31

Ruan YL, Llewellyn DJ, Furbank RT (2003) Suppression of sucrose synthase gene expression represses cotton fiber cell initiation, elongation and seed development. Plant Cell 15:952–964

Sairam RK, Srivastava GC (2002) Changes in antioxidant activity in sub-cellular fraction of tolerant and susceptible wheat genotypes in response to long term salt stress. Plant Sci 162:897–904

Sairam RK, Kumutha D, Ezhilmathi K, Deshmukh PS, Srivastava GC (2008) Physiology and biochemistry of waterlogging tolerance in plants. Biol Plant 52:401–412

Sairam RK, Kumutha D, Ezhilmathi K, Chinnusamy V, Meena RC (2009) Waterlogging induced oxidative stress and antioxidant enzymes activity in pigeon pea. Biol Plant 53:493–504

Sairam RK, Dharmar K, Lekshmy S, Chinnusam V (2011) Expression of antioxidant defense genes in mung bean (*Vigna radiata* L.) roots under water-logging is associated with hypoxia tolerance. Acta Physiol Plant 33:735–744

Sauter M (2013) Root responses to flooding. Curr Opin Plant Biol 16:282–286
Sculthorpe CD (1967) The biology of aquatic vascular plants. Edward Arnold, London
Setter TL, Waters I (2003) Review of prospects for germplasm improvement for waterlogging tolerance in wheat, barley and oats. Plant Soil 253:1–34
Setter TL, Waters I, Sharma SK, Singh KN, Kulshreshtha N, Yaduvanshi NPS, Ram PC, Singh BN, Rane J, McDonald G, Khabaz Saberi H, Biddulph TB, Wilson R, Barclay I, McLean R, Cakir M (2009) Review of wheat improvement for waterlogging tolerance in Australia and India: the importance of anaerobiosis and element toxicities associated with different soils. Ann Bot 103:221–235
Shiono K, Takahashi H, Colmer TD, Nakazono M (2008) Role of ethylene in acclimations to promote oxygen transport in roots of plants in waterlogged soils. Plant Sci 175:52–58
Shiono K, Ogawa S, Yamazaki S, Isoda H, Fujimura T, Nakazono M, Colmer TD (2011) Contrasting dynamics of radial O -loss barrier induction and aerenchyma formation in rice roots of two lengths. Ann Bot 107:89–99
Smethurst CF, Garnet T, Shabala S (2005) Nutrition and chlorophyll fluorescence responses of lucerne (*Medicago sativa*) to waterlogging subsequent recovery. Plant Soil 270:31–45
Srivastava JP, Gangey SK, Shahi JP (2007) Water logging resistance in maize in relation to growth, mineral composition and some biochemical parameters. Indian J Plant Physiol 12:28–33
Striker GG (2012) Flooding stress on plants: anatomical, morphological and physiological responses. In: Mworia J (ed) Botany. Intech, Rijeka, pp 3–28
Suralta RR, Yamauchi A (2008) Root growth, aerenchyma development, and oxygen transport in rice genotypes subjected to drought and waterlogging. Environ Exp Bot 64:5–82
Tan W, Liu J, Dai T, Jing Q, Cao W, Jiang D (2008) Alterations in photosynthesis and antioxidant enzyme activity in winter wheat subjected to post-anthesis water-logging. Photosynthetica 46 (1):21–27
Tang B, Xu S, Zou X, Zheng Y, Qiu F (2010) Changes of antioxidative enzymes and lipid peroxidation in leaves and roots of waterlogging-tolerant and waterlogging-sensitive maize genotypes at seedling stage. Agric Sci China 9:651–661
Tarekegne A, Bennie ATP, Labuschagne MT (2000) Effects of soil waterlogging on the concentration and uptake of selected nutrients in wheat genotypes differing in tolerance. In: The eleventh regional wheat workshop for Eastern, Central and Southern Africa. CIMMYT, Addis Abeba, pp 253–263
Vandoorne B, Descamps C, Mathieu AS, Van den Ende W, Vergauwen R, Javaux M, Lutts S (2014) Long term intermittent flooding stress affects plant growth and inulin synthesis of *Cichorium intybus* (var. sativum). Plant Soil 376:291–305
Vasellati V, Oesterheld M, Medan D, Loreti J (2001) Effects of flooding and drought on the anatomy of *Paspalum dilatatum*. Ann Bot 88:355–360
Vidoz ML, Loreti E, Mensuali A, Alpi A, Perata P (2010) Hormonal interplay during adventitious root formation in flooded tomato plants. Plant J 63:551–562
Voesenek LACJ, Benschop JJ, Bou J, Cox MCH, Groeneveld HW, Millenaar FF, Vreeburg RA, Peeters AJ (2003) Interactions between plant hormones regulate submergence-induced shoot elongation in the flooding tolerant dicot *Rumex palustris*. Ann Bot 91:205–211
Wang TW, Arteca RN (1992) Effect of low O_2 root stress on ethylene biosynthesis in tomato plants (*Lycopersicon esculentum* Mill. cv. Heinz 1350). Plant Physiol 98:97–100
Webb JA, Fletcher RA (1996) Paclobutrazol protects wheat seedlings from injury due to waterlogging. Plant Growth Regul 18:201–206
Wegner LH (2010) Oxygen transport in waterlogged plants. In: Mancuso S, Shabala S (eds) Waterlogging signalling and tolerance in plants. Springer, Berlin, pp 3–22
Wei W, Li D, Wang L, Ding X, Zhang Y, Gao Y, Zhang X (2013) Morpho-anatomical and physiological responses to waterlogging of sesame (*Sesamum indicum* L.) Plant Sci 208:102–111

Xu F, Wang X, Wu Q, Zhang X, Wang L (2012) Physiological responses differences of different genotype sesames to flooding stress. Adv J Food Sci Technol 4(6):352–356

Yaduvanshi NPS, Setter TL, Sharma SK, Singh KN, Kulshreshtha N (2010) Waterlogging effects on wheat yield, redox potential, manganese and iron in different soils of India. Paper presented at the 19th world congress of soil Science, Brisbane, Australia, 1–6 August, pp 45–48

Yang DQ, Yang JX, Hu YW (1994) Effects of S-3307 on some physiological characteristics of rape seedlings. Plant Physiol Commun 30:182–185

Yin D, Chen S, Chen F, Guan Z, Fang W (2009) Morphological and physiological responses of two chrysanthemum cultivars differing in their tolerance to waterlogging. Environ Exp Bot 67:87–93

Yiu JC, Liu CW, Fang DYT, Lai YS (2009) Waterlogging tolerance of Welsh onion (*Allium fistulosum* L.) enhanced by exogenous spermidine and spermine. Plant Physiol Biochem 47:710–716

Yordanova RY, Christov KN, Popova LP (2004) Antioxidative enzymes in barley plants subjected to soil flooding. Environ Exp Bot 51:93–101

Zaidi PH, Rafique S, Rai PK, Singh NN, Srinivasan G (2004) Tolerance to excess moisture in maize (*Zea mays* L.): susceptible crop stages and identification of tolerant genotypes. Field Crop Res 90:189–202

Zhang G, Tanakamaru K, Abe J, Morita S (2007) Influence of waterlogging on some anti-oxidative enzymatic activities of two barley genotypes differing in anoxia tolerance. Acta Physiol Plant 29:171–176

Zheng C, Jiang D, Liu F, Dai T, Jing Q, Cao W (2009) Effects of salt and waterlogging stresses and their combination on leaf photosynthesis, chloroplast ATP synthesis, and antioxidant capacity in wheat. Plant Sci 176:575–582

Zhou M (2010) Improvement of plant waterlogging tolerance. In: Mancuso S, Shabala S (eds) Waterlogging signalling and tolerance in plants. Springer, Berlin/Heidelberg, pp 197–213

Role of Reactive Oxygen Species in Water-Deficit Stress Response

11

Vishnu Varthini Nachimuthu, Balaji Aravindhan Pandian, and S. Robin

Abstract

Water deficit is the major limiting factor in crop production which has a significant influence by changing climate. With increasing occurrence of water-deficit stress in warming globe, it becomes inevitable to comprehend and dissect the role of oxidation-induced metabolic alterations. Reactive oxygen species (ROS), such as singlet oxygen, hydrogen peroxide, superoxide anion, and hydroxyl radicals which are produced in low level at intercellular sites especially in chloroplasts and peroxisomes, has had shown enhanced concentrations during water-deficit condition. Higher amount of ROS acts as toxic substance which can damage cellular components such as lipids, proteins, and DNA. Plants have evolved its antioxidant systems to restrict ROS accumulation as general stress signaling process. This oxidative stress plays a major role in plant water-deficit responses in terms of oxidative damage and related redox signaling.

Keywords

Reactive oxygen species • Drought stress • Oxidative damage • Cell signaling

V.V. Nachimuthu (✉) • B.A. Pandian • S. Robin
Centre of Excellence in Molecular Breeding, Tamil Nadu Agricultural University, Coimbatore, Tamil Nadu, India
e-mail: popvarun@gmail.com

M.I.R. Khan, N.A. Khan (eds.), *Reactive Oxygen Species and Antioxidant Systems in Plants: Role and Regulation under Abiotic Stress*,
DOI 10.1007/978-981-10-5254-5_11

Abbreviations

AA	Ascorbic acid
ABA	Abscisic acid
AOS	Active oxygen species
APX	Ascorbate peroxidase
ATP	Adenosine triphosphate
CAT	Catalase
GR	Glutathione reductase
GSH	Glutathione
GSSG/2GSH	Glutathione disulfide–glutathione couple
Ipt	Isopentenyl transferase
MAPK	Mitogen-activated protein kinases
RBOH	Respiratory-bound oxidase homolog
ROI	Reactive oxygen intermediates
ROS	Reactive oxygen species
RuBP	Ribulose-1,5-bisphosphate
RWC	Relative water content
SOD	Superoxide dismutase

11.1 Introduction

Reactive oxygen species (ROS) are chemicals formed as a toxic by-product of aerobic metabolism by partial reduction and excitation of atmospheric oxygen; they are also called as active oxygen species (AOS) and reactive oxygen intermediates (ROI). Four major forms of reactive oxygen species are singlet oxygen ($^1O_2$), superoxide radical (O_2^-), hydrogen peroxide (H_2O_2), and hydroxyl radical (OH^-). ROS is having an oxidative property which leads to cell death due to the oxidation of different cellular components (Mittler 2002). It is produced in exogenous as well as endogenous cellular systems. ROS are produced majorly from smoke, industrial pollutants, xenobiotics, and radiation, whereas the endogenous ROS are produced in cell organelles of both plants and animals. Taking into account the ROS accumulation in cells causing oxidative stress, it also involves in key signaling process with its capacity of site-specific production and scavenging to maintain balance at cellular level (Mittler et al. 2011).

Oxidative stress caused by ROS accumulation in cellular organelles is one of the important stress responses in various plant species. ROS with its site-specific nature has its production in different cellular components such as mitochondria, chloroplasts, peroxisomes, plasma membrane, and apoplast (Gill and Tuteja 2010). Light utilizing autotrophic plants has chloroplast where the existence of photosynthetic electron transport system favors for the production of different ROS. Apart from this site, mitochondria, apoplast, cytosol, and peroxisomes are the major producers of superoxide and hydrogen peroxide species. Cellular components have the capacity to scavenge these species through different scavenging systems such as superoxide dismutase, flavonoids, sugars, proline, ascorbate, glutathione, catalase, peroxidases, carotenoids, and α-tocopherol. Hence, disruption of delicate balance existing in this cyclic process of production and scavenging ROS that occurs constantly at cellular level leads to oxidation stress. This oxidative stress causes potential toxic effects of macromolecule denaturation including DNA, RNA, protein, lipid oxidation, and damage (Mittler 2016).

Various biotic and abiotic stress conditions in plants including drought, salinity, heat stress, low-temperature stress, nutrient deficiency, heavy metals, ultraviolet radiation, and pathogen attack disrupt cellular homeostasis thereby causing oxidative stress (Sharma et al. 2012; Khan and Khan 2014; Khan et al. 2014, 2015a, b, 2016a, b; Thao et al. 2015). In abiotic stress, ROS accumulation is used as a defensive mechanism to trigger programmed cell death (Levine et al. 1994). Drought stress, one of the major climate change impacts, affects plant metabolism by enhancing the production of ROS. When the plant faces water-deficit stress, the first response will be stomatal closure as the main drought tolerance mechanism to avoid water losses due to transpiration, but this will have a deleterious effect on photosynthesis with restricted CO_2 uptake, higher rate of photorespiration, and altered photosynthetic metabolism that leads to the production of ROS in different cell organelles (Mittler 2002). In the course of evolution, plants developed various ROS-scavenging mechanisms, but prolonged stress conditions stimulate higher ROS production which led to oxidation of cell organelles followed by cell death. This oxidative damage serves as a major limitation in plant performance under drought stress condition. Genotypes with better antioxidant capacity and tolerance to higher ROS level thrive well under water-deficit stress as tolerant cultivars (Mattos and Moretti 2015).

In addition to the by-products of aerobic metabolism, ROS plays a major role in plant signaling. Initially, it was identified primarily as by-products of metabolism, but recent studies have recognized ROS for its key role in plant signaling network (Mittler et al. 2011). Deviation from cellular homeostasis maintained by ROS production and scavenging process was utilized for signaling process. It plays a central role in systemic-acquired acclimation process through redox regulatory network (Dietz 2014). ROS signaling is mediated by various enzymes such as respiratory-bound oxidase homolog (RBOH), peroxidases, and other ROS through highly localized accumulation. Hence, with these different roles of ROS, it becomes necessary to integrate in the context of water-deficit stress to understand plant response and utilize it in developing stress-tolerant cultivars.

11.2 Water-Deficit Stress: An Overview

Several environmental stresses affect the plant growth and development which subsequently impede the productivity. Drought is one of the environmental stresses which drastically affects the plants and decreases crop productivity (Lambers et al. 2008).

Drought is nothing but prolonged water-deficit conditions due to nonavailability of rainfall and other water sources which results in soil water deficit and subsequent plant water deficit. The severity of drought is unpredictable as it depends on many factors such as occurrence and distribution of rainfall, evaporative demands, and moisture-storing capacity of soils (Wery et al. 1994). During water-deficit conditions, absorption of photosynthetically active radiations, radiation use efficiency, and harvest index are reduced based on the similarity (Farooq et al. 2009).

11.2.1 Factors Causing Drought

Till now various countries across the world are facing drought conditions threatening the food security of the nations. There are various factors found to be causing water-deficit conditions which are explained below.

Precipitation is condensation of atmospheric water vapor that falls under gravity; this is a major source of freshwater to the world and the only water source for 80 % of the agricultural land mass in the world. Precipitation is caused by three mechanisms: convective, stratiform, and orographic rainfall (Emmanouil and Anagnostou 2004; Dore et al. 2006). Any factor affecting these mechanisms such as wind and ridges of high-pressure areas aloft can prevent or restrict the development of thunderstorm activity or rainfall over one certain region, and temperature paves way to lesser precipitation which triggers water-deficit conditions.

The term El Niño refers to the large-scale ocean–atmosphere climate interaction linked to a periodic warming in sea surface temperatures across the central and east-central equatorial Pacific. Dry conditions than normal can be seen from December to February in south-central Africa which directly affects the monsoon of South Asia and Australia causing severe drought. Human interventions play a major role in affecting the water cycle of the earth, human activities such as deforestation, overutilization of groundwater, and pollution. Deforestation can influence the occurrence of dry conditions since it reduces forest's watershed potential. Growing construction and urbanization on many water bodies are closed which directly affects the water table. All these interventions affect the water sources other than rainfall and pave way to severe drought in the absence of rain (BBC 2016).

11.2.2 Effect of Water Deficit on Plants' Responses

Water is the prime component for the physiological process of the plants; hence, nonavailability of water affects the physiology of the plants which impacts various processes of plants (Fig. 11.1).

11.2.2.1 Crop Growth and Yield

Water-deficit stress severely limits crop growth especially in arid and semiarid regions of the world as it affects all stages of plant growth and development (Shamim et al. 2009). Reduced availability of water reduces the germination and seedling growth (Harris et al. 2002; Kaya et al. 2006). The turgor pressure of the cell is important for a cell to maintain rigidity for various cellular processes and growth and reproduction of plant cells affected by water-deficit stress which negatively impacts the turgor pressure. Interruption of water flow from xylem to other tissues and impaired mitosis affects the cell elongation, and expansion results in reduced height (Anjum et al. 2011; Naomi et al. 1998). As the stress affects the basic functions of the plant cell which results in hormonal imbalance and affects the production of various enzymes, this negatively impacts flowering and grain filling (Farooq et al. 2009). The leaf area was affected adversely in both the main shoot and tillers of all the varieties. Reduction in leaf area by water stress is an important cause of reduced crop yield through reduction in photosynthesis (Rucker et al. 1995). Prolonged water-deficit conditions will reduce the biomass of the plant; water-stressed plants will have less biomass compared to a normal plant.

Fig. 11.1 Effects of water-deficit stress in crop plants

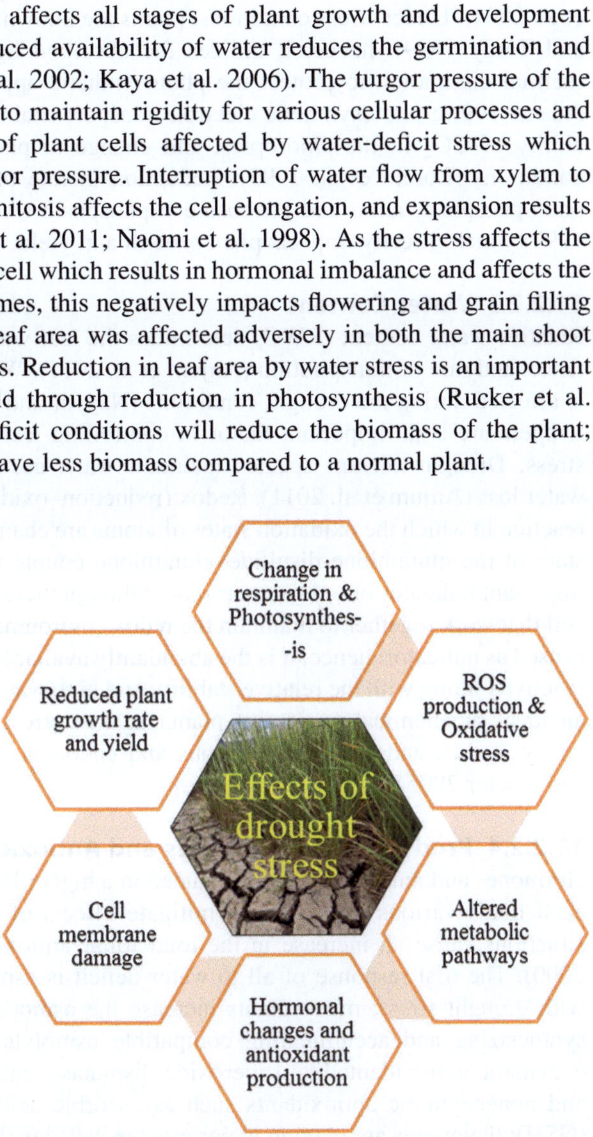

11.2.2.2 Respiration and Photosynthesis

Respiration and photosynthesis are the processes dreadfully affected during the water-deficit conditions. Consequent reduction of leaf expansion coupled with early leaf senescence impacts the photosynthetic machinery (Wahid and Rasul 2005). The rate photosynthesis often limits plant growth when soil water availability is reduced (Huang and Fu 2000). Stomatal closure is one of the first-line defense mechanisms deployed by the plants during drought; this reduces the CO_2 intake. Reduction in Rubisco quantity reduces the rate of photosynthesis; regeneration of RuBP has been shown to limit the photosynthetic capacity (Vu et al. 1999; Bota et al. 2004). Under stress conditions, plants are known to lose chlorophyll (Havaux and Tardy 1999); hence, it utilizes the absorbed light to other processes, like thermal dissipation to protect the photosynthetic apparatus (Deming-Adams and Adams 1996); photosynthetic electron transport chain is finely tuned to the availability of CO_2 in the chloroplast and change to photosystem II under drought conditions (Loreto et al. 1995). Reduction in rate of photosynthesis affects the photophosphorylation and adenosine triphosphate (ATP) synthesis (Tezara et al. 1999) which directly impacts plant growth and yield.

11.2.2.3 Water Content

Relative water content (RWC) determines the leaf water potential, stomatal resistance, and rate of transpiration (Anjum et al. 2011). Water content of the plant cell is affected during the drought conditions (Nayyar and Gupta 2006). Increased leaf temperature is an important factor in controlling leaf water status under drought stress. Drought-tolerant species maintain water-use efficiency by reducing the water loss (Anjum et al. 2011). Redox (reduction–oxidation reaction) is a chemical reaction in which the oxidation states of atoms are changed. In plant cells, the redox state of the glutathione disulfide–glutathione couple (GSSG/2GSH) serves as an important indicator of redox environment though there are many redox couples in a cell that work together to maintain the redox environment; the GSSG/2GSH couple is used as indicator; hence, it is the abundantly available redox couple in a cell. This reactivity along with the relative stability and high water solubility of GSH makes it an ideal biochemical to protect plants against stress including oxidative stress, heavy metals, and certain exogenous and endogenous organic chemicals (Foyer and Noctor 2005.)

11.2.2.4 Production of Hormones and Antioxidants

Hormones and antioxidants are produced in a higher level during stress conditions; each holds various properties to mitigate effects of drought stress. Most stress situations cause an increase in the total foliar antioxidant activity (Pastori et al. 2000). The first response of all to water deficit is osmotic adjustment, to counter with drought stress; many plants increase the osmotic potential of their cells by synthesizing and accumulating compatible osmolytes such as proline. Various enzymatic antioxidants like superoxide dismutase, catalase, ascorbate peroxidase, and nonenzymatic antioxidants such as ascorbic acid (AA) reduced glutathione (GSH); flavonoids are playing major roles in cell signaling and scavenging reactive oxygen species thus protecting the cells during water-deficit conditions. A brief discussion is given in the next section.

11.2.2.5 ROS Production

The effects of the water-deficit conditions create a downregulation in photosystem II which creates imbalance in light capture and utilization and then results in the production of reactive oxygen species such as singlet oxygen ($^1O_2$), superoxide radial (O_2^-), hydrogen peroxide (H_2O_2), and hydroxyl radical ($OH^•$). DNA nicking, amino acid and protein oxidation, and lipid peroxidation are potential for creating oxidative damage to the plant cells (Asada 1999; Johnson et al. 2003). ROS are always available in optimal levels which are increased during stress conditions; they also serve in signaling. A detailed discussion is given in the next section.

11.3 ROS Production and Its Scavenging Under Drought Stress

Cell disruption during abiotic stress conditions in plants is majorly due to oxidative damage created by relative oxygen species at various cellular compartments (Bowler et al. 1992). As a result of evolutions, plants developed various enzymatic and nonenzymatic mechanisms (scavengers) to fight against oxidative damage (Table 11.1). ROS response by different plants is explained below.

Table 11.1 ROS synthesis and scavenging in plants

S. no	ROS	Source	Sub cellular location	Oxidative stress nature	Scavengers
1	Superoxide	Atmospheric oxygen	Chloroplast, mitochondria, apoplast, peroxisomes	Dismutate enzymatically or nonenzymatically to hydrogen peroxide, reacts with proteins	Superoxide dismutase, ascorbate,
2	Hydroxyl radical	Hydrogen peroxide	Fenton reaction at active site of Rubisco	Most reactive ROS, reacts with all biological molecules and damage cellular components	Flavonoids, proline, ascorbate
3	Hydrogen peroxide	Superoxide radical	Chloroplast, mitochondria, peroxisome, apoplast, cytosol	Moderately reactive, oxidizes cysteine and methionine residues, reacts with DNA and proteins	Ascorbate peroxidase, catalase, peroxidase, flavonoids
4	Singlet oxygen	Atmospheric oxygen	Membrane, chloroplast, and mitochondria	Reacts with proteins, DNA G residue and lipids	Carotenoids and α-tocopherol

Adopted and modified from Mittler (2016)

Several antioxidant enzymes serve as active oxygen-scavenging systems, such as SOD, APX, GR, and CAT, and some low molecules of nonenzyme antioxidants, such as ascorbic acid (AsA) and reduced glutathione (GSH) (Bowler et al. 1992). Increase in ROS activity is reported in several stress conditions such as drought, chilling, and salinity stress. It plays a major role in stress signaling and triggers various biochemical activities. Increased activity of Halliwell–Asada–Foyer cycle, or the water–water cycle in chloroplasts, can maintain close to normal levels of PSII and PSI activity during stress reducing the inhibition of photosynthesis and decreasing ROS levels. Premature senescence flag leaf (psf) mutant in rice is reported to have higher leaf O_2^- generation rate and H_2O_2 content accumulation, and the decreased activities of SOD, CAT, and APX in the psf leaves prove that ROS activity is strongly responsible for the increased ROS level and the accelerated leaf senescence (Wang et al. 2016). The same phenomenon during prematured leaf senescence is also reported in sorghum (Djanaguiraman et al. 2010). Four rice varieties in which two are sensitive and two are resistant to chilling and drought stress were studied for ROS activity during stress period; the study resulted that tolerance to drought or chilling in rice is well associated with the enhanced capacity of antioxidative system under drought or chilling condition (Guo et al. 2006). Abscisic acid accumulation is found to trigger the increased generation of reactive oxygen species and upregulates the activities of antioxidant enzymes in maize leaves during water stress conditions (Jiang and Zhang 2002). A correlation between antioxidant enzyme activity and water stress or salinity tolerance was demonstrated by comparison of a tolerant cultivar with a sensitive cultivar in various plant species, such as rice and foxtail millet (*Setaria italica*) along with higher activities of antioxidant enzymes in tolerant cultivars. Aluminum toxicity is found to enhance the ROS production in black gram (Chowra et al. 2016). During water-deficit stress, cytokinin synthesis is limited along with root growth inhibition. A study showed that overexpressing isopentenyl transferase (*ipt*) gene has exhibited higher root growth during drought stress and found to be having lower levels of ROS production compared to wild type (Xu et al. 2016).

Plants evolved with various ROS-scavenging methods naturally. Modern science paves way to identify these mechanisms for improving systemic-acquired acclimation during stress conditions. Overexpression of transcription factors such as Zat10, Zat12, or JERF3 increased the expression of ROS-scavenging genes during drought stresses (Chen et al. 2012). Selenium application found partly alleviates HT-induced oxidative stress and negative impacts of HT on physiology, growth, and yield of grain sorghum (Djanaguiraman et al. 2010).

11.4 ROS as Signaling Molecules Under Drought Stress

ROS primarily obtained as the toxic by-product of aerobic metabolism causes oxidative stress by accumulating at a level that exceeds the detoxification ability by scavengers in different cellular compartments. It was initially considered as harmful to biological organisms by inducing oxidative stress in response to biotic and abiotic stresses. Later, many studies have established the role of ROS in signal transduction pathways (Mittler et al. 2011). Hence, ROS plays a dual role by inducing oxidative stress and signaling for defense or acclimation response under stress condition.

Drought or water-deficit stress in plants leads to oxidative stress by affecting the balance between ROS production and scavenging mechanism. This also acts as signaling molecules by promoting ROS as a signal intermediate in stomatal closure through abscisic acid which is accumulated in response to water stress in plants (Yan et al. 2007). ROS are used as signaling molecules because of its temporal and spatial control of ROS signals in various cellular components. Initial burst of ROS production mediated by RBOH (ROS-generating respiratory burst oxidase) could stimulate cell-to-cell communication which results in long-distance auto-propagating signals or ROS wave. In order to accomplish this role, ROS oxidative level should be maintained to avoid oxidative stress (Mittler et al. 2004).

This redox regulatory network requires various intermediate molecules such as mitogen-activated protein kinases (MAPK), calcium, and phytohormones that transduce external stress for genetic response. The role of MAPK family activated by ABA pathway has a well-established role in water-deficit stress. In rice, under water-deficit condition, two MAPK genes *OsMSRMK2* and *OsMAPK5* were activated in rice (Xiong and Yang 2003). In *Arabidopsis*, *AtMEKK1* and *AtMPK3* factors are activated under water-deficit condition (Mizoguchi et al. 1996). Overexpression of MAPK kinase like gene *DSM1* has increased dehydration and oxidative stress tolerance at seedling stage by scavenging ROS. Zong et al. (2009) have demonstrated higher osmotic stress tolerance by peroxidases in tobacco by overexpressing *ZmMPK7*, a maize MAPK gene. In cotton, two MAPK kinases (MAPKKs), *GhMKK1* and *GhMKK5*, have been characterized to be involved in stress resistance and ROS homeostasis (Zhang et al. 2012; Lu et al. 2013). *GhMKK1* was shown to increase drought tolerance in tobacco when it was overexpressed, whereas *GhMKK5* reduced tolerance level through upregulated expression of ROS-related and cell death marker genes and resulted in excessive accumulation of H_2O_2 and hypersensitive response (HR)-like cell death. Several studies have reported the role of *MPK6*, *MKK1*, and *MKKK20*, members of MAPK-mediated pathway in accumulating ROS under various abiotic stresses (Golldack et al. 2014). Thus, functional relation of MAPK hub to abiotic stress tolerance through ROS signaling pathways was revealed in various crops.

Reactive oxygen species and calcium (Ca^{2+}) interaction was well established in plants for various developmental processes and stress tolerance such as stomatal regulation and programmed cell death and abiotic stress tolerance such as water deficit and salinity. ROS and Ca^{2+} levels at cellular components have shown spatial and temporal coincidences in response to various stresses. Both the signaling networks have nitric oxide (NO), a prominent signaling molecule as a common reactive radical. Calcium-dependent protein kinase (CDPK), an important down-stream component regulator in Ca^{2+}-regulating pathways, has shown to be responsive to various abiotic stresses in different crops through ROS regulation. *OsDMI3*, a Ca^{2+}-/CaM-dependent protein kinase (CCaMK), is necessary for ABA-induced increases in the expression and the activities of SOD and CAT in rice (Shi et al. 2012). Functional characterization of *OsCPK4* by Campo et al. (2014) revealed that overexpression of this gene has shown increased tolerance to drought and salt stress by inhibiting membrane lipid peroxidation. RBOH-dependent singlet oxygen production acts as the most tightly regulated ROS-signaling hub directly linking ROS signaling with calcium signaling and protein phosphorylation events. It plays a significant role in stomatal closure, signaling under light and salinity stress, fiber growth in cotton, and peroxule outgrowth from peroxisomes (Dietz et al. 2016).

The role of various phytohormones such as ABA, ethylene, salicylic acid, jasmonates, etc. in signal transduction and its interaction with ROS-induced signals were observed for different biotic and abiotic stress tolerance. Among these hormones, ABA is an important hormone that plays a major role in drought stress signaling. Stomatal closure induced by ABA during water deficit is mediated by ROS, in particular H_2O_2, which is synthesized by plasma membrane-bound NADPH oxidase by elevating cytosolic Ca_2+ level through activation of channels. ABA and ROS-signaling pathway interaction was demonstrated in maize bZIP transcription factor ABP9, glutamate carboxypeptidase AMP1, and the ankyrin-repeat protein ITN1 for abiotic stress such as drought and salinity (Golldack et al. 2014). ABA-induced H_2O_2 production increased the expression and activity of *ZmMPK5* which is pivotal for ABA-based antioxidant defense. Its activation increases the expression and the activity of NADPH oxidase; thus, there is a positive correlation that exists for NADPH oxidase, H_2O_2, and *ZmMPK5* production in ABA signaling. A similar pattern of increased ROS accumulation in chloroplast induced by ABA was observed in tomato by Zhou et al. (2014). Thus, several studies have demonstrated ROS as secondary messengers in various phytohormone signaling pathways.

11.5 Conclusion and Future Perspectives

Previous understanding of toxic effects of ROS due to oxidative stress is appended with new fact by deciphering the key role of ROS as central player in plant signaling reactions through ROS wave. The results obtained in various recent studies set the stage for dissecting the pathway for systemic-acquired acclimation through ROS. It serves as a new target area in different crops to further improve drought tolerance to breed climate smart crop varieties. Further studies on metabolic alterations by ROS production and signaling may facilitate to address the complexity of plant response to drought stress.

References

Anjum SA, Xie XY, Wang LC, Saleem MF, Man C, Lei W (2011) Morphological, physiological and biochemical responses of plants to drought stress. Afr J Agric Res 6:2026–2032

Asada K (1999) The water-water cycle in chloroplasts: scavenging of active oxygens and dissipation of excess photons. Annu Rev Plant Physiol Plant Mol Biol 50:601–639

BBC – GCSE (2016) Bitesize: human activities causing drought. Bbc.co.uk. N.p

Bota J, Flexas J, Medrano H (2004) Is photosynthesis limited by decreased Rubisco activity and RuBP content under progressive water stress? New Phytol 162:671–681

Bowler C, Montagu MV, Inze D (1992) Superoxide dismutase and stress tolerance. Annu Rev Plant Biol 1:83–116

Campo S, Baldrich P, Messeguer J, Lalanne E, Coca M, San Segundo B (2014) Overexpression of a calcium-dependent protein kinase confers salt and drought tolerance in rice by preventing membrane lipid peroxidation. Plant Physiol 165:688–704

Chen L, Song Y, Li S, Zhang L, Zou C (2012) The role of WRKY transcription factors in plant abiotic stresses. Biochim Biophys Acta 1819:120–128

Chowra U, Yanase E, Koyama H, Panda SK (2016) Aluminium-induced excessive ROS causes cellular damage and metabolic shifts in black gram *Vigna mungo* (L.) Hepper. Protoplasma 254:1–10

Deming-Adams B, Adams WW III (1996) The role of xanthophylls cycle carotenoids in the protection of photosynthesis. Trends Plant Sci 1:21–26

Dietz KJ (2014) Redox regulation of transcription factors in plant stress acclimation and development. Antioxid Redox Signal 21:1356–1372

Dietz KJ, Mittler R, Noctor G (2016) Recent progress in understanding the role of reactive oxygen species in plant cell signaling. Plant Physiol 171:1535–1539

Djanaguiraman M, Prasad PVV, Seppanen M (2010) Selenium protects sorghum leaves from oxidative damage under high temperature stress by enhancing antioxidant defense system. Plant Physiol Biochem 12:999–1007

Dore M, Mousavi-Baygi RI, Smith J, Hall D, Fowler TW, Choularton (2006) A model of annual orographic precipitation and acid deposition and its application to Snowdonia. Atmos Environ 18:3316–3326

Emmanouil N, Anagnostou (2004) A convective/stratiform precipitation classification algorithm for volume scanning weather radar observations. Meteorol Appl. Cambridge University Press 4:291–300

Farooq M, Wahid A, Kobayashi N, Fujita D, Basra SMA (2009) Plant drought stress: effects, mechanisms and management. In: Sustainable agriculture. Springer/EDP sciences/INRA, pp 185–212

Foyer CH, Noctor G (2005) Oxidant and antioxidant signalling in plants: a re-evaluation of the concept of oxidative stress in a physiological context. Plant Cell Environ 8:1056–1071

Gill SS, Tuteja N (2010) Reactive oxygen species and antioxidant machinery in abiotic stress tolerance in crop plants. Plant Physiol Biochem 48:909–930

Golldack D, Li C, Mohan H, Probst N (2014) Tolerance to drought and salt stress in plants: unraveling the signaling networks. Front Plant Sci 5:151. http://doi.org/10.3389/fpls.2014.00151

Guo Z, Ou W, Lu S, Zhong Q (2006) Differential responses of antioxidative system to chilling and drought in four rice cultivars differing in sensitivity. Plant Physiol Biochem 11:828–836

Harris D, Tripathi RS, Joshi A (2002) On-farm seed priming to improve crop establishment and yield in dry direct-seeded rice. In: Pandey S, Mortimer M, Wade L, Tuong TP, Lopes K, Hardy B (eds) Direct seeding: research strategies and opportunities. International Rice Research Institute, Manila, pp 231–240

Havaux M, Tardy F (1999) Loss of chlorophyll with limited reduction of photosynthesis as an adaptive response of Syrian barley landraces to high-light and heat stress. Funct Plant Biol 26:569–578

Huang B, Fu J (2000) Photosynthesis, respiration, and carbon allocation of two cool-season perennial grasses in response to surface soil drying. Plant Soil 227:17–26

Jiang M, Zhang J (2002) Water stress-induced abscisic acid accumulation triggers the increased generation of reactive oxygen species and up-regulates the activities of antioxidant enzymes in maize leaves. J Exp Bot 379:2401–2410

Johnson SM, Doherty SJ, Croy RRD (2003) Biphasic superoxide generation in potato tubers. A self-amplifying response to stress. Plant Physiol 13:1440–1449

Kaya MD, Okçub G, Ataka M, Çıkılıc Y, Kolsarıcıa Ö (2006) Seed treatments to overcome salt and drought stress during germinationin sunflower (*Helianthus annuus* L.) Eur J Agron 24:291–295

Khan MIR, Khan NA (2014) Ethylene reverses photosynthetic inhibition by nickel and zinc in mustard through changes in PS II activity, photosynthetic-nitrogen use efficiency and antioxidant metabolism. Protoplasma 251:1007–1019

Khan MIR, Asgher M, Khan NA (2014) Alleviation of salt-induced photosynthesis and growth inhibition by salicylic acid involves glycinebetaine and ethylene in mungbean (*Vigna radiata* L.) Plant Physiol Biochem 80:67–74

Khan MIR, Nazir F, Asgher M, Per TS, Khan NA (2015a) Selenium and sulfur influence ethylene formation and alleviate cadmium-induced oxidative stress by improving proline and glutathione production in wheat. J Plant Physiol 178:9–18

Khan MIR, Asgher M, Fatma M, Per TS, Khan NA (2015b) Drought stress vis a vis plant functions in the era of climate change. Clim Chang Environ Sustain 3:13–25

Khan MIR, Iqbal N, Masood A, Mobin M, Anjum NA, Khan NA (2016a) Modulation and significance of nitrogen and sulfur metabolism in cadmium challenged plants. Plant Growth Regul 78:1–11

Khan MIR, Khan NA, Masood A, Per TS, Asgher M (2016b) Hydrogen peroxide alleviates nickel-inhibited photosynthetic responses through increase in use-efficiency of nitrogen and sulfur, and glutathione production in mustard. Front Plant Sci 7:44

Lambers H, Chapin III FS, Pons TL (2008) Plant water relations. In: Plant physiological ecology. Springer, New York, pp 163–223

Levine A, Tenhaken R, Dixon R, Lamb C (1994) H_2O_2 from the oxidative burst orchestrates the plant hypersensitive disease resistance response. Cell 79:583–593

Loreto F, Tricoli D, Di Marco G (1995) On the relationship between electron transport rate and photosynthesis in leaves of the C4 plant *Sorghum bicolor* exposed to water stress, temperature changes and carbon metabolism inhibition. Aust J Plant Physiol 22:885–892

Lu W, Chu X, Li Y, Wang C, Guo X (2013) Cotton GhMKK1 induces the tolerance of salt and drought stress, and mediates defence responses to pathogen infection in transgenic *Nicotiana benthamiana*. PLoS One 8:e68503

Mattos LM, Moretti CL (2015) Oxidative stress in plants under drought conditions and the role of different enzymes. Enzyme Eng 5:2

Mittler R (2002) Oxidative stress, antioxidants and stress tolerance. Trends Plant Sci 7:405–410

Mittler R (2016) ROS are good. Trends Plant Sci 10:623–630

Mittler R, Vanderauwera S, Gollery M, Van Breusegem F (2004) Reactive oxygen gene network of plants. Trends Plant Sci 9:490–498

Mittler R, Vanderauwera S, Suzuki N, Miller G, Tognetti VB, Vandepoele K, Gollery M, Shulaev V, Van Breusegem F (2011) ROS signaling: the new wave? Trends Plant Sci 16:300–309

Mizoguchi T, Irie K, Hirayama T, Hayashida N, Yamaguchi-Shinozaki K, Matsumoto K, Shinozaki K (1996) A gene encoding a mitogen-activated protein kinase kinase kinase is induced simultaneously with genes for a mitogen-activated protein kinase and an S6 ribosomal protein kinase by touch, cold, and water stress in *Arabidopsis thaliana*. Proc Natl Acad Sci 93:765–769

Nayyar H, Gupta D (2006) Differential sensitivity of C 3 and C 4 plants to water deficit stress: association with oxidative stress and antioxidants. Environ Exp Bot 58:106–113

Pastori G, Mullineaux P, Foyer CH (2000) Post transcriptional regulation prevents accumulation of glutathione reductase protein and activity in the bundle sheath cells of maize. Implication on the sensitivity of maize to low temperatures. Plant Physiol 122:667–675

Rucker KS, Kvien CK, Holbrook CC, Hook JE (1995) Identification of peanut genotypes with improved drought avoidance traits. Peanut Sci 24:14–18

Shamim A, Rashid A, Muhammad YA, Ashraf M, Ejaz AW (2009) Sunflower (*Helianthus annuus* L.) response to drought stress at germination and seedling growth stages. Pak J Bot 41:647–654

Sharma P, Jha AB, Dubey RS, Pessarakli M (2012) Reactive oxygen species, oxidative damage, and antioxidative defense mechanism in plants under stressful conditions. J Bot 50:2–18

Shi B, Ni L, Zhang A, Cao J, Zhang H, Qin T, Tan M, Zhang J, Jiang M (2012) OsDMI3 is a novel component of abscisic acid signaling in the induction of antioxidant defense in leaves of rice. Mol Plant 5:1359–1374

Tezara W, Mitchell VJ, Driscoll SD, Lawlor DW (1999) Water stress inhibits plant photosynthesis by decreasing coupling factor and ATP. Nature 401:914–917

Thao NP, Khan MIR, Thu NBA, Hoang XLT, Asgher M, Khan NA, Tran L-SP (2015) Role of ethylene and its crosstalk with other signaling molecules in plant responses to heavy metal stress. Plant Physiol 169:73–84

Vu JCV, Gesch RW, Allen LH, Boote KJ, Bowes G (1999) CO2 enrichment delays a rapid, drought induced decrease in Rubisco small subunit transcript abundance. J Plant Physiol 155:139

Wahid A, Rasul E (2005) Photosynthesis in leaf, stem, flower and fruit. In: Pessarakli M (ed) Handbook of photosynthesis, 2nd edn. CRC Press, Florida, pp 479–497

Wang F, Liu J, Chen M, Zhou L, Li Z, Zhao Q, Pan G, Cheng F (2016) Involvement of abscisic acid in PSII photodamage and D1 protein turnover for light-induced premature senescence of rice flag leaves. PLoS One 11:e0161203

Wery J, Silim SN, Knights EJ, Malhotra RS, Cousin R (1994) Screening techniques and sources and tolerance to extremes of moisture and air temperature in cool season food legumes. Euphytica 73:73–83

Xiong L, Yang Y (2003) Disease resistance and abiotic stress tolerance in rice are inversely modulated by an abscisic acid–inducible mitogen-activated protein kinase. Plant Cell 15:745–759

Xu Y, Burgess P, Zhang X, Huang B (2016) Enhancing cytokinin synthesis by overexpressing ipt alleviated drought inhibition of root growth through activating ROS-scavenging systems in *Agrostis stolonifera*. J Exp Bot 3:756–769

Yan J, Tsuichihara N, Etoh T, Iwai S (2007) Reactive oxygen species and nitric oxide are involved in ABA inhibition of stomatal opening. Plant Cell Environ 30:1320–1325

Zhang L, Li Y, Lu W, Meng F, Wu C-A, Guo X (2012) Cotton GhMKK5 affects disease resistance, induces HR-like cell death, and reduces the tolerance to salt and drought stress in transgenic Nicotiana benthamiana. J Exp Bot 63:3935–3951

Zhou J, Wang J, Li X, Xia X-J, Zhou Y-H, Shi K, Chen Z, Yu J-Q (2014) H2O2 mediates the crosstalk of brassinosteroid and abscisic acid in tomato responses to heat and oxidative stresses. J Exp Bot 65:4371–4383

Zong X-J, Li D-P, Gu L-K, Li D-Q, Liu L-X, Hu X-L (2009) Abscisic acid and hydrogen peroxide induce a novel maize group C MAP kinase gene, ZmMPK7, which is responsible for the removal of reactive oxygen species. Planta 229:485–495

12 Contribution of Glutathione in Heavy Metal Stress Tolerance in Plants

Mohd Asgher, Tasir S. Per, Shagufta Anjum, M. Iqbal R. Khan, Asim Masood, Susheel Verma, and Nafees A. Khan

Abstract

Plants are exposed to myriad of abiotic stresses such as salt stress, heavy metal, ozone, UV-B radiation, low and high temperatures, drought, and many more which are posing threats to agricultural soil. Among these abiotic stresses, the heavy metal pollutants, added by anthropogenic activities, had enormous impact on different ecosystems including plants. The increment of toxic heavy metals in the atmosphere is a serious problem to living organisms. Heavy metal induces excess generation of reactive oxygen species (ROS) which in turn is responsible for oxidative stress and impairs processes at physiological, biochemical, and molecular level resulting in loss of crop productivity. Plants possess well-organized antioxidant defense system to mitigate oxidative stress. The plant antioxidant system constitutes various antioxidant enzymes and metabolites. Reduced glutathione (GSH) is an important constituent of ascorbate (AsA)-GSH (AsA-GSH). Glutathione consists of Glu-Cys-Gly with GSH-related enzymes such as GSH reductase (GR), GSH peroxidase (GPX), and GSH

M. Asgher (✉) • S. Anjum • S. Verma
Centre for Biodiversity Studies, School of Biosciences and Biotechnology, Baba Ghulam Shah Badshah University, Rajouri 185234, India
e-mail: asghermohd@gmail.com

T.S. Per • A. Masood • N.A. Khan (✉)
Plant Physiology Section, Department of Botany, Aligarh Muslim University, Aligarh 202002, India
e-mail: naf9@lycos.com

M.I.R. Khan
Plant Physiology Section, Department of Botany, Aligarh Muslim University, Aligarh 202002, India

Crop and Environmental Sciences Division, International Rice Research Institute, Los Banos, DAPO Box 7777, Philippines

M.I.R. Khan, N.A. Khan (eds.), *Reactive Oxygen Species and Antioxidant Systems in Plants: Role and Regulation under Abiotic Stress*,
DOI 10.1007/978-981-10-5254-5_12

sulfotransferases (GSTs), involved in controlling the ROS concentrations in cell by protecting plants against oxidative stress induced by heavy metals. This chapter highlights (a) responses of plants to heavy metals and its consequences on photosynthesis and growth, (b) heavy metal stress detoxification and antioxidants, and (c) importance of glutathione and its biosynthesis, transport, and signaling under optimal and metal stress condition.

Keywords
Antioxidants • Glutathione • Metal stress • Tolerance

Abbreviations

Al	Aluminum
APX	Ascorbate peroxidase
As	Arsenic
AsA	Ascorbate
BSO	Buthionine sulfoximine
Ca	Calcium
CAT	Catalase
Cd	Cadmium
Cr	Chromium
Cu	Copper
Cys	Cysteine
Fe	Iron
GPX	Glutathione peroxidase
GR	Glutathione reductase
GSH	Reduced glutathione
GSSG	Oxidized glutathione
GSTs	Glutathione S-transferases
HMs	Heavy metals
K	Potassium
MAPK	Mitogen-activated protein kinase
Ni	Nickel
NO	Nitric oxide
Pb	Lead
PCs	Phytochelatins
PSII	Photosystem II
ROS	Reactive oxygen species
SNO	S-nitrosothiol
SOD	Superoxide dismutase
Zn	Zinc

12.1 Introduction

Rapid industrialization and urbanization significantly increased addition of unrestricted heavy metals (HMs) in agricultural soils that leads to a reduction in crop productivity and economy (Tran and Popova 2013; Gallego et al. 2012; Choppala et al. 2014; Thao et al. 2015; Kao 2015; Asgher et al. 2015a; Rizwan et al. 2016; Per et al. 2016). Due to the long half-life time and nondegradable nature, metals are critically toxic and are considered as a serious threat to the agricultural and environmental systems. Heavy metals caused crop yield loss worldwide having wider availability and high toxicity responsible for limiting plant metabolism and productivity (Nazar et al. 2012; Gallego et al. 2012; Choppala et al. 2014; Khan and Khan 2014; Khan et al. 2015; Kao 2015). As a primary response, HM induced production of reactive oxygen species (ROS) (Gill and Tuteja 2010; Sytar et al. 2013; Anjum et al. 2014; Khan and Khan 2014; Rizwan et al. 2016). In response to excessive ROS production, plants showed various strategies to overcome HM stress including antioxidants to mitigate ROS (Liu et al. 2008a, b; Ammar et al. 2008; Ali et al. 2011; Gill et al. 2011; Luo et al. 2011; Masood et al. 2012; Asgher et al. 2014; Khan and Khan 2014; Per et al. 2016).

A well-organized antioxidant network present in plants has a potential to mitigate the ill effects induced by oxidative stress. Enzymatic antioxidants are superoxide dismutase (SOD), catalase (CAT), ascorbate peroxidase (APX), and glutathione reductase (GR). Nonenzymatic antioxidants include ascorbate (AsA) and reduced glutathione (GSH) and constitute (AsA-GSH) a pathway that helps in reducing the level of ROS so as to maintain proper cellular homeostasis on exposure to HM stress (Noctor and Foyer 1998; Mahmood et al. 2010; Sofo et al. 2010; Anjum et al. 2012).

Glutathione have been assigned numerous roles in plants (Mullineaux and Rausch 2005; Noctor et al. 2011, 2012; Jozefczak et al. 2012; Diaz-Vivancos et al. 2015). Glutathione leads to the synthesis of metal-binding peptides, phytochelatins (PCs), and involved in heavy metal tolerance (Yadav 2010; Zhang et al. 2010; Anjum et al. 2012; Mohamed et al. 2012). Glutathione is a promising candidate for ROS detoxification including HMs such as, Cd, nickel (Ni), zinc (Zn), As and copper (Cu) (Wang et al. 2011a; Noctor et al. 2011; Anjum et al. 2012; Nakamura et al. 2013; Asgher et al. 2014; Khan and Khan 2014; Talukdar and Talukdar 2014; Asgher et al. 2015b; Khan et al. 2016). Similarly, mRNA level of genes participates in GSH synthesis enhanced in *Arabidopsis* under metal stress (Semane et al. 2007). GSH, a component of AsA-GSH cycle, removes excess H_2O_2 from the cell (Noctor and Foyer 1998); the ratio of GSH/GSSG can be maintained by GR under optimal growth conditions (Szalai et al. 2009). The present chapter highlights the involvement of GSH in HM stress tolerance in plants, detoxification mechanisms induced under HMs, and GSH participation in enhanced HM stress tolerance by maintaining redox homoeostasis. Glutathione overcomes the HM stresses by binding the HMs and keeping the cell free from metal toxicity (Table 12.1).

Table 12.1 A summary of GSH responses in different plants exposed to various heavy metals

S. no.	Heavy metals	Concentration	Response	Plant	References
1.	Ni	50,100, and 200 mg Ni kg^{-1} soil	+	*Brassica juncea*	Khan et al. (2016)
2.	Ni	200 mg Ni/Zn kg^{-1} soil	+	*Brassica juncea*	Khan and Khan (2014)
3.	Cd	50 μM	+	*Brassica juncea*	Per et al. (2016), Khan et al. (2016) and Asgher et al. (2014)
4.	Cd	200 mg Cd kg^{-1} soil	+	*Triticum aestivum*, *Brassica juncea*	Khan et al. (2015) and Masood et al. (2012)
5.	Cd	5 mM	–	*Oryza sativa*	Hsu and Kao (2004)
6.	Cd	50 μM	+	*Phragmites australis*	Pietrini et al. (2003)
7.	Cd	100 μM	+	*Lycium chinense*	Guan et al. (2015)
8.	As	25 μM	–	*Oryza sativa*	Kumar et al. (2014)
9.	As	0, 100, and 500 μM	+	*Pteris vittata*	Sakai et al. (2010)
10.	As	25 μM, 50 μM	+	*Oryza sativa*	Dixit et al. (2016)
11.	Zn	200 mg Zn kg^{-1} soil	+	*Brassica juncea*	Khan and Khan (2014)
12.	Zn	0.5 mM Zn	+	*Brassica oleracea*	Barrameda-Medina et al. (2014)
13.	Zn	50 μM	+	*Phaseolus vulgaris*	Cuypers et al. (2001)
14.	Cu	8 mM	–	*Zea mays*	Wang et al. (2011a)
15.	Cu	1, 10, 100, 1000 μM	–	*Silene cucubalus*	De Vos et al. (1992)
16.	Cu	100 μ M	+	*Oryza sativa*	Mostofa et al. (2015)
17.	Hg	40 μM	+	*Medicago sativa*	Zhou et al. (2008)
18.	Cr	100 μM	–	*Oryza sativa*	Qiu et al. (2013)
19.	Al	160 μM	+	*Oryza sativa*	Sharma and Dubey (2007)
20.	Al	100 μM	+	*Helianthus annuus*	Ruiz et al. (2006)
21.	Pb	100 and 500 mg L^{-1} Pb	+	*Iris lactea*	Yuan et al. (2015)

(+) and (–) signs indicate increase and decrease, respectively

12.2 Responses of Plants to Heavy Metals and Its Consequences on Photosynthesis and Growth

Heavy metal addition to agricultural soil has increased considerably due to the increasing anthropogenic activities. Heavy metals such as Cu, Zn, iron (Fe), manganese, molybdenum, and Ni are essential micronutrients and perform physiological and biochemical role. Higher levels of HMs such as Cd, Al, chromium (Cr), Pb, and As cause toxicities. The increase in higher range of HMs leads to growth inhibition by inducing toxicity symptoms in plants. Mechanisms of enzyme inhibition under HM stress involve binding of HMs to sulfhydryl groups (-SH) and lack of essential metal in the metal protein complex (Assche and Clijsters 1990). Lin et al. (2008) showed that As damaged DNA in *Vicia faba* plants' leaves and root tips. Cadmium damages DNA and replaces Zn ion in the Zn fingers and consequently interferes with the transcriptional mechanism (Sarkar 1995). Ghasemi et al. (2012) in *Zea mays* showed that excess Ni damages photosynthetic protein complexes, and the rate of Hill reaction decreased by increasing Ni concentration. Heavy metals trigger the formation of ROS such as singlet oxygen and hydroxyl radical, leading to injuries to macromolecules such as DNA, RNA, lipids, carbohydrates, and proteins (Lombardi and Sebastiani 2005; Yurekli and Porgali 2006; Emamverdian et al. 2015).

Heavy metal accumulation inhibits photosynthesis and growth (Ventrella et al. 2011; Per et al. 2016; Khan et al. 2016; Sheetal et al. 2016). They disrupt photosynthetic processes and resulted in numerous toxic effects, including photooxidative damage, and alter chloroplast structure and photosystem II (PSII) activity (Mys'liwa-Kurdziel et al. 2002; Ventrella et al. 2011; Khan and Khan 2014; Khan et al. 2016). Plant accumulates HMs from soil, and high level of HMs disrupts the metabolic process such as photosynthesis and growth (Khan et al. 2016). Numerous studies have demonstrated that PSII is the main target of HM stress (Patsikka et al. 2002; Sarvari 2005; Khan et al. 2015), and this inhibition is attributed due to dissociation of the oxygen-evolving complex proteins and to the displacement of the cofactors such as calcium (Ca), chloride, and manganese necessary for water splitting (Sersen et al. 1997; Patsikka et al. 2002), reduced transpiration, and photosynthesis resulting in inhibition of plant growth and alters seed germination, nutrient concentration in roots and leaves, and grain yield (Ashraf et al. 2016).

Heavy metals disrupt photosynthetic processes and protein associated with it and removal of important constituents such as magnesium and change chlorophyll role and structure (Kupper et al. 2003). Heavy metals such as Cd in *Brassica napus*, *Brassica juncea*, and *Gossypium hirsutum* (Wan et al. 2011; Masood et al. 2012; Asgher et al. 2014; Liu et al. 2014), Ni in *Brassica juncea* (Khan and Khan 2014), Zn in *Brassica juncea* (Khan and Khan 2014), As in *Myracrodruon urundeuva* (Gomes et al. 2014), and Cu in *Scenedesmus quadricauda* (Kupper et al. 2003) inhibit photosynthesis. Aluminum caused reduction of stomatal opening and photosynthesis (Vitorello et al. 2005), and high concentrations of Al inhibit photosynthesis (Ribeiro et al. 2013). The photosynthetic energy storage on exposure to mercury and Cu was inhibited; on the other hand, energy storage was insensitive to Pb; and these HMs inhibit PSII (Boucher and Carpentier 1999). The level of photosynthetic pigments and soluble proteins reduced only upon exposure to high Zn concentrations (Parlak and Yilmaz 2012). Higher Zn concentration resulted in significant reduction in the chlorophyll under all applied Zn concentrations (Li et al. 2013). Chromium toxicity caused ultrastructural changes in chloroplasts and reduced photosynthesis (Ali et al. 2013). Higher Pb levels also disrupt photosynthetic rate and affect water-use efficiency (Ahmad et al. 2011). Increasing concentration of Cd led to decreased growth, photosynthesis, and yield (Gill et al. 2011).

Heavy metal alters the mineral nutritional status of plants resulting in detrimental effect on photosynthetic, growth, and yield parameters. Aluminum exposure in *Theobahia* reduces mineral nutrient content except potassium which showed an increase (Ribeiro et al. 2013). Cadmium reduced Ca in *Brassica napus* (Wan et al. 2011). Chromium decreased the uptake of manganese in *Brassica oleracea* (Chatterjee and Chatterjee 2000) and in *Amaranthus viridis* (Liu et al. 2008a, b). Increase in the concentration of Pb leads to a decrease in the level of K and Cu in different parts of root, stem, and leaves (Ahmad et al. 2011). Lead exposure reduced the uptake of mineral ions in *Triticum aestivum* and *Spinacia oleracea* (Lamhamdi et al. 2013). Increasing Cd concentrations also interferes with mineral nutrient uptake which caused a reduction in growth attributes (Mohamed et al. 2012).

Cadmium inhibits growth in *Brassica juncea* and *Gossypium hirsutum* (Liu et al. 2014; Khan et al. 2016), As in *Myracrodruon urundeuva* (Gomes et al. 2014), and Ni in *Brassica juncea* (Khan and Khan 2014). *Vicia faba* under arsenate supplementation showed inhibition in shoot/root growth (Lin et al. 2008). Similar decrease in total dry mass was noted with increasing Al concentrations in growth medium in *Theobahia* (Ribeiro et al. 2013). Mercury concentrations in the nutrient solution reduced biomass in *Oryza sativa* (Du et al. 2005). Lead accumulation results in reduced growth in *Triticum aestivum* and *Spinacia oleracea* (Lamhamdi et al. 2013).

12.3 Heavy Metal Stress Detoxification and Antioxidants

Studies indicated the key role of antioxidant system that are employ the role of protection when plants are exposed to numerous HMs stress (Wang et al. 2011a; Talukdar and Talukdar 2014; Kao 2015). Plants develop various strategies to overcome these stresses by the production of various antioxidant systems (Harada et al. 2002; Gill and Tuteja 2010; Noctor et al. 2011; Masood et al. 2012; Asgher et al. 2014) which have a potential to alleviate harmful effects of stress. Lee et al. (2007) reported that genes of antioxidant system Cu/Zn-SOD and APX in *Festuca arundinacea* are tolerant to herbicide and metal stress.

Glutathione reductase has a potential to increase the GSH/GSSG ratio in plants, and this ratio decreased under stress conditions (Kao 2015). Glutathione reductase activity in *Brassica napus* leaves increased at low concentration of Cd; on the other hand, higher concentration of Cd decreased the activity of GR; while studying *Brassica juncea* leaves, they found that by changing the Cd concentration, GR activity remains unchanged (Nouairi et al. 2009). Gill et al. (2011) found that Cd causes a significant rise in the level of antioxidants in *Brassica juncea*. Higher activity of these antioxidants maintained a redox environment in the cell that might protect plants from harmful effects of metal stress, and plants behave normally (Gill and Tuteja 2010; Anjum et al. 2012). Xu et al. (2008) studied that transgenic

Arabidopsis having overexpressed antioxidant gene in *Hordeum vulgare* were tolerant to Cd stress. Lou et al. (2011) studied in *Lolium perenne* that HM stress induced expression of antioxidant genes. Glutathione S-transferases (GSTs) are ubiquitous enzymes in plants, and they are multifunctional proteins encoded by a large gene family. Overexpressing *PpGST* gene enhanced tolerance in tobacco exposed to HMs (Liu et al. 2013). Glutathione peroxidase (GPX), APX, and GST help in the reduction of H_2O_2 which prevents cellular damage (Anjum et al. 2012; Ozyigit et al. 2016).

12.4 Glutathione: A Non-enzymatic Antioxidant Biosynthesis, Transport, and Signaling Under Optimal and Metal Stress Conditions

Glutathione, a tripeptide antioxidant, is a nonprotein thiol found in plant tissues and plays an important role in abiotic stress such as HMs (Kao 2015) and at even low concentrations acts as redox buffer and is confined to plastids and cytosol (Wachter et al. 2005; Pasternak et al. 2008).

In the first step of GSH synthesis, cysteine (Cys) and glutamate (Glu) form γ-Glu-Cys, a reaction catalyzed by γ-glutamylcysteine synthetase (γ-ECS, GSH1). Further in the second step, glycine combines with Glu and Cys to form GSH, and the enzyme involved is GSH synthetase (Fig. 12.1). Research publications are available on the structural, functional, and regulatory role of GSH (Gill et al. 2013; Zechmann 2014). It was revealed that GSH is found in all important organelles such as cytosol, mitochondria, nucleus, and cytosol; and high GSH initiates the adjustment of redox in the cytosol responsible for a decreased signaling of oxidative stress with increased GSH (Diaz Vivancos et al. 2010, 2015).

In *Vicia faba* GSH and GSSG uptake studied in protoplasts and leaf tissues, GSSG predominates than GSH as far as uptake is concerned in this plant (Jamai et al. 1996). Accumulation or synthesis of GSH totally depends on the cytosolic GSH pool present in cells. Pasternak et al. (2008) reported in *Arabidopsis* that mutants which lack GSH synthetase encoding *GSH2* gene have no GSH and are not viable; cytosolic synthesis of GSH is important for plant growth. The regulation of *BjGT1*in *Brassica juncea* by Cd suggests a role of this GSH transporter on Cd exposure; these findings show that *BjGT1* mediates GSH uptake (Bogs et al. 2003). *OsGT1* transporter in *Oryza sativa* and *BjGT1* in *Brassica juncea BjGT1* are the transporter of GSH derivatives (Cagnac et al. 2004). GSH transporter genes *ZmGST27*, *ZmGT1*, and *ZmMRP1* encoding GST in *Zea mays* are key in the xenobiotic detoxification in plants (Pang et al. 2012). Xiang and Oliver (1998) conducted a research on *Arabidopsis* and conclude that supplementation of metals such as Cd and Cu enhanced genes of GSH synthesis as well as enzymes particularly GR. GSH involved in phytochelatins maintained proper homeostasis in cell and initiates signaling in plants (Srivalli and Khanna-Chopra 2008; Anjum et al. 2012). GSH provides tolerance to Fe deficiency and is also involved in nitric oxide (NO)-mediated signaling due to deficiency of Fe (Shanmugam et al. 2015).

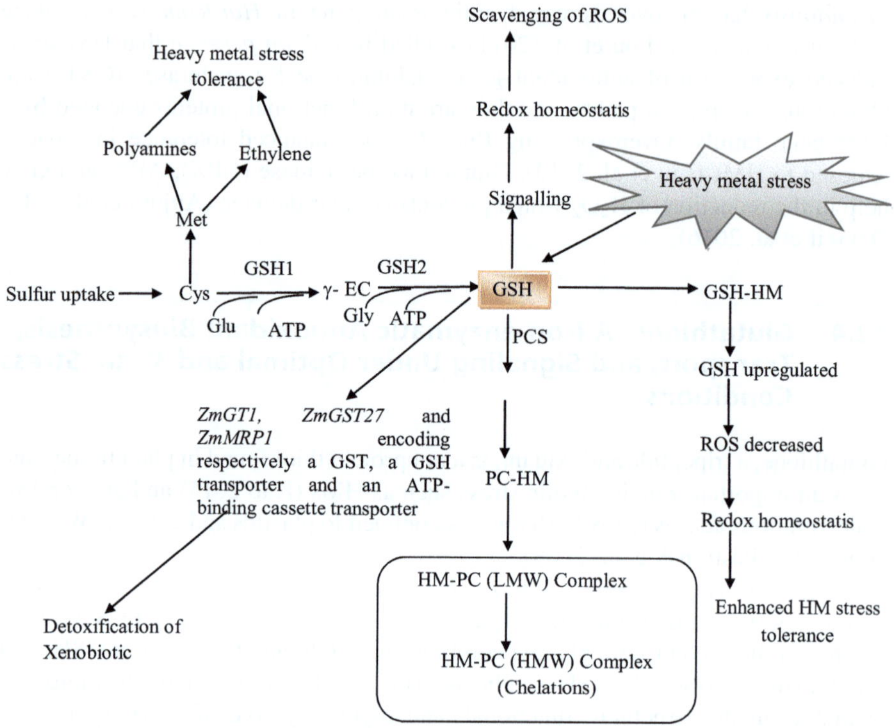

Fig. 12.1 An overview of GSH biosynthesis, signaling, detoxification of heavy metals/ xenobiotics, role of transporters, and scavenging of ROS in plants and heavy metal stress tolerance. *Cys* cysteine, *γ-EC* Glu-Cys synthetase, *GSH* reduced glutathione, *GSHS* glutathione synthetase, *GSTs* glutathione S-transferases, *HMW* high molecular weight, *LMW* low molecular weight, *Glu* glutamate, *Gly* glycine, *Met* methionine, *PCS* phytochelatin synthase, *PCs* phytochelatins

Transport of GSH into nucleus involved in nuclear GSH cycle helps in maintaining cell cycle (Vivancos et al. 2010; Zagorchev et al. 2013). GSH participates in imparting tolerance against numerous HMs by overexpressing *SAT* gene (Freeman and Salt 2007). Changing redox ratio are crucial for the signaling related to regulation of cell cycle other processes in cell when exposed to oxidative stress (Schafer and Buettner 2001). Glutathionylation, a modification of proteins, is involved in preventing Cys groups from oxidation or oxidative signaling (Foyer and Noctor 2005).

Glutathione induces ethylene biosynthetic enzymes by inducing ACS and ACO gene; GSH enhances tolerance in wild-type *Arabidopsis* plant but did not in ethylene signaling mutant showing a GSH-mediated resistance against stress through ethylene (Datta et al. 2015). GSH triggers the biosynthesis of phytohormones and other genes associated with signaling molecules revealed by translatome analysis (Cheng et al. 2015). GSH plays a key role in stress tolerance (Szalai et al. 2009) and maintains a signaling of NO pathways by retaining a redox state in plants. Report indicates that increased level of serine acetyl transferase in *Thlaspi* elevates GSH level and conferred Ni stress tolerance (Freeman et al. 2004).

12.5 Importance of Glutathione Under Heavy Metal Stress

Heavy metal toxicity induces ROS when its level increases the threshold value, imparting a severe injury to plants (Jozefczak et al. 2012). Increase of GSH was involved in enhancing Ni tolerance in *Thlaspi*, a hyperaccumulator of this plant (Freeman et al. 2004). An overview of biosynthesis of GSH, signaling, and detoxification of HMs and stress tolerance in plants has been depicted in Fig. 12.1. In Zn/Cd hyperaccumulator *Arabis paniculata*, the potential role of GSH and PCs in Cd uptake and complexation has also been reported (Zeng et al. 2009). Recent studies in plants show that GSH and its metabolizing enzymes such as GSTs, GPX, and other enzymes protect against oxidative stress (Hossain et al. 2012). Transgenic *Arabidopsis* plants with low GSH level show hypersensitivity to Cd stress due to the limited capacity of these plants to induce the formation of PCs (Xiang et al. 2001). Detoxification process involves complexation of metal with GSH, formation of PCs, and vacuolar sequestration. Stresses trigger a series of mitogen-activated protein kinase (MAPK) pathways of metal-detoxifying ligands (Sytar et al. 2013). Sulfate assimilatory enzymes and GSH biosynthetic enzymes were triggered by Cd (Asgher et al. 2014, 2015b; Khan et al. 2016); sulfur supplementation increases enzyme of sulfur assimilation, and GSH might reduce oxidative stress governed by Cd (Asgher et al. 2014, 2015b; Khan et al. 2015, 2016). Pietrini et al. (2003) conducted a research on *Phragmites australis* and concluded that GSH protects important photosynthetic enzymes under Cd exposure. Glutathione synthetase enzyme is involved in the biosynthesis of GSH; overexpressing this enzyme in plants is a better option for enhancing tolerance to Cd stress (Zhu et al. 1999). Glutathione upregulates the activities of antioxidant enzymes in *Lens culinaris* and conferred tolerance by scavenging excess ROS (Talukdar and Talukdar 2014). Glutathione synthesis in plants are key thiol which make a relief to on exposure to numerous HMs such as Cd, As and Hg (Hernandez et al. 2015). Khan et al. (2016) studied in *Brassica juncea* cultivars that higher GSH production helps in reversing the Ni stress more efficiently and protects photosynthesis, maintaining high plant dry mass. Loscos et al. (2008) studied that hGSH synthetase transcript levels and its activity did not show any enhancement in nodules of root on exposure to metal treatment; mRNA activity remains unchanged. Manipulation of GSH and/or PC content is involved in detoxifying the metal stress in plant cells (Noctor et al. 2011; Iqbal et al. 2012). The synthesis of PCs via GSH participates in reducing ROS by chelation. PCs form low and high molecular weight complex in vacuoles by combining with sulfide and bind maximum HM in vacuole and reduce its concentration in cytosol and ensure the cell is free from HM stress (Iqbal et al. 2012; Fig. 12.1). Plants such as *Arabidopsis* mutants indicate a GR inhibition than wild type on exposure to Hg, with depleted GSH level, and GSH/GSSG ratio remains unchanged (Sobrino-Plata et al. 2014).It has been shown that GSH inhibition by its inhibitors makes worse effect on plants by increasing accumulation of ROS and reducing contents of NO and S-nitrosothiol (SNO) in tomato roots. Additionally, exogenous GSH alleviated Cd toxicity by minimizing ROS accumulation and increasing contents of NO and SNO, and antioxidants that eventually reduced stress, and modulated the cellular redox balance through maintaining increased GSH:GSSG and AsA:DHA ratio (Hasan et al. 2016). More interestingly, it was found that application of NaCl alleviates Cd-induced growth inhibition in the halophyte *Sesuvium portulacastrum* by modulating GSH content resulting in protecting redox balance and photosynthesis (Wali et al. 2016).

Exogenous GSH supplementation increased photosynthetic attributes and growth and uptake of mineral nutrient and decreased Cd uptake and ultimately increased yield (Cai et al. 2010; Cao et al. 2015). Cadmium and Cu inhibit growth and chlorophyll content, and growth was promoted by the application of GSH (Son et al. 2014). Sodium nitroprusside and GSH reduced the Cu acquisition in *Oryza sativa* via inducing antioxidants enzyme and glyoxalase systems (Mostofo et al. 2014). Chromium toxicity was overcome in *Oryza sativa* by improving mineral nutrient uptake and growth and by reducing the damage to cell structure (Qiu et al. 2013). Exposure of *Iris lactea* to 150 mg L^{-1} GSH and Cys minimizes the ill effect of Pb and shows an increment in tolerance (Yuan et al. 2015). Glutathione addition reduced oxidative stress induced by Cd by minimizing Cd accumulation in plant parts maintained proper structure of thyakoid membranes and overall chloroplasts (Wang et al. 2011b). Cotton seedlings' (28 days old) growth was resorted under Pb toxicity (500 μM) when seedlings were treated with 50 μM GSH. GSH also maintained photosynthetic features along with reduced MDA, H_2O_2, and Pb contents in cotton leaves. Exogenously sourced GSH improved antioxidant enzyme activity (CAT, SOD, POD, GR, and APX) under Pb toxicity (Khan et al. 2016).

12.6 Conclusion and Future Prospects

Glutathione performs the scavenging of ROS, providing tolerance to HM-exposed plants. Work on GSH role in enhancing stress tolerance is available in literature; however, literature regarding the mechanism of GSH regulation with HMs is meager. Introducing the GSH gene in plants through genetic engineering is a feasible strategy to overcome HM stress in plants. Research/reports on HMs induced influence on GSH synthesis; its redox and related enzymes of its biosynthesis are lacking. The present review literatures cover the role of GSH as an important redox buffer to maintain a reducing environment in the cell. Despite a lot of studies on thiol biosynthesis, proper maintenance of the redox state and the gap between biochemical and molecular pathways are not studied. Engineering the GSH pathway enzymes against stress is a better strategy to overcome the stress. The present studies overview a role of GSH as a redox buffer to keep the environment reducing. These reports would be important for imparting tolerance by enhancing the tolerance and minimizing the environmental effects of HMs. The production of thiols and sulfur metabolism need special attention in the near future for the improvement of agricultural productivity under stressful environment. Transgenic research will improve stress tolerance by incorporating/engineering one of the thiols or its enzymes to combat ROS accumulation.

Acknowledgments First author is thankful to SERB, Department of Science and Technology, New Delhi, India, in the form of National Post-Doctoral Fellowship (N-PDF; PDF/2016/000591).

References

Ahmad MSA, Ashraf M, Tabassam Q et al (2011) Lead (Pb)-induced regulation of growth, photosynthesis, and mineral nutrition in maize (*Zea mays* L.) plants at early growth stages. Biol Trace Elem Res 144:1229–1239

Ali S, Bai P, Zeng F et al (2011) The ecotoxicological and interactive effects of chromium and aluminum on growth, oxidative damage and antioxidant enzymes on two barley genotypes differing in Al tolerance. Environ Exp Bot 70:185–191

Ali S, Farooq MA, Hussain S et al (2013) Alleviation of chromium toxicity by hydrogen sulfide in barley. Environ Toxicol Chem 32:2234–2239

Ammar WB, Nouairi I, Zarrouk M et al (2008) Antioxidative response to cadmium in roots and leaves of tomato plants. Biol Plant 52:727–731

Anjum NA, Ahmad I, Mohmood I et al (2012) Modulation of glutathione and its related enzymes in plants' responses to toxic metals and metalloids – a review. Environ Exp Bot 75:307–324

Anjum NA, Aref IM, Duarte AC et al (2014) Glutathione and proline can coordinately make plants withstand the joint attack of metal (loid) and salinity stresses. Front Plant Sci 5. doi:10.3389/fpls.2014.00662

Asgher M, Khan NA, Khan MIR et al (2014) Ethylene production is associated with alleviation of cadmium-induced oxidative stress by sulfur in mustard types differing in ethylene sensitivity. Ecotoxicol Environ Saf 106:54–61

Asgher M, Khan MIR, Anjum NA et al (2015a) Minimising toxicity of cadmium in plants – role of plant growth regulators. Protoplasma 252:399–413

Asgher M, Khan MIR, Fatma M et al (2015b) Potentiality of ethylene in sulfur-mediated counteracting adverse effects of cadmium in plants. In: Chakraborty U, Chakraborty BN (eds) Abiotic stresses in crop plants. CAB International, UK, pp 136–163

Ashraf MY, Roohi M, Iqbal Z et al (2016) Cadmium (Cd) and lead (Pb) induced changes in growth, some biochemical attributes and mineral accumulation in two cultivars of mung bean (*Vigna radiata* L.) Wilczek. Commun Soil Sci Plant Anal. doi:10.1080/00103624.2015.1118117

Assche FV, Clijsters H (1990) Effects of metals on enzyme activity in plants. Plant Cell Environ 13:195–206

Barrameda-Medina Y, Montesinos-Pereira D, Romero L et al (2014) Role of GSH homeostasis under Zn toxicity in plants with different Zn tolerance. Plant Sci 227:110–121

Bogs J, Bourbouloux A, Cagnac O et al (2003) Functional characterization and expression analysis of a glutathione transporter, BjGT1, from *Brassica juncea*: evidence for regulation by heavy metal exposure. Plant Cell Environ 26:1703–1711

Boucher N, Carpentier R (1999) Hg^{2+}, Cu^{2+}, and Pb^{2+}-induced changes in photosystem II photochemical yield and energy storage in isolated thylakoid membranes: a study using simultaneous fluorescence and photoacoustic measurements. Photosynth Res 59:167–174

Cagnac O, Bourbouloux A, Chakrabarty D et al (2004) AtOPT6 transports glutathione derivatives and is induced by primisulfuron. Plant Physiol 135:1378–1138

Cai Y, Lin L, Cheng W et al (2010) Genotypic dependent effect of exogenous glutathione on Cd-induced changes in cadmium and mineral uptake and accumulation in rice seedlings (*Oryza sativa*). Plant Soil Environ 56:524–533

Cao F, Cai Y, Liu L et al (2015) Differences in photosynthesis, yield and grain cadmium accumulation as affected by exogenous cadmium and glutathione in the two rice genotypes. Plant Growth Regul 75:715–723

Chatterjee J, Chatterjee C (2000) Phytotoxicity of cobalt, chromium and copper in cauliflower. Environ Pollut 109:69–74

Cheng MC, Ko K, Chang WL et al (2015) Increased glutathione contributes to stress tolerance and global translational changes in Arabidopsis. Plant J 83:926–939

Choppala G, Saifullah BN et al (2014) Cellular mechanisms in higher plants governing tolerance to cadmium toxicity. Crit Rev Plant Sci 33:374–391

Cuypers A, Vangronsveld J, Clijsters H (2001) The redox status of plant cells (AsA and GSH) is sensitive to zinc imposed oxidative stress in roots and primary leaves of *Phaseolus vulgaris*. Plant Physiol Biochem 39:657–664

Datta R, Kumar D, Sultana A et al (2015) Glutathione regulates 1-aminocyclopropane-1-carboxylate synthase transcription via WRKY33 and 1 aminocyclopropane-1-carboxylate oxidase by modulating messenger RNA stability to induce ethylene synthesis during stress. Plant Physiol 169:2963–2981

De Vos CHR, Vonk MJ, Vooijs R et al (1992) Glutathione depletion due to copper-induced phytochelatin synthesis causes oxidative stress in *Silene cucubalus*. Plant Physiol 98:853–858

Diaz Vivancos P, Dong Y, Ziegler K et al (2010) Recruitment of glutathione into the nucleus during cell proliferation adjusts whole-cell redox homeostasis in Arabidopsis thaliana and lowers the oxidative defence shield. Plant J 64:825–838

Diaz-Vivancos P, Simone AD, Kiddle G et al (2015) Glutathione–linking cell proliferation to oxidative stress. Free Radic Biol Med 89:1154–1164

Dixit G, Singh AP, Kumar A et al (2016) Reduced arsenic accumulation in rice (*Oryza sativa* L.) shoot involves sulfur mediated improved thiol metabolism, antioxidant system and altered arsenic transporters. Plant Physiol Biochem 99:86–96

Du X, Zhu YG, Liu WJ et al (2005) Uptake of mercury (Hg) by seedlings of rice (*Oryza sativa* L.) grown in solution culture and interactions with arsenate uptake. Environ Exp Bot 54:1–7

Emamverdian A, Ding Y, Mokhberdoran F et al (2015) Heavy metal stress and some mechanisms of plant defense response. Sci World J. org/10.1155/2015/756120

Foyer CH, Noctor G (2005) Redox homeostasis and antioxidant signaling: a metabolic interface between stress perception and physiological responses. Plant Cell 17:1866–1875

Freeman JL, Salt DE (2007) The metal tolerance profile of *Thlaspi goesingense* is mimicked in *Arabidopsis thaliana* heterologously expressing serine acetyl-transferase. BMC Plant Biol 7:63

Freeman JL, Persans MW, Nieman K et al (2004) Increased glutathione biosynthesis plays a role in nickel tolerance in *Thlaspi* nickel hyperaccumulators. Plant Cell 16:2176–2191

Gallego SM, Pena LB, Barcia RA et al (2012) Unravelling cadmium toxicity and tolerance in plants: insight into regulatory mechanisms. Environ Exp Bot 83:33–46

Ghasemi F, Heidari R, Jameii R et al (2012) Effects of Ni^{2+} toxicity on Hill reaction and membrane functionality in maize. J Stress Physiol Biochem 8:55–61

Gill SS, Tuteja N (2010) Reactive oxygen species and antioxidant machinery in abiotic stress tolerance in crop plants. Plant Physiol Biochem 48:909–930

Gill SS, Khan NA, Tuteja N (2011) Differential cadmium stress tolerance in five Indian mustard (*Brassica juncea* L.) cultivars: an evaluation of the role of antioxidant machinery. Plant Signal Behav 6:293–300

Gill SS, Anjum NA, Hasanuzzaman M et al (2013) Glutathione and glutathione reductase: a boon in disguise for plant abiotic stress defense operations. Plant Physiol Biochem 70:204–212

Gomes MP, Soaresb AM, Garcia QS (2014) Phosphorous and sulfur nutrition modulate antioxidant defenses in *Myracrodruon urundeuva* plants exposed to arsenic. J Hazard Mater 276:97–104

Guan C, Ji J, Jia C et al (2015) A GSHS-like gene from *Lycium chinense* may be regulated by cadmium-induced endogenous salicylic acid and overexpression of this gene enhances tolerance to cadmium stress in Arabidopsis. Plant Cell Rep 34:871–884

Harada E, Yamaguchi Y, Koizumi N et al (2002) Cadmium stress induces production of thiol compounds and transcripts for enzymes involved in sulfur assimilation pathways in Arabidopsis. J Plant Physiol 159:445–448

Hasan MK, Liu C, Wang F et al (2016) Glutathione-mediated regulation of nitric oxide, S--nitrosothiol and redox homeostasis confers cadmium tolerance by inducing transcription factors and stress response genes in tomato. Chemosphere 161:536–545

Hernández LE, Sobrino-Plata J, Montero-Palmero MB et al (2015) Contribution of glutathione to the control of cellular redox homeostasis under toxic metal and metalloid stress. J Exp Bot. doi:10.1093/jxb/erv063

Hossain MA, Piyatida P, da Silva JAT et al (2012) Molecular mechanism of heavy metal toxicity and tolerance in plants: central role of glutathione in detoxification of reactive oxygen species and methylglyoxal and in heavy metal chelation. J Bot. org/10.1155/2012/872875

Hsu TY, Kao CH (2004) Cadmium toxicity is reduced by nitric oxide in rice leaves. Plant Growth Regul 42:227–238

Iqbal N, Khan NA, Khan MIR et al (2012) Sulfur in the alleviation of cadmium-induced oxidative stress in plants. In: Ahmad P, Prasad MNV (eds) Environmental adaptations and stress tolerance of plants in the era of climate change. Springer-Verlag, New York, pp 429–446

Jamai A, Tommasini R, Martinoia E et al (1996) Characterization of glutathione uptake in broad bean leaf protoplasts. Plant Physiol 111(4):1145–1152

Jozefczak M, Remans T, Vangronsveld J et al (2012) Glutathione is a key player in metal induced oxidative stress defenses. Int J Mol Sci 13:3145–3175

Kao CH (2015) Role of glutathione in abiotic stress tolerance of rice plants. J Taiwan Agric Res 164:167–176

Khan NA, Asgher M, Per TS, Masood A, Fatma M, Khan MIR (2016) Ethylene potentiates sulfur-mediated reversal of cadmium inhibited photosynthetic responses in mustard. Front Plant Sci 7:1628

Khan MIR, Khan NA (2014) Ethylene reverses photosynthetic inhibition by nickel and zinc in mustard through changes in PSII activity, photosynthetic nitrogen use efficiency, and antioxidant metabolism. Protoplasma 251:1007–1019

Khan MIR, Nazir F, Asgher M et al (2015) Selenium and sulfur influence ethylene formation and alleviate cadmium-induced oxidative stress by improving proline and glutathione production in wheat. J Plant Physiol 173:9–18

Khan MIR, Khan NA, Masood A et al (2016) Hydrogen peroxide alleviates nickel-inhibited photosynthetic responses through increase in use-efficiency of nitrogen and sulfur, and glutathione production in mustard. Front Plant Sci 7:44. doi:10.3389/fpls.2016.00044

Khan M, Daud MK, Basharat A, Khan MJ, Azizullah A, Muhammad N, Muhammad N, Rehman Z, Zhu SJ (2016) Alleviation of lead-induced physiological, metabolic, and ultramorphological changes in leaves of upland cotton through glutathione. Environ Sci Pollut Res 23(9):8431–8440

Kumar A, Singh RP, Singh PK et al (2014) Selenium ameliorates arsenic induced oxidative stress through modulation of antioxidant enzymes and thiols in rice (*Oryza sativa* L.) Ecotoxicology 23:1153–1163

Küpper H, Šetlík I, Šetliková E et al (2003) Copper-induced inhibition of photosynthesis: limiting steps of in vivo copper chlorophyll formation in *Scenedesmus quadricauda*. Funct Plant Biol 30:1187–1196

Lamhamdi M, El Galiou O, Bakrim A et al (2013) Effect of lead stress on mineral content and growth of wheat (*Triticum aestivum*) and spinach (*Spinacia oleracea*) seedlings. Saudi J Biol Sci 20:29–36

Lee SH, Ahsan N, Lee KW et al (2007) Simultaneous overexpression of both Cu-Zn superoxide dismutase and ascorbate peroxidase in transgenic tall fescue plants confers increased tolerance to a wide range of abiotic stresses. Plant Physiol 164:1626–1638

Li X, Yang Y, Jia L et al (2013) Zinc-induced oxidative damage, antioxidant enzyme response and proline metabolism in roots and leaves of wheat plants. Ecotoxicol Environ Saf 89:150–157

Lin A, Zhang X, Zhu YG et al (2008) Arsenate-induced toxicity: effects on antioxidative enzymes and DNA damage in *Vicia faba*. Environ Toxicol Chem 27:413–419

Liu D, Zou J, Wang M et al (2008a) Hexavalent chromium uptake and its effects on mineral uptake, antioxidant defence system and photosynthesis in *Amaranthus viridis* L. Bioresour Technol 99:2628–2636

Liu D, Li TQ, Jin XF et al (2008b) Lead induced changes in the growth and antioxidant metabolism of the lead accumulating and non-accumulating ecotypes of *Sedum alfredii*. J Integr Plant Biol 50:129–140

Liu D, Liu Y, Rao J et al (2013) Overexpression of the glutathione S-transferase gene from *Pyrus pyrifolia* fruit improves tolerance to abiotic stress in transgenic tobacco plants. Mol Biol 47:515–523

Liu L, Sun H, Chen J et al (2014) Effects of cadmium (Cd) on seedling growth traits and photosynthesis parameters in cotton (*Gossypium hirsutum* L.) Plants Omics J 7:284–290

Lombardi L, Sebastiani L (2005) Copper toxicity in *Prunus cerasifera*: growth and antioxidant enzymes responses of in vitro grown plants. Plant Sci 168:797–802

Loscos J, Matamoros MA, Becana M (2008) Ascorbate and homoglutathione metabolism in common bean nodules under stress conditions and during natural senescence. Plant Physiol 146:1282–1292

Luo H, Li H, Zhang X et al (2011) Antioxidant responses and gene expression in perennial ryegrass (*Lolium perenne* L.) under cadmium stress. Ecotoxicology 20:770–778

Mahmood Q, Ahmad R, Kwak SS et al (2010) Ascorbate and glutathione: protectors of plants in oxidative stress. In: Anjum NA, Umar S, Chan MT (eds) Ascorbate-glutathione pathway and stress tolerance in plants. Springer, Netherland, pp 209–230

Masood A, Iqbal N, Khan NA (2012) Role of ethylene in alleviation of cadmium-induced photosynthetic capacity inhibition by sulfur in mustard. Plant Cell Environ 35:524–533

Mohamed AA, Castagna A, Ranieri A et al (2012) Cadmium tolerance in *Brassica juncea* roots and shoots is affected by antioxidant status and phytochelatin biosynthesis. Plant Physiol Biochem 57:15–22

Mostofa MG, Hossain MA, Fujita M et al (2015) Physiological and biochemical mechanisms associated with trehalose-induced copper-stress tolerance in rice. Sci Rep 5:11433. doi:10.1038/srep11433

Mostofo MG, Seraj ZI, Fujita M (2014) Exogenous sodium nitroprusside and glutathione alleviate copper toxicity by reducing copper uptake and oxidative damage in rice (*Oryza sativa* L.) seedlings. Protoplasma 251:1373–1386

Mullineaux PM, Rausch T (2005) Glutathione, photosynthesis and the redox regulation of stress-responsive gene expression. Photosynth Res 86:459–474

Myśliwa-Kurdziel B, Prasad MNV, Strzałka K (2002) Heavy metal influence on the light phase of photosynthesis physiology and biochemistry of metal toxicity and tolerance in plants. Springer, Netherland, pp 229–255

Nakamura S, Suzui N, Nagasaka et al (2013) Application of glutathione to roots selectively inhibits cadmium transport from roots to shoots in oilseed rape. J Exp Bot 64:1073–1081

Nazar R, Iqbal N, Masood A et al (2012) Cadmium toxicity in plants and role of mineral nutrients in its alleviation. Am J Plant Sci 3:1476–1489

Noctor G, Foyer CH (1998) Ascorbate and glutathione: keeping active oxygen under control. Annu Rev Plant Physiol Plant Mol Biol 49:249–279

Noctor G, Queval G, Mhamdi A et al (2011) Glutathione. In: The Arabidopsis book. Am Soc Plant Biol:1–32

Noctor G, Mhamdi A, Chaouch S et al (2012) Glutathione in plants: an integrated overview. Plant Cell Environ 35:454–484

Nouairi I, Ammar WB, Youssef NB et al (2009) Antioxidant defense system in leaves of Indian mustard (*Brassica juncea*) and rape (*Brassica napus*) under cadmium stress. Acta Physiol Plant 31:237–247

Ozyigit II, Filiz E, Vatansever R et al (2016) Identification and comparative analysis of H_2O_2-scavenging enzymes (ascorbate peroxidase and glutathione peroxidase) in selected plants employing bioinformatics approaches. Front Plant Sci 7:301. doi:10.3389/fpls.2016.00301

Pang S, Duan L, Liu Z et al (2012) Co-induction of a glutathione-S-transferase, a glutathione transporter and an ABC transporter in maize by xenobiotics. PLoS One 7:e40712

Parlak KU, Yilmaz DD (2012) Response of antioxidant defences to Zn stress in three duckweed species. Ecotoxicol Environ Safe 85:52–58

Pasternak M, Lim B, Wirtz M et al (2008) Restricting glutathione biosynthesis to the cytosol is sufficient for normal plant development. Plant J 53:999–1012

Pätsikkä E, Kairavuo M, Šeršen F et al (2002) Excess copper predisposes photosystem II to photoinhibition in vivo by outcompeting iron and causing decrease in leaf chlorophyll. Plant Physiol 129:1359–1367

Per TS, Khan S, Asgher M et al (2016) Photosynthetic and growth responses of two mustard cultivars differing in phytocystatin activity under cadmium stress. Photosynthetica. doi:10.1007/s11099-016-0205-y

Pietrini F, Iannelli MA, Pasqualini S et al (2003) Interaction of cadmium with glutathione and photosynthesis in developing leaves and chloroplasts of *Phragmites australis* (Cav.) Trin. ex Steudel. Plant Physiol 133:829–837

Qiu B, Zeng F, Cai S et al (2013) Alleviation of chromium toxicity in rice seedlings by applying exogenous glutathione. J Plant Physiol 170:772–779

Ribeiro MAQ, de Almeida AAF, Mielke MS et al (2013) Aluminum effects on growth, photosynthesis, and mineral nutrition of Cacao genotypes. J Plant Nutr 36:1161–1179

Rizwan M, Ali S, Adrees M et al (2016) Cadmium stress in rice: toxic effects, tolerance mechanisms, and management: a critical review. Environ Sci Pollut Res 23:17859–17879

Ruiz JM, Rivero RM, Boron LR (2006) Increases synthesis of glutathione in sunflower plants subjected to aluminum stress. Plant Soil 279:25–30

Sakai Y, Watanabe T, Wasaki J et al (2010) Influence of arsenic stress on synthesis and localization of low-molecular-weight thiols in *Pteris vittata*. Environ Pollut 158:3663–3669

Sarkar B (1995) Metal replacement in DNA-binding zinc finger proteins and its relevance to mutagenicity and carcinogenicity through free radical generation. Nutrition 11:646–649

Sárvári É (2005) Effects of heavy metals on chlorophyll-protein complexes in higher plants: causes and consequences. Handb Photosyn 2:865–888

Schafer FQ, Buettner GR (2001) Redox environment of the cell as viewed through the redox state of the glutathione disulfide/glutathione couple. Free Rad Biol Med 30:1191–1212

Semane B, Cuypers A, Smeets K et al (2007) Cadmium responses in *Arabidopsis thaliana*: glutathione metabolism and antioxidative defence system. Physiol Plant 129:519–528

Šeršeň F, Král'ová K, Bumbálová A et al (1997) The effect of Cu (II) ions bound with tridentate Schiff base ligands upon the photosynthetic apparatus. J Plant Physiol 151:299–305

Shanmugam V, Wang YW, Tsednee M et al (2015) Glutathione plays an essential role in nitric oxide-mediated iron-deficiency signaling and iron-deficiency tolerance in Arabidopsis. Plant J 84:464–477

Sharma P, Dubey RS (2007) Involvement of oxidative stress and role of antioxidative defense system in growing rice seedlings exposed to toxic concentrations of aluminum. Plant Cell Rep 26:2027–2038

Sheetal KR, Singh D, Anand A et al (2016) Heavy metal accumulation and effects on growth, biomass and physiological processes in mustard. Indian J Plant Physiol 21:219–223

Sobrino-Plata J, Meyssen D, Cuypers A et al (2014) Glutathione is a key antioxidant metabolite to cope with mercury and cadmium stress. Plant Soil 377:369–381

Sofo A, Cicco N, Paraggio M et al (2010) Regulation of the ascorbate–glutathione cycle in plants under drought stress. In: Anjum NA, Umar S, Chan MT (eds) Ascorbate-glutathione pathway and stress tolerance in plants. Springer, Dordrecht, pp 137–190

Son JA, Narayanankutty DP, Roh KS (2014) Influence of exogenous application of glutathione on rubisco and rubisco activase in heavy metal-stressed tobacco plant grown in vitro. Saudi J Biol Sci 21:89–97

Srivalli S, Khanna-Chopra R (2008) Role of glutathione in abiotic stress tolerance. In: Khan NA, Singh S, Umar S (eds) Sulfur assimilation and abiotic stress in plants. Springer, Berlin, pp 207–225

Sytar O, Kumar A, Latowski D et al (2013) Heavy metal-induced oxidative damage, defense reactions, and detoxification mechanisms in plants. Acta Physiol Plant 35:985–999

Szalai G, Kellos T, Galiba G et al (2009) Glutathione as an antioxidant and regulatory molecule in plants under abiotic stress conditions. J Plant Growth Regul 28:66–80

Talukdar D, Talukdar T (2014) Coordinated response of sulfate transport, cysteine biosynthesis, and glutathione-mediated antioxidant defense in lentil (*Lens culinaris* Medik.) genotypes exposed to arsenic. Protoplasma 251:839–855

Thao NP, Khan MIR, Thu NBA et al (2015) Role of ethylene and its cross talk with other signaling molecules in plant responses to heavy metal stress. Plant Physiol 169:73–84

Tran TA, Popova LP (2013) Functions and toxicity of cadmium in plants: recent advances and future prospects. Turk J Bot 37:1–13

Ventrella A, Catucci L, Placido T et al (2011) Biomaterials based on photosynthetic membranes as potential sensors for herbicides. Biosens Bioelectron 26:4747–4752

Vitorello VA, Capaldi FR, Stefanuto VA (2005) Recent advances in aluminum toxicity and resistance in higher plants. Braz J Plant Physiol 17:129–143

Vivancos PD, Wolff T, Markovic J et al (2010) A nuclear glutathione cycle within the cell cycle. Biochem J 431:169–178

Wachter A, Wolf S, Steininger H et al (2005) Differential targeting of GSH1 and GSH2 is achieved by multiple transcription initiation: implications for the compartmentation of glutathione biosynthesis in the Brassicaceae. Plant J 41:15–30

Wali M, Gunse' B, Llugany M et al (2016) High salinity helps the halophyte *Sesuvium portulacastrum* in defense against Cd toxicity by maintaining redox balance and photosynthesis. Planta 244:333–346

Wan G, Najeeb U, Jilani G et al (2011) Calcium invigorates the cadmium-stressed Brassica napus L plants by strengthening their photosynthetic system. Environ Sci Pollut Res 18:1478–1486

Wang ZY, Nie HL, Tie S et al (2011a) Effects of excess copper on the oxidative stress in roots of maize. Afr J Agric Res 6:4998–5004

Wang CF, Cai Y et al (2011b) Modulation of exogenous glutathione in ultrastructure and photosynthetic performance against Cd stress in the two barley genotypes differing in Cd tolerance. Biol Trace Elem Res 144:1275–1288

Xiang C, Oliver DJ (1998) Glutathione metabolic genes co-ordinately respond to heavy metals and jasmonic acid in Arabidopsis. Plant Cell 10:1539–1550

Xiang C, Werner BL, E'Lise MC et al (2001) The biological functions of glutathione revisited in Arabidopsis transgenic plants with altered glutathione levels. Plant Physiol 126:564–574

Xu WF, Shi WM, Ueda A et al (2008) Mechanisms of salt tolerance in transgenic *Arabidopsis thaliana* carrying a peroxisomal ascorbate peroxidase gene from barley. Pedosphere 18:486–495

Yadav SK (2010) Heavy metals toxicity in plants: an overview on the role of glutathione and phytochelatins in heavy metal stress tolerance of plants. South Afr J Bot 76:167–179

Yuan H, Zhang Y, Huang S et al (2015) Effects of exogenous glutathione and cysteine on growth, lead accumulation, and tolerance of *Iris lactea* var. *chinensis*. Environ Sci Pollut Res 22:2808–2816

Yurekli F, Porgali ZB (2006) The effects of excessive exposure to copper in bean plants. Acta Biol Cracov Ser Bot 48:7–13

Zagorchev L, Charlotte E, Seal CE et al (2013) A central role for thiols in plant tolerance to abiotic stress. Int J Mol Sci 14:7405–7432

Zechmann B (2014) Compartment-specific importance of glutathione during abiotic and biotic stress. Front Plant Sci 5:566

Zeng X, Ma LQ, Qiu R et al (2009) Responses of non-protein thiols to Cd exposure in Cd hyperaccumulator *Arabis paniculata Franch.* Environ Exp Bot 66:242–248

Zhang ZC, Chen BX, Qiu BS (2010) Phytochelatin synthesis plays a similar role in shoots of the cadmium hyperaccumulator *Sedum alfredii* as in non-resistant plants. Plant Cell Environ 33:1248–1255

Zhu YL, Pilon-Smits EA, Jouanin L et al (1999) Overexpression of glutathione synthetase in Indian mustard enhances cadmium accumulation and tolerance. Plant Physiol 119:73–80

Production of Antioxidant and Oxidant Metabolites in Tomato Plants Infected with *Verticillium dahliae* Under Saline Conditions

13

Murat Dikilitas, Nurcan Yucel, and Sibel Dervis

Abstract

The objective of this research was to study the levels of antioxidant and oxidant metabolites such as total protein, total soluble sugars, proline, peroxidase, catalase, malondialdehyde, anthocyanin and phenol contents in tomato plants (*Lycopersicon esculentum* Mill. cv. Ailsa Craig) inoculated with *Verticillium dahliae* under various NaCl concentrations (50, 100, 200 and 250 mmol l^{-1}). *V. dahliae* alone resulted in increases in oxidant (malondialdehyde and phenol) and antioxidant metabolites (proline, anthocyanin and sugar) as well as in antioxidant enzymes (catalase and peroxidase). However, the pathogenic effect of *V. dahliae* increased with the increase of NaCl stress and resulted in depletion of antioxidant metabolites and enzymes after 150 mmol l^{-1} NaCl level, while oxidant levels showed an increased trend. Results suggest that *V. dahliae* could be able to survive under high salt stress conditions and cause combined stress that affect the resistance of the resistant cultivars and reduce crop yield.

Keywords

Antioxidant • Oxidant metabolites • Salt stress • Tolerance • Tomato

M. Dikilitas (✉) • N. Yucel
Department of Plant Protection, Faculty of Agriculture, Harran University, 63200, Sanliurfa, Turkey
e-mail: m.dikilitas@gmail.com

S. Dervis
Department of Plant Protection, Faculty of Agriculture, Mustafa Kemal University, Hatay, Antakya, Turkey

M.I.R. Khan, N.A. Khan (eds.), *Reactive Oxygen Species and Antioxidant Systems in Plants: Role and Regulation under Abiotic Stress*,
DOI 10.1007/978-981-10-5254-5_13

13.1 Introduction

Salinity is one of the world's most serious environmental issues affecting crop growth, productivity and quality. Although salinity existed before human activities, the problem is now increasing at a very fast rate due to anthropogenic activities. With the increase of salinity over the years, it has now been estimated that more than 50% of agricultural soils would be under the threat of salinity by the mid of this century (Hasanuzzaman et al. 2014). Salinity could affect the plant growth via high concentration of toxic ions, ion imbalance and negative water potential (Dikilitas and Karakas 2012). High concentration of sodium chloride is known to be toxicant for humans, animals, plants and microorganisms, and it enters the metabolism of higher organisms and plants through food chain and irrigation, respectively. Increases in Na^+ and Cl^- content in plant tissue can increase the oxidative stress and cause deterioration in cell wall and cell components (Agami 2014). Excess accumulation of NaCl in soil may induce various visible symptoms such as chlorosis, blight and growth reduction in plants. As a result, crop and quality reductions in plants are inevitable.

Apart from salinity, pathogens (especially soilborne) may cause detrimental effects on crop plants. For example, one of the most devastating soil fungi which affects a great number of plants including crop plants, shrubs and ornamental plants and even trees is a wilt fungus *Verticillium dahliae* Kleb. It is a systemic pathogen that causes wilt in plants. The fungus enters the root through wounds that damages the vascular system or grows between the cells of the meristem to gain access the immature xylem vessel elements (Pegg 1989). The fungus is able to enter the root of young plants or moderately resistant plants via secretion of the cell wall-degrading enzymes and alters the physiology of infected plants, accelerates their senescence and reduces yield (Goicoechea 2006). Under abiotic stress conditions, gaining of entrance by the pathogen becomes easy through the increased secretion of lytic enzymes or reduced defence barriers of the host (Turco et al. 1999; Dikilitas and Karakas 2012). Any additional stress during pathogenesis would result in additional production of reactive oxygen species (ROS) even if the host becomes resistant. Cells, in general, produce superoxide, hydrogen peroxide and hydroxyl radicals during interactions with abiotic or biotic stress agents. In almost all interactions, the production of ROS is rapid and transient. The ROS production and increased activity of enzymes are strongly connected with the induction of defence responses and may play a significant role for the activation of genes encoding protective proteins (Lamb and Dixon 1997). However, enhanced production of ROS causes oxidative damage to macromolecules such as pigments, proteins, nucleic acids, etc., and leads to lipid peroxidation (Apel and Hirt 2004). For example, Von Tiedemann (1997) stated that the generation of ROS assisted the colonization of bean leaves by *Botrytis cinerea*.

Although fungi result in oxidative stress in plants, fungi themselves get also stressed by the abiotic stress factors such as salinity and drought. Under these circumstances, if pathogens tolerate or adapt to saline conditions, the additive negative effects on crop plants are inevitable (Dikilitas and Karakas 2014). In some cases, salinity may lead to stimulation of pathogen growth and dissemination (Turco et al. 1999; You et al. 2011). There are only a few reports that salt negatively affect the pathogen growth (Elmer 2002). In majority of cases, negative effects of salinity and pathogens on crop plants have been reported. Salt interacts with soil pathogens in the vicinity of root area. Therefore, there are very complex interactions among plant, pathogen and salt. For example, a nutrient solution with an electrical conductivity of 20 mS cm^{-1} increased the severity of *Fusarium oxysporum* f.sp. *vasinfectum* wilt of cotton. A decrease in the diameter of xylem vessels was a normal pathological response of the infected plants grown under no salt conditions (Turco et al. 2002). However, when salinity was involved, further decreases in xylem vessels were evident but the number of vessels at this time increased to conduct the water in vessel elements. This, however, greatly favoured the infection process of tracheovascular pathogen (Turco et al. 2002). Similarly, Saadamand et al. (2008) stated that 8-week-old pistachio seedlings exhibited higher symptoms of wilt disease and root/shoot colonization when exposed to salt stress in *V. dahliae*-infested soil. Again, You et al. (2011) reported that 40 mmol l^{-1} NaCl with 1 mmol l^{-1} $CaSO_4$ stimulated the growth and development of *Macrophomina phaseolina* which eventually increased disease development and severity. Plants infected with *M. phaseolina* showed increased tissue concentrations of Na^+ and Cl^- ions but decreased K^+ concentration.

In saline conditions, avirulent isolates also become virulent. Regragui et al. (2003) stated that tomato plants var. Marmande showed great susceptibility under 80 mmol l^{-1} NaCl stress to *Verticillium albo-atrum*. With the effect of salt, avirulent isolate gained a new pathogenic aptitude which exhibited increased carboxymethylcellulose activity along with the slight increase in mycelial growth under salt stress.

Since the salinity problem is ever increasing and threatening the non-salinized areas, the interactions between salinity and biotic agents from now on are inevitable. Therefore, the problem of salinity or wilt disease would become even more acute when plants are exposed to both NaCl and the disease (Dikilitas 2003; Dikilitas and Karakas 2012). Although plants develop a protective system against oxidative stress, the combined effects of stress would endanger even the resistance mechanisms of the resistant cultivars. The consequence would not only be reduction in plant production, but also quality would deteriorate more quickly.

Various enzymes such as superoxide dismutase (SOD), catalase (CAT), guaiacol peroxidase (POD) and ascorbate peroxidase (APX) are involved to protect the integrity of cell wall and cell components. POD, in special, play important roles in lignin and suberin formation and act as free radical scavengers and protect cells from their oxidative toxicity (Leon et al. 1995). However, excess accumulation of phenols may cause disorders in cell membranes and result in lipid peroxidation. Generally, adaptation of plants to stress involves the increase of intracellular solutes such as proline, total soluble sugars (TSS) and total soluble proteins. However, the combined stress may not give enough time to plants to accumulate those osmoprotectants and soluble macromolecules to reduce the severity of stress. Therefore, the combined pathogenic attack and salinity stress may result in rapid disease progress and necrosis (Dikilitas 2003). In this study, the levels of solutes in leaves of inoculated tomato plants under various NaCl stress conditions were measured and compared with those of plants growing under salinity stress or pathogen threat alone. By this way, we aimed to determine the resistance of the moderately resistant tomato cultivar Ailsa Craig if the cultivation would be sustained under the effect of combined stress.

13.2 Materials and Methods

The experiment was conducted in a greenhouse of Harran University, Faculty of Agriculture, S. Urfa, Turkey. Tomato seeds (*Lycopersicon esculentum* Mill. cv. Ailsa Craig) were first surface sterilized with 1% NaOCl for 5 min, rinsed thoroughly with deionized water, then soaked overnight in running tap water at room temperature and germinated in trays of sterilized moist sand in the greenhouse conditions at 25 ± 2 °C. Following 6 weeks, the seedlings were uprooted, and the soil particles were washed off under running tap water. The seedlings were then inoculated for 10 min with the conidia (1×10^6 ml^{-1}) of *Verticillium dahliae* isolated from tomato with a root dip inoculation method (Dikilitas 2003). The inoculated seedlings were transplanted into 16-cm diameter plastic pots containing local field soil, compost and perlite (1:1:1). For control group, distilled water was used instead of fungal conidial solution for the same period of time. Inoculated or non-inoculated groups of tomato seedlings were irrigated with 100 ml of various concentrations of NaCl (0, 50, 100, 200 and 250 mmol l^{-1}) every other day in the first week. To avoid from the osmotic shock, 50 mmol l^{-1} increment of NaCl was ensured in every irrigation until the highest concentration of NaCl (250 mmol l^{-1}) was reached. The plants were then irrigated on demand with respective salt concentrations. After 3 weeks of inoculation and salt treatment, leaves from treated and control plants were collected for biochemical analysis.

The enzyme extract was prepared by homogenizing leaf samples (0.5 g fresh weight, FW) in a prechilled mortar and pestle containing 4 ml of 50 mmol l^{-1} potassium phosphate buffer (pH 7.0) with 1 mmol l^{-1} EDTA for determination of POD and CAT activities. The homogenate was centrifuged at 13,000 g for 10 min at 4 °C. The supernatant was stored at −80 °C until used for the biochemical analyses.

POD activity was measured with guaiacol as the substrate in a total volume of 3 ml. The reaction mixture consisted of 50 mmol l^{-1} potassium phosphate buffer (pH 6.5), 13 mmol l^{-1} guaiacol, 5 mmol l^{-1} H_2O_2 and enzyme extract. Enzyme activity was then calculated as the rate of increase of absorbance due to oxidation of guaiacol per minute and expressed as $\Delta OD_{470\ nm}$ min^{-1} mg^{-1} protein (Chance and Maehly 1995; Shahbazi et al. 2010).

CAT activity was measured by monitoring the disappearance of H_2O_2 through measuring the decrease in absorbance at 240 nm of a reaction mixture consisted of 0.1 ml enzyme extract and 2.8 ml potassium phosphate buffer (0.1 mol l^{-1}, pH 7.0) containing 1 mmol l^{-1} Na_2EDTA. The biochemical reaction was started by adding

0.1 ml of 10 mmol l^{-1} H_2O_2. Samples without H_2O_2 were used as blank. Enzyme activity was expressed as $\Delta OD_{240\ nm}$ min^{-1} mg^{-1} protein (Chance and Maehly 1995; Shahbazi et al. 2010).

Anthocyanin contents of leaves were determined according to the method of Mancinelli (1990). A 1 g leaf sample was extracted in 3 ml methanol-HCl (1% HCl, v/v). The samples were then left at 4 °C in the fridge for 48 h. Then the extract was filtered through two layers of cheesecloth, and the anthocyanin content was measured by spectrophotometrically at 530 and 657 nm wavelengths. The results were expressed as $\Delta A_{(530–657)}$ g^{-1} FW.

Phenol content of leaves was determined according to Singleton and Rossi (1965) with slight modifications (Bolat et al. 2014) by using Folin-Ciocalteu solution. A 0.5 g of leaf tissue was ground in liquid nitrogen with a mortar and pestle. It was then extracted with 5 ml of 80% methanol and centrifuged at 5000 g for 10 min. To determine the amount of total phenols, a 100 μl of the supernatant was mixed with 1.8 ml Folin-Ciocalteu solution (ten times diluted), shaken vigorously and then kept for 5 min at room temperature. Then 1.2 ml of 20% Na_2CO_3 was added, and the volume was made up to 6 ml with distilled water. The mixture was incubated for 1 h at room temperature, and the absorption of developed blue colour was measured at 765 nm against methanol blank. The content of total soluble phenols were expressed as mg gallic acid equivalent g^{-1} leaf FW. The standard curve was prepared from a stock solution of gallic acid (4 mg ml^{-1}) in which the curve was prepared by dissolving various concentrations of gallic acid (0.02–1.8 mg ml^{-1}) in 80% methanol. The standard solution of gallic acid was reacted as stated above.

Proline content of the leaves was measured with acid ninhydrin reagent according to Bates et al. (1973). Proline was extracted from 0.5 g of leaf sample by grinding in 10 ml of 3% sulphosalicylic acid, and the mixture was then centrifuged at 10,000 g for 10 min. Two millilitres of the supernatant was then added into test tubes to which 2 ml of freshly prepared acid ninhydrin solution, and 2 ml of acetic acid glacial was mixed. The tubes were placed in a water bath for 1 h at 90 °C, and the reaction was terminated in an ice bath. The mixture was then extracted with 5 ml of toluene and vortexed for 15 s. After allowing to stand at room temperature for 20 min in darkness to separate the toluene and aqueous phase, the toluene phase was then carefully collected into clean test tubes, and the absorbance of the fraction was read at 520 nm against a toluene blank in a UV spectrophotometer. Proline content in the sample was expressed as μg g^{-1} FW. The standard curve was prepared by employing L-proline.

The level of lipid peroxidation in leaf tissue was measured in terms of malondialdehyde (MDA) content, which was determined according to the method of Sairam and Saxena (2000) with slight modifications. A 0.5 g leaf tissue was grounded with a pestle and mortar in 5 ml of 0.1% trichloroacetic acid (TCA), and the homogenate was centrifuged at 10,000 g for 5 min. To 1 mL of the supernatant, 4 ml of 20% v/v TCA containing 0.5% v/v thiobarbituric acid (TBA) was added. The homogenate was heated at 95 °C for 30 min and then quickly cooled on ice. The mixture was centrifuged again at 10,000 g for 5 min, and the absorbance of the supernatant was measured at 532 and 600 nm. MDA content of the leaves was expressed as nmol g^{-1} FW using the coefficient of absorbance of 155 mmol l^{-1} cm^{-1}.

$$\mathrm{MDA}\left(\mathrm{nmo}\ g^{-1}\right) = \frac{\text{Extract volume (ml)} \times \left[(A_{532} - A_{600})/\left(155\mathrm{mM}^{-1}\mathrm{cm}^{-1}\right)\right]}{\text{sample amount (g)}} \times 10^3$$

For the determination of total protein, a 0.5 g leaf was grounded in 50 mmol l^{-1} phosphate buffer (pH 7.0) with quartz sand at 4 °C using prechilled mortar and pestle. The extract was centrifuged (10,000 g for 10 min at 4 °C), and the supernatant was used for soluble protein quantification. Protein concentration was determined according to Coomassie Brilliant Blue G250 method using bovine serum albumin as a standard at 595 nm colorimetric wavelength (Bradford 1976).

Total sugar contents were estimated with phenol-sulphuric acid method at 490 nm according to Hodge and Hofreiter (1962). A standard curve was prepared using D-Glucose over the range of 10–100 μg, and distilled water was used as a control. A 1 ml of 5% (w/v) phenol solution (Sigma) was added to 1 ml carbohydrate solution (containing 10–100 μg carbohydrate). The solution prepared for protein determination was also used for carbohydrate determination. The solution was mixed and added to 5 ml concentrated sulphuric acid (analytic reagent). The tubes were initially incubated at room temperature for 10 min and placed in a water bath at 30 °C for 20 min. The absorbance at 490 nm was then recorded. The amount of carbohydrate in a sample is expressed in glucose equivalents.

All biochemical analyses were made with a UV spectrophotometer (Shimadzu UV-1700).

Statistical analysis was made with the analysis of variance (ANOVA). The significance of difference among means was compared by using Duncan's multiple range test.

13.3 Results and Discussion

In the past, salt stress had little importance as it was specific to coastal regions. However, the issue is now more problematic than ever in areas where excess irrigation or irrigation with marginal quality of water was used. Increase of shortage of fresh water and decrease in agricultural land increased the pressure on using marginal quality of water. Therefore, previously as non-saline characterized areas became saline with the wrong agricultural applications. In these areas, inhabiting soil pathogens also become affected either positively or not affected at all or tolerate the salinity stress much better than crop plants do in terms of survival rate and longevity. Previously, *V. dahliae* and saline irrigation water was reported on potato. The disease expression and fungal colonization were higher in plants under both pathogen and salinity stress (Nachmias et al. 1993). Similarly, Howell et al. (1994) reported that lucerne cultivars growing in areas where soil salinity was a problem could undergo an increase in susceptibility to *Verticillium albo-atrum* and suffer a greater yield loss due to combined stress.

In our study, the results showed that the growth of fungus was delayed, while metabolites necessary for the growth were not affected (data not shown). At this growth stage, crop plants may weaken and loose ability to synthesize defence metabolites against the attacking pathogen and salinity stress. Moreover, salt stress may weaken the genetic resistance of plants to disease by inducing changes in susceptibility of root tissue to pathogens (Daami-Remadi et al. 2009). For example, Ragazzi and Moricca (1995) reported that a moderately resistant cotton cultivar to *Fusarium oxysporum* f.sp. *vasinfectum* showed susceptibility under water stress by displaying a fairly high percent infected leaf area. Daami-Remadi et al. (2009) stated that salinity stress (2–10 g of NaCl l^{-1}) in *Fusarium*-inoculated tomato plants enhanced the severity of *Fusarium* wilt disease and resulted in a significant increase in the leaf damage index. The combined effect of the pathogen *Fusarium oxysporum* f. sp. *lycopersici* and salinity was also reflected onto fruit fresh weight, which was significantly reduced by 40.

Increased NaCl concentrations under *V. dahliae* negatively affected the response of tomato. It decreased antioxidant activity by decreasing antioxidant enzyme levels after a certain level (150 mmol l^{-1} NaCl). Even low concentrations of salt could have a devastating effect on crop plants when combined with *V. dahliae*. High salt concentrations, on the other hand, retard the development of fungal mycelia, but it would not always kill the fungus completely. Only, the longer incubation period would be needed for the fungi to complete their life cycle. For example, Daami-Remadi et al. (2009) showed that the growth of *Fusarium oxysporum* f.sp. *lycopersici* completed its growth in 35 days in vitro conditions, while the same fungus completed its growth in 62 days in 10 g l^{-1} NaCl conditions.

In this study, the combined effect of *V. dahliae* and increased concentrations of NaCl (0–250 mmol l^{-1}) resulted in decreases in antioxidant and increase in oxidant metabolites, Table 13.1. All living organisms respond to unfavourable conditions by accumulating low molecular weight compounds to protect the cells from invading organisms or stress agents. In general, the production of low molecular compounds such as amino acids, sugars, proteins and carbohydrates has been produced as a response to stress. Defence enzymes such as POD, CAT, glucosamine S-transferase (GST) and ascorbate peroxidase (APX) are important enzymes in modulating the defence responses. POD, which use phenolics as the electron donors, play a part of defence mechanisms against pathogens at the cell wall membrane (Hiraga et al. 2001; Kuzniak et al. 2011). Antioxidant enzymes such as POD and CAT increased upon increase of NaCl levels; however, the increase of antioxidant enzymes in response to NaCl decreased after 150 mmol l^{-1} NaCl level, Table 13.1. Although the involvement of *V. dahliae* resulted in further increases in antioxidant enzyme levels, however, the increase in enzyme levels ended at 150 mmol l^{-1} NaCl level. Then, higher NaCl concentrations alone or in combination with *V. dahliae* led to decreases in POD and CAT levels. At 250 mmol l^{-1} NaCl level, both POD and CAT activities were minimized. It is clear that salinity and disease interactions played important roles in reducing antioxidant enzymes

Table 13.1 Biochemical responses of tomato plants inoculated with *V. dahliae* under various concentrations of NaCl salinity

NaCl (mmol l^{-1})	*V. dahliae*	POD $\Delta OD_{470\ nm}$ min^{-1} mg^{-1} protein	CAT $\Delta OD_{240\ nm}$ min^{-1} mg^{-1} protein	Proline (μg g^{-1}FW)	Anthocyanin ΔA (530–657) g^{-1} FW	Protein (μg g $^{-1}$ FW)	Sugar (mg g $^{-1}$ FW)
0	–	0.3	0.11	12 ± 2	0.1	2 ± 0.5	6 ± 1
0	+	**0.4***	0.12	**19 ± 3**	0.1	2.1 ± o.6	8 ± 2
50	–	**1.1**	**0.14**	**34 ± 4**	0.2	2.3 ± 0.5	**12 ± 2**
50	+	**1.4**	**0.15**	**45 ± 6**	0.2	2.4 ± 0.5	**15 ± 2**
100	–	**1.5**	**0.18**	**120 ± 7**	**0.24**	2.5 ± 0.5	**18 ± 3**
100	+	**1.8**	**0.21**	**125 ± 8**	**0.23**	2.8 ± 0.4	**23 ± 3**
150	–	**2.4**	**0.15**	**115 ± 8**	0.2	**3.5 ± 0.4**	**25 ± 3**
150	+	**2.3**	**0.13**	**119 ± 12**	0.2	**3.8 ± 0.6**	**30 ± 5**
200	–	**2.1**	**0.13**	**80 ± 10**	0.1	2.1 ± 0.3	**23 ± 4**
200	+	**1.8**	0.11	**45 ± 8**	0.1	1.8 ± 0.3	**28 ± 4**
250	–	**0.5**	0.10	**50 ± 8**	0.1	1 ± 0.3	10 ± 2
250	+	0.3	0.10	**34 ± 6**	0.05	0.8 ± 0.3	7 ± 2

*The figures in bold are significantly different from those of control plants at $p < 0.05$ level of significance. Results were expressed as mean ± standard error

after. This clearly showed that the fungus acted as a significant pathogen under saline conditions and reduced the antioxidant potential of plants by diminishing the antioxidant enzymes. Similar trends were also observed with those of proline, anthocyanin, total soluble protein and sugar contents. Increased NaCl levels resulted in accumulation of proline, protein and sugar contents, Table 13.1. A sixfold increase was evident in proline accumulation at 100 mmol l^{-1} NaCl level. Control value of tomato for proline accumulation was 12 μg g^{-1} FW, while the inoculation of *V. dahliae* resulted in further increase in proline accumulation (19 μg g^{-1} FW) in tomato plants. Increase in NaCl concentrations caused much higher accumulation in proline contents. When the pathogenic effect of *V. dahliae* was combined with the increase of NaCl, a further increase was evident. Similar trends were also observed with protein and sugar contents. However, a breaking point was reached at 150 mmol l^{-1} NaCl level. Above 150 mmol l^{-1} NaCl, a decline in proline, protein and sugar contents was observed. Involvement of *V. dahliae* resulted in further decline in these metabolites. Levels of antioxidant metabolites at 200 and 250 mmol l^{-1} NaCl conditions were much lower than those of plants grown in no salt or no fungus conditions, Table 13.1. The significance of the osmoprotective function of proline accumulation has been reported in many plant studies. Proline serves as a storage sink for carbon and nitrogen, stabilizes subcellular structures and buffers cellular redox potential under stress (Chinnusamy et al. 2005). The role of proline in scavenging ROS and protecting the cells from ROS-induced damaging effects has been discussed in many studies (Jahantigh et al. 2016). In this study, decline in proline contents showed a similar trend as in the case of antioxidant enzymes. For example, Abass (2016) stated that the decline in growth parameters was accompanied with an increase of free proline and H_2O_2 production in callus exposed to either culture filtrate of *Aspergillus* sp. or high NaCl (205.34 mmol l^{-1}) concentration. Increases in free amino acids and proline and protein content were also reported in fungi under salinity stress. El-Abyad et al. (1992) reported that *Fusarium solani* and *Sclerotia rolfsii* showed a great tolerance to salinity by accumulating free amino acids under salinity stress. Therefore, any increases in proline, protein and free amino acid contents not only arise from the plants, but the infecting fungi could contribute to the increase of those metabolites in plants if sampling was made only from plants. Kuzniak et al. (2011) stated that the increased proline accumulation in callus of *Mesembryanthemum crystallinum* restricted the proliferation of *Botrytis cinerea*. They stated that *B. cinerea* easily colonized the callus surface and intercellular spaces of halophyte *M. crystallinum* and resulted in increases in MDA and antioxidant enzymes. Penetration was 67% lower in NaCl-adapted tissues than in the inoculated control tissue. The non-salt-adapted control callus provided a more favourable environment for the fungus. This was ensured with the enhanced antioxidant potential of ascorbate, ascorbate peroxidase as well as guaiacol and syringaldazine peroxidases. Although biochemical and genetic background of this phenomenon need to be further evaluated, increased antioxidant levels before fungal inoculation enhanced the capacity to withstand the pathogen infection. It is important to note that here, the restricted pathogen colonization was in salt-adapted callus of halophyte *M. crystallinum*. However, this case was different in salt-adapted callus of glycophyte *Medicago sativa* that the colonization and infection by *Verticillium albo-atrum* were severe (Dikilitas 2003). Involvement of pathogen under NaCl conditions may also have negative effect on water transportation system in tomato plants creating drought conditions along with the toxicity and nutrient imbalance in plants. Turco et al. (2002) stated that flow in the lymph system in cotton plants was slow due to reduced diameter of xylem vessels under *Fusarium oxysporum* f.sp. *vasinfectum* infection. However, as a response to disease, the number of xylem vessel elements was greatly increased to conduct the sap more efficiently. Under saline stress, due to enhanced growth of mycelium and sporulation, the colonization by the fungus was much severe. Similar findings were also made by Ragazzi et al. (1996) on chlamydospore viability and *Fusarium* virulence. Again, Turco et al. (1999) stated that the greater enzymatic activity such as pectate lyase and polygalacturonase activity by *Fusarium* was evident in saline-enriched media. This clearly showed that why soil pathogens were so severe in plants under saline conditions. Tomato plants grown in the presence of NaCl showed an increased trend in terms of protein accumulation, Table 13.1. This increase was higher in leaves grown at 100 and 150 mmol l^{-1} NaCl when compared to control.

Under saline stress, pathogens also get stressed, but its tolerance to salinity may differ from those of crop plants and tolerate much higher concentrations of NaCl than crop plants do. Under such circumstances, pathogens might still become virulent as in normal growth conditions. Some authors reported that the mechanism of the increased virulence of the fungus under salinity could be related to accumulation of melanin, which is one of the most important pigments in fungi. This pigment is involved for the penetration of host cell wall by enhancing the rigidity of the mycelium of the fungus (Gessler et al. 2014). It also plays crucial roles for the survival of fungi in harsh conditions and protects the fungal cell from the hydrolytic enzymes produced by other microorganisms or by the plant roots.

In some cases, salinity does not cause any stress on fungus growth. For example, Al-Sadi et al. (2010) reported that the development of *Pythium* spp. was unaffected at 5 dS m^{-1} salinity level, while this level was stressful for cucumber plants. In fact, IC_{50} values of *Pythium* spp. for growth were varied from 23 to 62 dS m^{-1}, with an average of 46 dS m^{-1}. Although oospore production was more sensitive to salinity and no spore was produced above 20 dS m^{-1}, this level, however, was very toxic to crop plants. Similarly, Attaby (2001) reported that the growth of *Penicillium chrysogenum* was maximum at 10% NaCl and tolerated NaCl concentration up to 18%. Similar findings were also made by Mert and Emekci (1987) on *Aspergillus flavus* and *P. chrysogenum*. Omori et al. (1995) also reported that a mutant TK-2 of *Saccharomyces* was tolerant to 18% NaCl.

Pathogenicity under saline conditions is also quicker when compared to normal pathogenicity conditions. Since the disease has been firmly established, yield becomes much lower. For example, Triky-Dotan et al. (2005) reported that disease onset was earlier when irrigation was made with saline water in tomato plants inoculated with *Fusarium oxysporum* f.sp. *radices-lycopersici*. One of the main reasons has been attributed to quick degradation of cellulose to

Table 13.2 Biochemical changes in stress metabolites in tomato plants inoculated with *V. dahliae* under various concentrations of NaCl salinity

NaCl (mmol l^{-1})	*V. dahliae*	Total phenol (mg g^{-1} FW)	MDA (nmol g^{-1} FW)
0	–	0.2	9 ± 2
0	+	0.24	11 ± 1
50	–	**0.66**	**13 ± 2**
50	+	**0.69**	**12 ± 2**
100	–	**1.2**	**14 ± 3**
100	+	**1.3**	**14 ± 3**
150	–	**1.8**	**16 ± 4**
150	+	**1.9**	**17 ± 4**
200	–	**2.2**	**18 ± 4**
200	+	**1.2**	**18 ± 5**
250	–	**1.1**	**22 ± 5**
250	+	**1.0**	**23 ± 4**

*The figures in bold are significantly different from those of control plants at $p < 0.05$ level of significance. Results were expressed as mean ± standard error

carboxymethylcellulose which seems to have a significant role in the progress of pathogenesis as provides a source of rich contents of carbohydrates to support the growth of wilt fungus in the host tissue (Turco et al. 1999). As shown in various studies, adequate and balanced mineral nutrition in many crops prevent crop plants from stress. In particular, potassium and chloride fertility have been effective in reducing crop injury from diseases. By this way, KCl or $CaCl_2$ applications may not only result in higher yields but also delay the disease onset. Although application of fertilization is not a substitute for fungicides, it is an important component in the integrated pest management.

When oxidant metabolites of tomato plants were assessed under various NaCl conditions with *V. dahliae* or not, the oxidant metabolites such as total phenol and MDA contents gradually increased as with the increase of NaCl. MDA contents were almost doubled towards the concentration of 200 mmol l^{-1} NaCl and continued to increase with further increases of NaCl concentrations. Involvement of *V. dahliae* also contributed to further increase in MDA in all NaCl levels (Table 13.2). Similar trend was also observed in phenol accumulation.

Salt stress is often associated with the increased generation of ROS such as H_2O_2 and superoxide radical anion (O_2^-) and lipid peroxidation (Kuzniak et al. 2011). Levels of ROS in plant cells are normally controlled by protective antioxidant activity. However, under the combined stress, ROS production can increase, and protective activity may then become inadequate. Williamson et al. (2007) suggested that the accumulation of ROS during *B. cinerea* infection resulted in accumulation of lipid peroxidation and depletion of antioxidants. In the present study, phenol content was found to be high in salinity or fungi-treated plants. There was a direct correlation between foliar phenol and MDA content.

Phenolic substances as secondary metabolites are involved in disease resistance in plants through inhibition of fungal extracellular enzyme by restricting mycelial growth as well as spore germination and oxidative phosphorylation and make defence strategies against biotic stress or plant diseases necessary for lignin biosynthesis which is one of the fundamental components of cell wall. However, high concentration of phenol could be toxic to plant cell metabolism. For example, Shahbazi et al. (2010) stated that potato plants inoculated with *Alternaria solani* accumulated phenol up to 4 days after inoculation. After the establishment of defence barriers, phenol levels decreased significantly. The less aggressive strain of *A. solani* elicited more phenol than the more aggressive isolate proving that the accumulation of phenol was related to the defence systems of plants. Although the antioxidant activity of phenols has been well documented, there is a lack of information on their toxicity. The experimental evidences suggest that the interaction between tannins and proteins or DNA may lead to the formation of both soluble and insoluble aggregates (Labieniec and Gabryelak 2006). Besides, gallic acid was described as having a cytotoxic effect on isolated hepatocytes (Nakagawa and Tayama 1995). Labieniec and Gabryelak (2006) stated that the phenolic acids at concentrations of 30 and 60 μmol l^{-1} caused DNA-based oxidation. They stated that the polyphenolic compounds such as tannic acid, ellagic acid and gallic acid had cytotoxic and oxidative properties in digestive gland cells of the freshwater mussel *Unio tumidus*. Higher concentrations (above 15 μmol l^{-1}) caused oxidative protein damage and DNA damage.

13.4 Conclusion and Future Prospects

Previously, scientists directed their attention to either the interactions between plants and abiotic stress or that between plants and biotic stress agents; however, it is now clear that the interactions among salinity, pathogens and plants are quite realistic and plants get affected by the both stress factors. On the other hand, the pathogen gets affected by salinity stress; however, pathogen under salinity stress becomes more aggressive if it gets excited by the stimulation of salt. In this case, more aggressive races of the pathogens might spread to non-saline or non-stressed areas and threat even the resistant cultivars. The presence of Na^+ ions in the nutrient or irrigation solution may increase the pathogenicity through the increased virulence or decreased resistance of the host. Mohammadi et al. (2007) reported that the interaction between soil salinity and pistachio (*Pistachio vera*) infection caused by *V. dahliae* resulted in a severe disease progression when compared to either of those stress agents. Salt stress significantly increased rootstock shoot and root colonization by *V. dahliae*. They also stated that salinity and *V. dahliae* interaction increased the concentrations of Na, K and Cl ion levels but decreased the K/Na ratio.

The idea of developing tomato cultivars with resistance to both *V. dahliae* and salinity has been criticized by breeders and scientists initially; however, the need for selection of new varieties with resistance to both pathogens and various environmental stresses is now more widely recognized (You et al. 2011; Dikilitas and Karakas 2012).

High concentration of NaCl may enhance the plant susceptibility to pathogen attack as well as reducing the pathogenic potential of the attacking microorganisms; however, low concentrations of NaCl, on the other hand, could have a stimulatory effect on plant pathogens. At this stage, any ingredients applied for nullifying the toxicity of NaCl such as $CaSO_4$ or KCl would increase the stimulation of the fungus. Boumaaza et al. (2015) reported that the mycelial growth of isolates of *Botrytis cinerea* was stimulated under salinity. The conidial production was stimulated by calcium or sodium salts. Only 300 mmol l^{-1} NaCl concentrations had inhibiting effect on *B. cinerea*. Therefore, it important to note that plants both under the threat of salt and pathogen attack should be carefully evaluated.

It is crucially important to increase K+ ion levels in host plants to maintain the resistance of plants under the threat of abiotic, biotic or their combined stresses. You et al. (2011) suggested that the disease and salinity combination reduced K^+ concentration, which slowed down the synthesis of defence enzymes and further synthesis of cellular metabolites. For example, hypocotyl tissues containing less K^+ and more Na^+ encouraged the spread of *M. phaseolina* following hypocotyl inoculation.

Growers should avoid using saline water even with low concentration of salt in seedling and flowering stages. They should also reduce cultural practices which increase salinity levels such as fertigation and excess irrigation in critical periods.

References

Abass MH (2016) Responses of date palm (*Phoenix dactylifera* L.) callus to biotic and abiotic stresses. Emir J Food Agric 28:66–74

Agami RA (2014) Applications of ascorbic acid or proline increase resistance to salt stress in barley seedlings. Biol Plant 58:341–347

Al-Sadi AM, Al-Masoudi RS, Al-Habsi N, Al-Said FA, Al-Rawahy SA, Ahmed M, Deadman L (2010) Effect of salinity on Pythium damping-off of cucumber and on the tolerance of *Pythium aphanidermatum*. Plant Pathol 59:112–120

Apel K, Hirt H (2004) Reactive oxygen species: metabolism, oxidative stress, and signal transduction. Annu Rev Plant Biol 55:373–399

Attaby HSH (2001) Influence of salinity stress on the growth, biochemical changes, and response to gamma irradiation of *Penicillium chrysogenum*. Pak J Biol Sci 4:703–706

Bates L, Waldren RP, Teare ID (1973) Rapid determination of free proline for water-stress studies. Plant Soil 39:205–207

Bolat I, Dikilitas M, Ercisli S, Ikinci A, Tonkaz T (2014) The effect of water stress on some morphological, physiological, biochemical characteristics and bud success on apple and quince rootstocks. Sci World J: 769732

Boumaaza B, Benkhelifa M, Belkhoudja M (2015) Effects of two salts compounds on mycelial growth, sporulation, and spore germination of six isolates of *Botrytis cinerea* in the Western North of Algeria. Int J Microbiol: 572626

Bradford MM (1976) A rapid and sensitive method for the quantitation of microgram quantities of protein utilizing the principle of protein-dye binding. Anal Biochem 72:248–254

Chance B, Maehly AC (1995) Assay of catalase and peroxidases. Methods Enzymol 2:764–775

Chinnusamy V, Jagendorf A, Zhu JK (2005) Understanding and improving salt tolerance in plants. Crop Sci 45:437–448

Daami-Remadi M, Souissi A, Houn BE, Mansour M, Nasraoui B (2009) Salinity effects on fusarium wilt severity and tomato growth. Dyn Soil Dyn Plant 3:61–69

Dikilitas M (2003) Effect of salinity & its interactions with *Verticillium albo-atrum* on the disease development in tomato (*Lycopersicon esculentum* Mill.) and lucerne (*Medicago sativa* & M. *media*) plants. Ph.D. Thesis, University of Wales, Swansea

Dikilitas M, Karakas S (2012) In: Ashraf M (ed) Behaviour of plant pathogens for crops under stress during the determination of physiological, biochemical, and molecular approaches for salt stress tolerance, Chapter 16. Crop production for agricultural improvement. Springer, Heidelberg, pp 417–441

Dikilitas M, Karakas S (2014) Crop plants under saline-adapted fungal pathogens: an overview. Chapter 8. In: Emerging technologies and management of crop stress tolerance, volume II A sustainable approach. Elsevier Academic Press, London, pp 173–185

EL-Abyad MS, Abu-Taleb AM, Khalil MS (1992) Impact of salinity stress on soil-borne fungi of sugar beet. Plant Soil 143:75–83

Elmer WH (2002) Influence of inoculum density of *Fusarium oxysporum* f.sp. *cyclaminis* and sodium chloride on cyclamen and the development of fusarium wilt. Plant Dis 86:389–393

Gessler NN, Egorova AS, Belozerskaya TA (2014) Melanin pigments of fungi under extreme environmental conditions (review). Appl Biochem Microbiol 50:105–113

Goicoechea N (2006) Verticillium-induced wilt in pepper: physiological disorders and perspectives for controlling the disease. Plant Pathol J 5:258–265

Hasanuzzaman M, Nahar K, Alam Md M, Bhowmik PC, Hossain Md A, Rahman MM, Prasad MNV, Ozturk M, Fujita M (2014) Potential use of halophytes to remediate saline soils. BioMed Res Int. Article ID 589341

Hiraga S, Sasaki K, Ito H, Ohashi Y, Matsui HA (2001) Large family of class III plant peroxidases. Plant Cell Physiol 42:462–468

Hodge JE, Hofreiter BT (1962) In: Whistler RL, Be Miller JN (eds) Methods in carbohydrate chemistry. Academic, New York

Howell AB, Francois L, Erwin DC (1994) Interactive effect of salinity and *Verticillium albo-atrum* on Verticillium wilt disease severity and yield of two alfalfa cultivars. Field Crop Res 37:247–251

Jahantigh O, Najafi F, Badi HN, Khavari-Nejad RA, Sanjarian F (2016) Changes in antioxidant enzymes activities and proline, total phenol and anthocyanin contents in *Hyssopus officinalis* L. plants under salt stress. Acta Biol Hung 67:195–204

Kuźniak E, Gabara B, Skłodowska M, Libik-Konieczny M, Miszalski Z (2011) Effects of NaCl on the response of Mesembryanthemum crystallinum callus to *Botrytis cinerea* infection. Biol Plant 55:423–430

Labieniec M, Gabryelak T (2006) Oxidatively modified proteins and DNA in digestive gland cells of the fresh-water mussel *Unio tumidus* in the presence of tannic acid and its derivatives. Mutat Res Genet Toxicol Environ Mutagen 603:48–55

Lamb C, Dixon RA (1997) The oxidative burst in plant disease resistance. Annu Rev Plant Physiol Plant Mol Biol 48:251–275

Leon L, Lawton MA, Raskin I (1995) Hydrogen peroxide stimulates salicylic acid biosynthesis in tobacco. Plant Physiol 108:1673–1678

Mancinelli AL (1990) Interaction between light quality and light quantity in the photoregulation of anthocyanin production. Plant Physiol 92:1191–1195

Mert HH, Ekmekci S (1987) The effect of salinity and osmotic pressure of the medium on the growth, sporulation, and changes, in the total organic acid content of *Aspergillus flavus* and *Penicillium chrysogenum*. Mycopathology 100:85–89

Mohammadi AH, Banihashemi Z, Maftoun M (2007) Interaction between salinity stress and verticillium wilt disease in three pistachio rootstocks in a calcareous soil. J Plant Nutr 30:241–252

Nachmias A, Kaufman Z, Livescu L, Tsror L, Meiri A, Caligari PDS (1993) Effects of salinity and its interactions with disease incidence on potatoes grown in hot climates. Phytoparasitica 21:245–255

Nakagawa Y, Tayama S (1995) Cytotoxicity of propyl gallate and related compounds in rat hepatocytes. Arch Toxicol 69:200–208

Omori T, Ogawa K, Shimoda M (1995) Breeding of high glycerol-producing shochu yeast (*Saccharomyces cerevisiae*) with acquired salt tolerance. J Ferment Bioeng 79:560–565

Pegg GF (1989) Pathogenesis in vascular diseases of plants. In: Ayres PG (ed) Effects of disease on the physiology of the growing plant. Cambridge University Press, New York, pp 149–177

Ragazzi A, Moricca S (1995) Infection of cotton by *Fusarium oxysporum* f.sp. *vasinfectum* as affected by water stress. Phytoparasitica 23:315–321

Ragazzi A, Broggio M, Soldi R, Moricca S, Dellavalle I, Sabatini S (1996) Biometry and pathogenicity *Fusarium oxysporum* f.sp. *vasinfectum* on a Na_2SO_4 enriched medium. J Plant Dis Prote 103:272–278

Regragui A, Rahouti M, Et Lahlou H (2003) Effect of saline stress on *Verticillium albo-atrum*: pathogenicity and production of cellulolytic enzymes in vitro. Cryptogam Mycol 24:167–174

Saadatmand AR, Banishashemi Z, Sepaskhah AR, Maftoun M (2008) Soil salinity and water stress and their effect on susceptibility to Verticillium wilt disease, ion composition and growth of pistachio. J Phytopathol 156:287–292

Sairam RK, Saxena DC (2000) Oxidative stress and antioxidants in wheat genotypes: possible mechanism of water stress tolerance. J Agron Crop Sci 184:55–61

Shahbazi H, Aminian H, Sahebani N, Halterman DA (2010) Biochemical evaluation of resistance responses of potato to different isolates of *Alternaria solani*. Phytopathology 100:454–459

Singleton VL, Rossi JA (1965) Colorimetry of total phenolics with phosphomolybdic-phosphotungstic acid reagents. Am J Enol Vitic 16:144–158

Triky-Dotan S, Yermiyahu U, Katan J, Gamliel A (2005) Development of crown and root rot disease of tomato under irrigation with saline water. Phytopathology 95:1438–1444

Turco E, Broggio M, Ragazzi A (1999) Enzymes produced by *Fusarium oxysporum* f.sp. *vasinfectum* compared on saline and a non-saline substrate. J Plant Disease Prot 106:333–341

Turco E, Naldini D, Ragazzi A (2002) Disease incidence and vessel anatomy in cotton plants infected with *Fusarium oxysporum* f.sp. *vasinfectum* under salinity stress. J Plant Dis Prot 109:15–24

Von Tiedemann A (1997) Evidence for a primary role of active oxygen species in induction of host cell death during infection of leaves with *Botrytis cinerea*. Physiol Mol Plant Pathol 50:151–166

Williamson B, Tudzynski B, Tudzynski P, van Kan JAL (2007) Botrytis cinerea: the cause of grey mould disease. Mol Plant Pathol 8:561–580

You M, Colmer TD, Barbetti MJ (2011) Salinity drives host reaction in *Phaseolus vulgaris* (common bean) to *Macrophomina phaseolina*. Func Plant Pathol 38:984–992 [illegible] (19[illegible]) [illegible] interaction with [illegible] tomato [illegible]. Physiol [illegible]

Nakagawa [illegible] (19[illegible]) [illegible] and [illegible] metabolism [illegible]

Olivera [illegible], Davies [illegible] (19[illegible]) Breeding of [illegible] with improved [illegible] tolerance. [illegible] 75:[illegible]

[illegible] (19[illegible]) [illegible] diseases [illegible]. In: [illegible] physiology of [illegible] plant. Cambridge University Press, New York, pp [illegible]

[illegible] A, [illegible] (19[illegible]) Induction of [illegible] affected by water stress. Phytochemistry 23:315–32[illegible]

[illegible] A, [illegible] M, [illegible] (19[illegible]) Salinity and [illegible] 107:[illegible]

[illegible] M, [illegible] (2000) Effect of salinity [illegible] pathogenicity and production of [illegible]. Cryptogam Mycol 21:[illegible]

[illegible], Hutchinson [illegible] (19[illegible]) Salinity and water stress and their effect on susceptibility to [illegible] with disease [illegible] and growth [illegible]. Phytopathology [illegible]

[illegible] (20[illegible]) [illegible]

[illegible]

[illegible]

[illegible]

[illegible] E, [illegible] M, [illegible] (19[illegible]) [illegible] Plant Dis Rep [illegible]

[illegible] (20[illegible]) Disease incidence and [illegible] plants [illegible]

MIX
Papier aus verantwortungsvollen Quellen
Paper from responsible sources
FSC® C105338

If you have any concerns about our products,
you can contact us on
ProductSafety@springernature.com

In case Publisher is established outside the EU,
the EU authorized representative is:
Springer Nature Customer Service Center GmbH
Europaplatz 3, 69115 Heidelberg, Germany

Printed by Libri Plureos GmbH
in Hamburg, Germany